Modern American **Environmentalists**

A BIOGRAPHICAL ENCYCLOPEDIA

Edited by GEORGE A. CEVASCO *and* RICHARD P. HARMOND

Foreword by EVERETT I. MENDELSOHN

THE JOHNS HOPKINS UNIVERSITY PRESS *Baltimore*

Printed in the United States of America on acid-free paper
9 8 7 6 5 4 3 2 1

The Johns Hopkins University Press
2715 North Charles Street
Baltimore, Maryland 21218-4363
www.press.jhu.edu

Library of Congress Control Number: 2008937528

A catalog record for this book is available from the British Library.

ISBN 13: 978-0-8018-9152-6
ISBN 10: 0-8018-9152-3

Special discounts are available for bulk purchases of this book. For more information, please contact Special Sales at 410-516-6936 or specialsales@press.jhu.edu.

The Johns Hopkins University Press uses environmentally friendly book materials, including recycled text paper that is composed of at least 30 percent post-consumer waste, whenever possible. All of our book papers are acid-free, and our jackets and covers are printed on paper with recycled content.

To George A. Cevasco,
Violet and William Harmond,
and Elizabeth DeSantis

With special thanks
to William L. Keogan of the
St. John's University Library
and the reference staff at
the Bay Shore–Brightwaters
Public Library

CONTENTS

FOREWORD

An earlier American sense of nature was shaped in large measure by the frontier—an "out there" that seemed to expand before the fledgling nation. There was a sense of awe and celebration: "O beautiful for spacious skies, for amber waves of grain, for purple mountain majesties above the fruited plain!" This frontier seemed endless and promising. Nature was bountiful and attractive. It drew the settlers across the territories and helped shape the values of the emerging society. A different view was more cautious, for beyond the frontier lay a seemingly inhospitable wilderness. Here, nature must be subdued and resources captured and exploited. The frontier's inhabitants, the Native Americans, were viewed as savages who needed to be subjugated. This contrasting view was also a formative element of the values of the new nation, of its pioneers, settlers, conquerors.

The twentieth century, which inherited a nation that stretched "from sea to shining sea," saw the rise of an industrial giant. That rise came with a price: plentiful resources that had seemed inexhaustible were being stretched; rivers were being tamed but also polluted; the endless frontier was now crowded with people; and the family farm was replaced by production that had become industrialized and guided by a new set of values established by agribusiness.

As the "endless frontier" seemed to close, a new sense of propriety of the land developed; a fear of the loss of the beauties of the land reinforced that earlier ethos of protection of the land and the great open spaces. The beauties of nature became something to be preserved and the as-yet unpopulated/underpopulated spaces something to be conserved. The first national parks and wilderness areas were developed in the West, and a movement led by the first environmentalists campaigned for a new national commitment to save large segments of the vanishing wilderness.

The story of what came next is told in this volume, giving place to the many individuals who took part in the multitude of efforts to establish and then practice a new set of values—a set of values regularly pulled in the different directions that were part of the life of this now-mature nation. On one side, values were tugged by constantly growing new industries and the steady expansion of urban and suburban habitats; on the other side, individuals worked vigorously defending the ideals of the necessity to protect the natural world—both the physical landscape and the numerous species of plants and animals whose survival was being threatened.

Many of the individuals encountered in these pages called attention to a set of problems not easily recognized earlier: the products of human habitation. For many, the symbolic turning point was the 1962 publication of Rachel Carson's *Silent Spring.* The book has become a classic of environmental prose. At the time, its mixture of evocative portraits of nature and descriptions of the deep intrusions of chemical pesticides made it a highly political challenge not only to the chemical industry but also to governing bodies at both the state and national levels. The congressional hearings it launched became the starting point for a flow of regulatory legislation and remedial efforts. The four and a half decades since *Silent Spring* have witnessed a resurgent but also very new and much-strengthened environmental movement. Perhaps the single most important new perspective has been the move from a focus on single challenges—for example, DDT—to the recognition of the much broader and more pervasive challenges brought by "climate change" and "global warming"—these the result of the cumulative effects of normal life in an industrial society whose wide use of fuels and energy sources has a significant enough impact to alter the very climate of the whole globe.

In 2007, the award of the Nobel Peace Prize to Albert Gore provided full international acknowledgment of the challenges global warming had brought and his role in sharply defining the threat and what must be done to deal with it. Former Vice President Gore's book *An Inconvenient Truth: The Planetary Emergency of Global Warming and What We Can Do About It* (2006) brackets the period since Carson's warning. In that time, a myriad of environmental protection efforts and protests saw the striking growth of a new and almost universal recognition that a problem of great magnitude faced us and that there were steps that could and should be taken to confront the problem.

The rich collection of entries in this volume forms a valuable map of what was done—politically, socially, and scientifically—to deal with a dangerously changing environment and the individuals who engaged in these efforts.

EVERETT I. MENDELSOHN

ACKNOWLEDGMENTS

A good book, it has been said, is one that is opened with expectation and closed with profit. We trust that such an observation is worthy of *Modern American Environmentalists*, a work that has had a long gestation. It began with the faith that Dr. Vincent Burke, Life Sciences Editor at the Johns Hopkins University Press, placed in the project. To him we are especially grateful for the support and suggestions he extended as this book took shape.

We also extend our thanks to the 88 contributors to this volume for their expertise and their efforts. These colleagues, for the most part active and retired university faculty, government officials, scientists, naturalists, conservationists, and ecologists, gave generously of their time and specialized knowledge.

Our gratitude is also extended to our advisers: Dr. Everett I. Mendelsohn, Professor of the History of Science, Harvard University; Dr. Keir Sterling, Command Historian, U.S. Army Combined Arms Support, Fort Lee, Virginia; and Professor Arthur Sherman, Acquisition Librarian, St. John's University. Their perceptive judgments contributed significantly to the overall quality of this book.

This volume could not have progressed without the cooperation of many other individuals to whom we also express our gratitude for their indispensable assistance in bringing this project to completion. We acknowledge the support, practical advice, and sound recommendations we received from the Research Staff in the St. John's University Library. Two erudite librarians come quickly to mind, William L. Keogan and Anthony Todman. For superior secretarial service and helping out in multiple ways, we are indebted to Jane Liguori, Lana Umali, and Anna Marie Mannuzza.

Without the help of all the aforementioned, *Modern American Environmentalists* would not have come into being as it has. We have the pleasure of formally expressing our gratitude and indebtedness to each of them.

INTRODUCTION

When Earth Day was first celebrated in April 1970, it had become obvious to many that we had to better deal with and show respect for our natural surroundings. The term *environmentalism* had come into general use, and every sensible person had become something of an environmentalist. No longer could we go on fouling our air, destroying the land and its resources, and polluting our streams, lakes, rivers, and oceans.

A more limited meaning of *environmentalist* suggests an individual closely concerned with the protection of the physical world. By extension the term has come to mean one dedicated to ecological issues, an advocate for the preservation, restoration, and enhancement of the natural environment.

To subsume all environmentalists under a single definition is problematic. There are, for example, advocates and promoters as well as activists and even radicals. On the one hand, there are those who recommend a biocentric philosophy and preach the need for cooperation between humanity and nature; on the other hand, there are those who demand what they dub "ecodefense" and go to such extremes as blowing up dams and incapacitating bulldozers. Indeed, it is not a simple task to categorize or delineate those we label as environmentalists. Likewise, it is not a simple matter to determine when and how the modern environmental movement began.

It is widely accepted, however, that the seemingly sudden appearance of a broadly popular environmentalism was a consequence of Rachel Carson's *Silent Spring*, published in 1962. As powerful as Carson's message was—and continues to be—there are obviously many other individuals and books responsible for the attention given to ecological matters. Historically considered, the roots of environmentalism first took hold in the middle of the nineteenth century, with concern for the environment growing exponentially over the past one-hundred-plus years since then. The men and women most responsible for the environmental movement in the United States during the twentieth century are acknowledged in this volume.

Modern American Environmentalists: A Biographical Encyclopedia is intended for individuals drawn to, and solicitous of, one of the most important and challenging problems of our time. It provides personal and professional information on the lives of those who contributed most to our understanding of the world in which we live and what must be done to preserve and improve our natural surroundings. This volume fills a gap in the literature, focusing on the accomplishments of a dedicated group of individuals who have enlarged and heightened our ecological knowledge.

To a large extent, this encyclopedia complements a good number of books on conservation and ecology that have been published recently. Unlike one of our previous volumes published in 1997, *Biographical Dictionary of American and Canadian Naturalists and Environmentalists* (with Keir B. Sterling and Lorne F. Hammond), which, as its title indicates, contains rather brief, dictionary-like entries, *Modern American Environmentalists* has longer biographical and critical essays. The *Biographical Dictionary*, additionally, lists Canadian figures; this volume does not. The former also lists many colonial figures; the latter does not. All the environmentalists selected for inclusion here made noteworthy contributions in the twentieth century and are native to the United States. Moreover, they have not been previously profiled as they are in this book. Each entry covers why or at least how each subject became interested in and involved with environmental causes, his or her specific contributions, and the subject's long-term influence or legacy. Broadly considered, *Modern American Environmentalists* is rather like a sequel and an update of our 1997 work.

Our selection of entrants for the volume is diverse, distinct, and broad. Chief among the entrants are those who have added to our appreciation of the beauty and vulnerability of our natural resources, our fields and forests, streams, lakes and rivers, our food supply and the very air we breathe. They are outstanding biologists, zoologists, agriculturists, botanists, ornithologists, apiculturists, ichthyologists, wildlife management specialists, park planners, museum scientists, and administrators. Among those profiled are numerous artists, poets, novelists, political figures, philosophers, photographers, and educators deeply involved with the physical and aesthetic qualities of life.

Every effort has been made to focus primarily on the most important men and women who have worked in a diversified range of environmental disciplines and activities over the last one hundred years. This encyclopedia, accordingly, covers a vast domain, but it would be presumptuous to suggest that the biographies and accomplishments of all who have made us aware of ecological matters could be encompassed between the covers of any one book. It may even be argued that certain individuals who should have been included have somehow been omitted. To such charges, we confess our culpability.

Prominent among the better-known environmentalists are Rachel Carson, Dian Fossey, Joseph W. Krutch, Stewart Udall, René J. Dubos, and Barry H. Lopez. Among the lesser known are many who proved to be very successful in various fields of endeavor but whose contributions to environmentalism have not been fully acknowledged. Charles A. Lindbergh, Thomas Merton, Edmund S. Muskie, Gary S. Snyder, and John Steinbeck, are just a few names that come to mind quickly.

A judgment was made to keep profiles under 3,000 words to allow for the inclusion of some 139 entries. We did not impose a rigid formula for individual entries, but obviously they have commonalities. First, each one specifies why an individual merits inclusion. Biographical facts about birth, family, and education are noted, along with why or how he or she became involved in environmental, conservation, or preservation causes. The subject's noteworthy activities are discussed and analyzed. Such matters as books written, legislation responsible for, and public or private organizations founded are covered in detail. A paragraph or two on the subject's long-term influence or legacy brings an entry to a close. Supplementing each entry is an up-to-date bibliography meant to encourage further reading and research.

The entries are not of equal length. Nor should they be. Some individuals are deserving of more attention than others, and we left the length up to the writer of the piece. Since an encyclopedia is only as good as its contributors make it, we sought and obtained entries from leading scholars and specialists working in environmental studies. This volume, consequently, is a cooperative venture on the part of 88 environmentalists, virtually all of whom have university or scientific affiliations and dedicate themselves to the subject. To them we extend our gratitude for their time and expertise.

Modern American Environmentalists, we trust, will prove to be a valuable reference tool to all who support, advocate, and promote environmental causes.

Modern American **Environmentalists**

Abbey, Edward (January 29, 1927–March 14, 1989). An essayist and novelist, Abbey was born and raised on a small farm in Home, Pennsylvania (near Indiana, Pennsylvania). His father, Paul Revere Abbey, a logger by trade and a registered Socialist and Wobbly organizer, taught him at an early age "to hate injustice, to defy the powerful, and to speak for the voiceless." His mother, Mildred Postlethwaite Abbey, was a schoolteacher. During the summer of 1944, Abbey undertook a three-month excursion to the West: he hitchhiked to Seattle and, by bus, hitchhiking, and boxcar, made his way to Arizona (where he spent a night in the Flagstaff city jail for vagrancy) and back to Pennsylvania. Soon after graduating from high school in 1945, he joined the U.S. Army and was stationed in Italy as a rifleman for the next two years. Abbey has written that the experience made him an anarchist.

The three-month journey to the West left a lasting impression on Abbey, and within a year of his return to the States, he moved to the outskirts of Albuquerque and became a student at the University of New Mexico, where he received a B.A. in philosophy in 1951. As he would continue to do up to the last year of his life, Abbey began to spend as much time as he could exploring the deserts and canyons of the region, growing more fascinated with the beauty and solitude of the West.

During his college days, Abbey began to write profusely, starting a private journal and writing fiction. In 1951, Abbey published a story, "Some Implications of Anarchy," in the university literary journal the *Thunderbird*, which Abbey also edited. The story marked the beginning of his lifelong exploration of anarchy as an alternative form of government. That issue of the journal aroused the anger of local civic and religious authorities because of a quote (taken from Voltaire but ironically attributed to Louisa May Alcott) that appeared on the issue's cover: "Man will never be free until the last king is strangled with the entrails of the last priest." Most of the copies were seized, and the journal's editors and supporters were reprimanded.

After graduation, Abbey spent the following year studying at the University of Edinburgh in Scotland as a Fulbright fellow and traveling throughout Europe. The following dozen years of his life were marked by restlessness, an inability or unwillingness to settle down. He had married a girl he had met as a student at New Mexico in the summer of 1950, but the marriage lasted only a year and a half. In December 1952, shortly after his return from Edinburgh, he married Rita Deanin, a marriage that would produce a son and a daughter. In the fall of 1953, with ambitions to become a professor of philosophy, he entered graduate school at Yale University, but dissatisfied (and, as he has said, baffled by a required course in symbolic logic), he stayed there only a matter of weeks and then took a New Jersey factory job. Although Abbey had developed a love for the West, as an aspiring writer his access to the publishing world was greater in New York. And Rita, an artist, felt that residence in New York would be more beneficial to her career as well. For two other brief periods, in 1962–1963 and 1964–1965, Abbey would attempt to live and work in the East. Living mostly in Hoboken, he worked at several occupations including writing technical manuals for General Electric (which, he claims, fired him for staring out the window too much) and doing social work in Brooklyn. Both of these periods ended with a feeling of resentment toward the technocratic bustle of city life and an escape to the open spaces of the Southwest.

Abbey returned to Albuquerque in 1954 and began work on an M.A. in philosophy at the University of New Mexico. That same year, he published his first novel, *Jonathan Troy*, an undistinguished fictional manifestation of his academic interests in anarchic government. Two years later, he became a father; completed his master's thesis, "Anarchism and the Morality of Violence"; published his second novel, *The Brave Cowboy* (which would be filmed as *Lonely Are the Brave* in 1962); and while Rita and his infant son spent the summer in New Jersey, began the first of many terms of employment by the National Park Service (NPS) as a park ranger at Utah's Arches National Monument. Until 1975, with brief interruptions, Abbey spent his subsequent springs and summers working for the NPS as a ranger and fire lookout and the balance of the year exploring the canyons, mountains, and rivers of the desert Southwest. In addition to Arches, Abbey's employment with the NPS included tours at Death Valley National Monument, Sunset Crater, Lassen Volcanic National Park, Lee's Ferry, Petrified Forest, Organ Pipe National Monument, Coronado National Forest, Grand Canyon National Park (North Rim), and Glacier National Park. From 1972 through 1975, he was manager of an Aravaipa Canyon wildlife preserve.

Abbey returned to Arches for a second season of work in 1957, and by October, he had accumulated four volumes of notes and drawings compiled over the last two years. (He would return once again to Arches in the early 1960s, but, more developed by that time, the park had lost much of its magic for Abbey.) Shortly after the tourist season ended in late September, Abbey entered Wallace Stegner's creative writing workshop at Stanford University on a fellowship; but by April 1958 he had decided to leave the program and take a job as a firefighter at the Gila National Forest in New Mexico. Meanwhile, he continued to write fiction. Drawing literary inspiration from, among others, B. Traven, Thomas Wolfe, Knut Hamsun, John Steinbeck, and Louis-Ferdinand Céline, almost all of his fiction is centered around the conflict between idealistic individuals and the faceless, tyrannical forces of industry, military, and government. Abbey's protagonists are, much like the author, anarchists by political persuasion and carry vital connections to the land (usually the desert Southwest) but are usually defeated in the end by the established order.

Garth McCann, in his monograph *Edward Abbey* (1977), suggests that Abbey's relationships with women have followed a similar pattern: "[H]e is an idealist who believes in love and marriage. . . . On the other hand, he occasionally shows contempt for women. He finds them irresistible, yet he cannot abide monogamy or the feeling that he is not free. Hence his numerous and erratic loves." Abbey and Rita divorced in August 1965, and in October of the same year, he married Judy Pepper, who died of leukemia on July 4, 1970, leaving Abbey with a second daughter. He married Rene Downing in 1973, was divorced some years later, and married Clarke Cartwright, with whom he had two sons.

By the mid-1960s, Abbey had taken up permanent residence in the West. Drawing from the journals he had kept at Arches in 1956 and 1957, he began to write his first book of nonfiction, *Desert Solitaire: A Season in the Wilderness*. Published in January 1968, the book initially received little attention and was soon out of print. Abbey was disappointed but, given past reception of his work, not surprised. In 1972, the book was, in Abbey's words, "exhumed and resurrected in paperback, in which it has enjoyed a modest but persistent life." A composite account of his three seasons employed as a park ranger at Arches, *Desert Solitaire* exalts individualism, simplification, solitude, and the stark beauty of the desert and is a frequently bitter criticism of a society dominated by materialism. Abbey moves seamlessly from eloquent descriptions of the land he loves to adventure stories to philosophical reflection, punctuated by his ever-present acerbic wit. Echoing Henry David Thoreau's reasons for living at Walden Pond, Abbey writes that he went to Arches "not only to evade for a while the clamor and filth and confusion of the cultural apparatus but also to confront, immediately and directly if it's possible, the bare bones of existence, the elemental and fundamental, the bedrock which sustains us." The book is a serious plea for saving wild places, a call for absolute adherence to the Wilderness Act of 1964, and a demand for the restoration of balance between civilization and wilderness, both of which are necessary for the survival of the human species.

In 1975 Abbey published what is perhaps his most popular book, *The Monkey Wrench Gang*, a comic novel about the adventures of four ecological saboteurs. The Gang operates in the Southwest, disabling bulldozers, pulling up survey stakes, burning down billboards, blowing up dams, and carrying out other illegal activities aimed at slowing the industrialization of the West. Publicized largely through word of mouth, *The Monkey Wrench Gang* has sold in excess of 500,000 copies. The novel, both in spirit and through its tactical suggestions, bore a direct influence on the emergence of the radical environmental organization Earth First!—which has been known to commit acts of sabotage similar to those described in the novel. Abbey, who held a lifelong distrust of large organizations, affiliated himself with Earth First! He published articles in its magazine and spoke at public Earth First!–sponsored demonstrations against development. Perhaps the most well known of these demonstrations consisted of a speech he delivered at Glen Canyon Dam (the object of much vituperation throughout the author's work) after fellow protesters unfurled a 300-foot plastic "crack" down the face of the dam. Abbey's work has also influenced more mainstream environmental groups. Peter Wild writes that Abbey motivated the Sierra Club to take a tougher stance and inspired public figures such as Robert Redford "to mount their own campaigns to save wilderness, though using somewhat more restrained methods than Abbey seems to suggest." According to Charles Little, Abbey's agenda for the road-building policies in the national parks, laid out in his essay "Polemic: Industrial Tourism and the National Parks" in *Desert Solitaire*, "became the basis for the recommendations of the staid old Conservation Foundation in their report, *National Parks for the Future*."

By the mid-1970s, Abbey was residing near Tucson and working full-time as a writer and speaker. For brief periods, Abbey had taught freshman composition at Western Carolina University and creative writing at the University of Utah. In 1981 he began teaching writing courses at the University of Arizona, where he eventually received tenure. After *Desert Solitaire*, Abbey was in high demand as a writer—his essays have been published in such diverse periodicals as *Rolling Stone*, *Reader's Digest*, *Harper's*, *Architectural Digest*, *Playboy*, *Sierra*, and the *New York Times Magazine*. As a journalist/correspondent, Abbey made trips to Australia and Alaska. The bulk of his periodical pieces are collected in the books *The Journey Home* (1977), *Abbey's Road* (1979), *Down the River* (1982), and *One Life at a Time, Please* (1988). Beginning in 1970, Abbey wrote the texts for four oversized photography books, *Appalachian Wilderness* (with Eliot Porter; 1970), *Slickrock* (with Philip Hyde; 1971), *The Hidden Canyon* (with John Blaustein; 1977), and *Desert Images: An American Landscape* (with David Muench; 1979), and a Time-Life book, *Cactus Country* (with Ernst Haas; 1973). Despite the glossy format of these books, their texts are, like his other nonfiction, autobiographical ("personal histories" is the term he has used to describe his nonfiction) and devoted to describing and praising the beauty of unspoiled lands and castigating humankind's "destructive development."

Abbey was described by his friend and fellow writer Edward Hoagland as possessing an "uncut grayish beard, slow speech, earnest eyes, red-dog-road shuffle, raw height and build, and jean jacket or shabby brown tweed." Another writer and friend, Barry Lopez, writes that Abbey was characterized by an ingenuous shyness, so at odds with the public image of a bold iconoclast. Numerous critics and reviewers have noted that some of Abbey's work is blatantly sexist, xenophobic, and politically inconsistent. To all of these charges, Abbey would be the first to agree: "If there's anyone still present whom I've failed to insult, I apologize," he once stated. A few reviewers of Abbey's work have been disturbed by the author's highly opinionated style on the grounds that it is detrimental to the serious cause of environmentalism. Wendell Berry, in "A Few Words in Favor of Edward Abbey," responds to this charge by stating that Abbey's subjectivity is not a liability but rather a key component in his importance as a writer: "My defense . . . begins with the fact that I want him to argue with, as I want to argue with Thoreau, another writer full of cranky opinions and strong feelings." Critical discussions of Abbey's work abound with comparisons of the author to Thoreau. Certainly Abbey, whose writings dramatize the inseparability of personal freedom and wilderness, is among the most important writers with environmental concerns in the latter half of the twentieth century.

In 1988, Abbey published the novel *The Fool's Progress*, which alternates narratively between a dying middle-aged man's journey from the West to the East and flashbacks to his younger days. The novel would be the last of the author's books published during his lifetime. Two other books were published posthumously: *Hayduke Lives!* (1990), a sequel to *The Monkey Wrench Gang*, and *A Voice Crying in the Wilderness* (1990), a collection of fragments and aphorisms. Abbey died of internal bleeding caused by a circulatory disorder. As he had requested in written instructions, a few close friends wrapped his body in his sleeping bag and buried it in an unmarked grave in the desert.

BRYAN L. MOORE

Cahalan, James M. *Edward Abbey: A Life*. Tucson, AZ, 2001.
Hoagland, Edward. "Standing Tough in the Desert." *New York Times Book Review*, May 7, 1989.
Loeffler, Jack. *Adventures with Ed: A Portrait of Abbey*. Albuquerque, NM, 2001.
Powell, Lawrence Clark. "A Singular Ranger." *Westways*, April 2, 1976.
Wild, Peter. *Pioneer Conservationists of Western America*. Missoula, MT, 1979.

Adams, Ansel (February 20, 1902–April 22, 1984). A photographer, teacher, writer, and conservationist, Adams was one of the most loved and most widely known personalities of the American conservation movement. Perhaps more than any other one person, Adams is responsible for creating the image of the American wilderness for the American public in the twentieth century. His photographic principles and techniques shaped the American perception and expectation of nature, the wilderness, and the national parks. He was most fortunate in being able to combine his art, first a hobby and then a livelihood, with his passion, the American wilderness and its conservation, to create his life's vocation.

Adams was born in San Francisco, California, into a moderately prosperous family with ties to insurance, lumber, and chemicals. His family home was on the dunes beyond the Golden Gate in the years before the bridge was built, and he grew up in a setting of natural beauty. Adams spent much of his hyperactive childhood roaming and exploring the hills and shores of the Bay Area, and he was witness at an early age to the effects of development on—and wanton destruction of—his natural playgrounds. His parents, Charles Hitchcock and Olive Bray Adams, gave ample latitude to his interests and allowed him to leave formal education with an eighth-grade diploma. The rest of Adams's education consisted of home tutoring by his father and his own explorations of the phenomena of life. Adams credits his father, a shy and seemingly conventional businessman who was fond of the pleasures of nature and amateur photography, for nurturing the "internal spark" that developed into the mature passions of this conservationist.

Adams received his first camera, a Kodak Box Brownie, as a gift from his parents during the family's first visit to Yosemite National Park in 1916. It was one of those rare moments that a person can name as the time and place of the origin of his life's work. Summer trips to Yosemite became a regular occurrence for Adams, and it was in Yosemite that he learned to hike, camp, climb, and take the photographs that were to serve as a visual diary of his mountain trips. It was also in Yosemite that Adams developed an intuition of a spiritual bond between humans and nature. He contracted influenza in 1919 at his San Francisco home and had difficulty overcoming the physical and emotional aftereffects. He returned to Yosemite and there recovered his health. Adams was convinced that it was the natural environment of Yosemite that enabled him to permanently overcome the emotional trauma he had suffered.

It was in Yosemite that Adams met Virginia Best, the daughter of a painter and park concessionaire, whom he married in 1929. Virginia inherited her father's Yosemite studio at his death, and in this way Adams's domestic life became linked to the park.

Adams's long association with the Sierra Club also began in Yosemite. He joined the club in 1919 and became summer custodian of its park headquarters, LeConte Memorial Lodge, in 1920; in 1928 he became the official photographer of its annual "outing." Adams's involvement with the club deepened with time, and in 1930 he was appointed the assistant manager of the yearly outings and began to contribute articles as well as photographs to the *Sierra Club Bulletin.* Virginia was also active in the club and became a member of the Sierra's board of directors in 1931. Adams took her place on the board in 1934 when she stepped down to devote time to their first child, and he continued to serve in that capacity for thirty-seven years.

Adams supported his family through commercial photography but continued to define and refine his personal conception of photography as an art form. *Parmelian Prints of the High Sierra* (1926), Adams's first published book, and his other nature pictures made him known to the Sierra's general membership. The value of these wilderness images to conservation issues was recognized, and in 1936 the Sierra Club sent Adams, armed with a portfolio of photographs of the Kings River country, to Washington, DC, to lobby Congress on behalf of the proposed Kings River National Park. This initial effort bore no fruit, but success came after publication of *The Sierra Nevada and the John Muir Trail* (1938). A copy given to Secretary of the Interior Harold Ickes was forwarded to the White House, where it greatly impressed President Franklin Delano Roosevelt. With Roosevelt's enthusiastic backing, Kings River became a national park in 1940.

Secretary Ickes attempted further use of Adams's talents in 1941 with a proposal to make a series of photographs of the national parks that were to be enlarged to mural size and hung at the Department of the Interior. The project was discontinued because of the outbreak of World War II, but Adams completed the project on his own initiative after the war. *My Camera in the National Parks* and *The National Parks and Monuments* (both 1950) came from this project, during which Adams also rethought his understanding of the meaning of conservation and the wilderness. He developed a distinction between recreation and a true wilderness experience, and he recognized that the intrinsic value of undisturbed nature was imperiled by intrusive activities or developments such as dams, which tame or modify a pristine environment. He desired that others come to realize that economic necessities can be counterbalanced by spiritual values of equal necessity. In his essay "The Meaning of the National Parks" (from *My Camera in the National Parks*), Adams argued that humankind had advanced to a state sufficient to consider resources other than materialistic and that these other resources "are, in fact, the symbols of spiritual life—a vast impersonal pantheism—transcending the confused myths and prescriptions that are presumed to clarify ethical and moral conduct. The clear realities of Nature seen with the inner eye of the spirit reveal the ultimate echo of God." Adams's God was in nature, and it was that God who had healed the sick youth in 1919. Adams was instrumental in changing Sierra Club wilderness policy from one favoring access for recreation to one of respect for the values of undisturbed wilderness. In 1957 he called for a halt to any further developments in the national parks.

Adams urged a nonconfrontational approach to conservation based on the use of reason in the conflict of values. His willingness to compromise led him to support nuclear power stations as the lesser of necessary evils—a stance that was severely criticized by many in the conservation movement—but he was unwilling to compromise when avoidable and irrevocable damage threatened the wilderness. He resigned from the Sierra board of directors in 1971, partly in disagreement over its policies, yet remained active in conservation activities. He devoted much of his time to the Wilderness Society because he approved of its strong concept of the wilderness. He helped start the Big Sur Foundation in 1977 to protect California's Big Sur coast, and he campaigned for a revamping of national park policy to restore the parks' deteriorated condition. His national stature was such that he was able to take his causes in person to Presidents Gerald Ford, Jimmy Carter, and Ronald Reagan—the last meeting "negative from the start."

Each of Adams's seven portfolios and thirty books shaped the American perception of the wilderness in its own way. Some of these are *Yosemite and the High Sierra* (1948), *Death Valley* (1954), *The Islands of Hawaii* (1958), *Yosemite Valley* (1959), *The Tetons and the Yellowstone* (1970), and *Photographs of the Southwest* (1976). *This Is the American Earth* (1960), produced in cooperation with Nancy Newhall, was called by U.S. Supreme Court Justice William O. Douglas "one of the great statements of conservation."

Adams was active in a variety of conservation groups. He was a trustee of the Foundation of Environmental Design, and he was president in 1956–1957 of Trustees for Conservation. He received the Sierra Club's John Muir Award in 1963 and the Conservation Service Award of the Interior Department in 1968. President Carter presented Adams with the Presidential Medal of Freedom in 1980.

Adams died in Carmel, California. He was survived by Virginia and their two children, Michael and Anne (Mrs. Ken Helms).

MICHAEL D. NICHOLS

Adams, Ansel. *Ansel Adams: An Autobiography.* Boston, MA, 1985.
"Adams, Ansel." *Current Biography Yearbook.* 1977.
Cahn, Robert. "Ansel Adams, Environmentalist." *Sierra,* May–June 1979, 31–49.

Archibald, George (July 13, 1946–). As a twenty-year-old college student working a summer job in the wilderness of western Canada, Archibald first encountered wild cranes in 1966. The experience proved pivotal in defining the course of his work and life. Archibald felt an instant fascination with the large birds' beauty and grace and their loud, unusual voices.

Learning of their highly threatened existence throughout the world, he became determined to dedicate himself to the study and preservation of the world's cranes. Cofounder and former director of the International Crane Foundation (ICF), started in 1972 in central Wisconsin, Archibald is now recognized as the world's leading scientific authority on cranes and as a pioneer in the captive breeding of endangered species. His work has taken him to countries around the world, and he has received awards and honors worldwide, including the United Nations Global 500 Honour for Environmental Achievement, the Netherlands' prestigious Order of the Ark, and the World Wildlife Fund Gold Medal. In 2006, he was honored by the Indianapolis Zoo as the first recipient of the $100,000 Indianapolis Prize, the largest monetary award ever bestowed on an individual for conservation of an animal species. Archibald is a member of the Wisconsin Conservation Hall of Fame and a recipient of the Wisconsin Historical Society's 2007 Aldo Leopold Award for Distinction in Environment and Conservation.

Archibald was born in New Glasgow, Nova Scotia, Canada. The second of six children of Donald and Lettie Archibald, both trained schoolteachers, he and his three brothers and two sisters grew up in the rural countryside near New Glasgow on a succession of small farms. Living in such a setting, young Archibald had ample opportunity to develop an interest in the natural world. From early childhood he was particularly interested in birds—domestic and wild—that lived on and around the local farms. Though the rest of the family did not share his interest, his parents supported it. By the time he was ten years old, he had made pets of several species of wild ducks, a Canada goose, and various others birds, constructing fenced-in pens and shelters made from old barn wood to house them.

Archibald attended a one-room schoolhouse through sixth grade, at which point he transferred to the larger local "consolidated" school. His love of nature and birds continued through his adolescence, and while in high school he volunteered at the local wildlife agency, assisting in a program to hatch ring-necked pheasants. Nonetheless, even as he graduated from high school in 1964, his interest in birds had never extended beyond the level of a beloved hobby. Following graduation, Archibald enrolled in nearby Acadia University. When he transferred the following year to Dalhousie University in Halifax, Nova Scotia, it was for the purpose of preparing to study medicine and become a physician. He and several friends planned to eventually open a medical clinic together—a plan that was never to be realized.

Seeking temporary work for the summer of 1966, Archibald learned of a game farm in the western province of Alberta, which greatly appealed to the naturalist in him. He wrote to the director of the farm, Al Oeming, and was offered employment for the summers of both 1966 and 1967. Excited as he was about the opportunity, he could never have imagined the central role it was to play in helping him to identify his true vocation, and his proper place in the world.

Archibald had always taken great pleasure in the natural world and its inhabitants, yet he had little experience with true wilderness. In fact, he viewed it as a somewhat forbidding entity, a place where a person could easily become lost or come to harm. His perspective took a radical turn, however, during that first summer in the wilds of Alberta. As never before, he felt encouraged to explore and discover the richness and beauty of the North American wilderness, and he began to comprehend the wealth and texture of life that it supports. Gradually becoming more and more at ease in venturing into the wilderness surrounding him, he was soon rewarded with the most important result of his new perspective: his introduction to wild cranes.

The first time Archibald witnessed cranes in the wild, he was immediately drawn to the large, beautiful birds, members of a distinct taxonomic family (Gruidae) whose loud, spectacular calls, elaborate mating displays, and habit of ground nesting set them apart from similar species, such as herons, egrets, storks, and flamingos.

During the course of his employment at the game farm, Archibald also gradually came to understand the concept and techniques of natural resource conservation. He was especially moved on learning of the perilous status of all fifteen crane species worldwide, owing to the degradation and destruction of the world's wetlands. Encouraged by Oeming, who himself had a special admiration for the unique character of cranes, Archibald's interest and concern for the future of the birds intensified.

By the time he returned to Dalhousie in the fall of 1966, Archibald had seriously begun to question the plans he had previously mapped out for his future, although he continued to pursue his medical studies, nonetheless. It wasn't until November 1967 that he finally made a decision to change the course of his education and, thereby, his life. Inspired by an article that appeared at that time in *National Geographic* magazine concerning the

Cornell Lab of Ornithology, in Ithaca, New York, Archibald decided to make a visit to the world-renowned ornithological facility. While there, he was engaged in conversation by William Dilger, an associate professor employed at the lab. Dilger quickly detected the passion for the plight of cranes that had been growing in Archibald since the summer of 1966. He urged the young man to make plans to enroll in graduate school at Cornell University and to pursue the study of ornithology, a suggestion that Archibald immediately embraced.

Archibald received his B.S. degree from Dalhousie in 1968. Not long afterward he left Canada behind and traveled to Cornell University to begin work on his Ph.D. His doctoral research focused on the behavior and evolution of cranes. He completed his degree in 1971.

While at Cornell, Archibald made the acquaintance of another graduate student, Ronald Sauey, who shared his deep interest and strong commitment to the study and preservation of endangered species of birds. The two became good friends and began serious discussions about organizing an effective means by which to save the world's cranes.

On his graduation from Cornell, Archibald accepted an offer to work on an elite, yearlong conservation project of international scope—eight months to be spent studying cranes in Japan, followed by four months on a similar project in Australia. The project would provide him an excellent chance to observe crane populations living in two extremely different environmental contexts—Japan featuring winter conditions in a highly populous country, Australia featuring conditions of extreme heat in an unpopulated region. By the time Archibald was ready to leave for Japan in January 1972, he and Sauey had outlined a plan to create a nonprofit foundation that would support the world's first scientific center devoted to the study and preservation of cranes. In particular, they hoped to initiate an unprecedented program of captive breeding and the subsequent reintroduction of cranes into the wild. The facility was to be developed on the site of a vacant farm owned by Sauey's father, located in central Wisconsin near the town of Baraboo. As Archibald left for Japan, Sauey had already begun devising ways to convert a former horse barn into housing for cranes.

Archibald's work in Japan focused on a flock of red-crowned cranes wintering on the island of Hokkaido. Contrary to the belief that the birds bred in Siberia, he discovered the flock nesting in a marsh on the north end of the island—an area largely targeted for development. Archibald quickly launched a public awareness campaign to save the breeding grounds. As a result, a large portion of the marsh was preserved, and Archibald established himself as a champion of crane conservation in the estimation of the people of Japan.

His work in Australia was similarly gratifying, resulting in the confirmed presence of a flock of eastern sarus cranes. This was particularly fortuitous in that the species is believed to have since become extinct within its original Asian range.

Archibald returned to the United States at the end of 1972, as planned, and joined Sauey in their creditable endeavor. In March 1973, the articles of incorporation for the ICF were submitted to the Wisconsin secretary of state, with Archibald named as the foundation's director.

The foundation made steady progress over the next several years. By 1976, fourteen crane species had been acquired (donated primarily by zoos, animal parks, and natural resource agencies), and hooded cranes were successfully bred in captivity at ICF for the first time in the world. That same year, Archibald embarked on what was to become a regular schedule of international travel when he visited Russia for the first time to study the particularly rare Siberian crane and to establish a working relationship with researchers of that country. In 1979, Archibald traveled to China and negotiated a collaborative program of crane conservation between ICF and the Chinese Academy of Sciences. By 1983, ICF had grown to the point of requiring a move to larger quarters a few miles north of the Sauey farm.

Tragically, Sauey passed away in 1987, at the age of thirty-nine. But he had lived to see the foundation gain recognition as the world's center for the study and preservation of cranes.

Through the support of ICF, Archibald continued his work and travels through the 1990s, conducting research and collaborating with colleagues throughout the world. He has personally studied eight species of cranes in Australia, Bhutan, China, Iran, India, Japan, Korea, Russia, and the United States. In addition, he has organized over 900 researchers, in more than sixty nations, working for the study and preservation of all fifteen species of cranes. Of special note are his recent efforts focusing on the Demilitarized Zone (DMZ) between North and South Korea, where endangered cranes and other species of wildlife exist with only a minimal human presence. In 2005, CNN founder Ted Turner gave his enthusiastic support to a proposal developed by Archibald to declare the 2.5-mile-wide, 155-mile-long

DMZ a World Heritage site. Such an action would ensure the protection of native plant and animal species of the site, along with its history.

Under Archibald's direction, ICF has come to claim the world's largest and most complete collection of cranes. All fifteen of the world's species of cranes have been bred successfully at ICF, and a flock of migratory whooping cranes has been established, which nests in central Wisconsin and winters in central Florida. The foundation has more than 7,000 members in the United States and other countries and attracts over 30,000 visitors annually.

Archibald pioneered many techniques for the rearing of cranes by humans but will probably be best remembered for the relationship he began in 1976 with "Tex," a nine-year-old female whooping crane who refused to mate. To induce the proper hormonal state to allow her to be artificially inseminated, Archibald imitated the courtship dance of a male crane—flapping his arms, leaping, jumping, calling, and even tossing sticks. After several years of repeating the springtime ritual, his efforts finally paid off in 1982, when Tex laid her first fertile egg at age fifteen. The egg hatched, producing a male that was named "Gee Whiz," and the event made national news. Archibald was invited to appear on the *Johnny Carson Show* to tell the story of Tex and Gee Whiz and show a film of his courtship of Tex. Sadly, the night he left for Los Angeles, Tex was killed by a raccoon. Archibald, nevertheless, went ahead with the appearance, and the poignant story he told on national television touched the audience deeply. Public awareness of ICF skyrocketed, donations reached an all-time high, and Archibald gained celebrity as the man who had danced with a crane.

Archibald resigned as director of ICF in November 2001 but continues to spend roughly half his time working to safeguard the future of the world's cranes. He is an ICF trustee and serves as director of the (Switzerland-based) World Conservation Union Working Group on Cranes and as a member of the recovery teams for whooping and Siberian cranes. Archibald lives on a farm near Baraboo, Wisconsin, with his wife, Kyoko, where he still pursues his boyhood hobby of raising a variety of wild and domestic birds.

MICHAELENE BROWN

Ackerman, Jennifer. "Cranes." *National Geographic*, April 2004.

Archibald, George. "Cranes: Litmus Birds of the State of the Earth." Lecture, Chautauqua Institution, Chautauqua, NY, 2002.

"Archibald, Who Danced with Cranes, Is Retiring." *Capital Times* (Madison, WI), July 14, 2000.

Katz, Barbara. *So Cranes May Dance.* Chicago, IL, 1993.

Atwood, Wallace Walter (October 1, 1872–July 24, 1949). The son of a planing mill owner, Atwood, a geologist, physiographer, geographer, and university president, was born and raised in Chicago, Illinois. After graduating from Chicago's West Division High School, he entered the recently established University of Chicago in December 1892. His interest in the environment developed after taking a summer field course in the Devil's Lake Driftless Area of Wisconsin from his professor, geographer-geologist Rollin D. Salisbury. This experience led to his early commitment to a career of systematic field study as a geographer-geologist.

During his graduate years at Chicago, Atwood held a variety of part-time positions working for the New Jersey Geological Survey, the Wisconsin Natural History Survey, and the U.S. Geological Survey. He also taught at precollegiate and teacher-training schools associated with the University of Chicago. It was at this time that he met and worked under the direction of noted educational reformer John Dewey. In 1901 he was named instructor of physiography and general geology by the University of Chicago, and in 1903 he received his Ph.D. from that institution.

Atwood conducted his own field study courses and continued to work with Salisbury, his mentor and friend. His first publication, written jointly with Salisbury, was published in 1897 and dealt with the geography and geology of the Devil's Lake region of Wisconsin. In 1909 Atwood was commissioned by the U.S. Geological Survey to make an in-depth study of the San Juan Mountains of Colorado.

Atwood became one of the foremost experts on the landscapes and landforms of the Rocky Mountain region. His principal scientific publication was a substantial monograph on the San Juan Mountains ("Physiography and Quaternary Geology of the San Juan Mountains, Colorado," U.S. Geological Survey Professional Paper No. 166, 1932, with K. F. Mather.) He published a variety of other technical papers, along with his son Wallace Jr., on other subregions of the Rocky Mountains. His general findings and approach to landform studies were published in his most famous book, *The Physiographic Provinces of North America*, published in Boston by Ginn and Company in 1940.

In 1913 Atwood left Chicago to take a position as professor of physiography at Harvard University. He revitalized the teaching of physical geography at Harvard by emphasizing the importance of field excursions and summer field courses. He attracted many students with, and his leadership in, field studies and his innovative teaching. According to one geographer, "Atwood himself was an inspiring teacher in field and classroom, notable for his personal interest in students . . . and for his celebrated technique of sketching landforms on the blackboard using both hands, simultaneously illustrating structure and surface features." As early as 1919 he developed an interest in using motion pictures to teach geographic concepts. He also was a pioneer in museum education and developed the Atwood Celestial Sphere, a forerunner of the modern planetarium.

Atwood was very concerned with environmental education in the public schools and authored a variety of textbooks, workbooks, study guides, and wall maps. His first text, *New Geography—Book Two* (1920), sold over 10 million copies, and estimates have been made that this textbook and related materials had been purchased for use in some 27,000 school districts and that they had been used by nearly 50 million students.

In early 1920 Atwood was offered and accepted the presidency of Clark University in Worcester, Massachusetts. He was hired at Clark to develop both undergraduate and graduate programs in geography and related fields. In 1921 he established the Graduate School of Geography at Clark, the second Ph.D.-granting geography department in America.

Aside from his technical publications and considerable ability in program development as an academic administrator, Atwood made a major impact on the development of geography as an academic discipline. He saw geography as a way of linking people with their natural environment. "Man does not conquer nature," he wrote. "He may discover the laws of nature and accomplish better and better adjustments to natural conditions on earth." Accordingly, he was also deeply committed to the wise use and protection of natural resources. His early research in the western United States for the U.S. Geological Survey provided insights into a variety of issues such as management of grazing lands, mining, and timber and water resources. He was an early advocate of the National Park System. In 1925 he was instrumental in establishing the internationally circulated journal *Economic Geography* to appeal to scholars who were concerned with the intelligent utilization of the world's resources.

Atwood made many lasting contributions to geographical and environmental science. As an active researcher, teacher, and administrator, he was able to encourage others to come to a better understanding of the complex natural systems on our planet.

ROBERT P. LARKIN

Koelsch, William K. "Wallace Walter Atwood." *Geographers Biobibliographical Studies* 3 (1979): 13–18.

Larkin, Robert P., and Gary L. Peters. *Biographical Dictionary of Geography.* Westport, CT, 1993.

Austin, Mary Hunter (September 9, 1868–August 13, 1934). A naturist, anthropologist, mystic, transcendentalist, feminist, conservationist, and ecologist, Austin was born in Carlinville, Illinois. After early years spent in the Midwest, she chose to live the rest of her life in the Southwest (southern California, New Mexico, Arizona).

As a child, Mary Hunter took advantage of her parents having the Chautauqua Literary and Scientific Circle meet at their home. The Chautauqua books and pamphlets were available to her, and she especially devoured Hugh Miller's *Old Red Sandstone* (1841), which introduced her to the early paleontological and geological studies of the earth. In fact, this was the first book young Mary bought when she had saved enough money. Because of the literary nature of her milieu, Mary was familiar with the writings of early naturalists John Burroughs, Henry David Thoreau, and Ralph Waldo Emerson. As a child and young adult, Mary knew Charles Robertson, a neighbor who was famous for his studies of the interrelation of plants and insects. Both the Chautauqua readings and Robertson probably influenced Mary to major in science in college; she graduated in 1888 from Blackburn College.

After Mary's graduation, she went with her family to homestead at Tejon in the San Joaquin Valley of California. In 1888, this area was a vast desert landscape little known to those outside its Sierra Nevada mountains and its arroyos. It was here that Mary first defined herself as a "naturist." In her autobiography *Earth Horizons* (1932), she writes of these years of exploration as learning "beauty-in-the wild" and seeking the "spirit to be evoked out of the land." She spent her hours outdoors, roaming the desert and keeping careful observations of the landscape and its creatures. But even with her scientific training, she had no language or terms to know what she was seeing, and this caused her considerable anxiety. Then she met General Edward Fitzgerald Beale, the owner of Tejon Ranch. Because of his position and contacts, he supplied Mary with government papers, agricultural reports, geological reports, and botanical surveys of the area. Finally, Mary learned the identifications of what she observed. Also, General Beale introduced Mary to the shepherds who roamed the desert; these individuals provided Mary with additional great quantities of information about the land and its activities. About this time, too, Mary began spending time with the Paiute Indians who lived in the area. From them she learned that all life is interrelated and that art is living and interpreting the common aspects. This interrelationship philosophy dominated the rest of Mary's life. She also learned the concept of being spiritually connected to the environment, which became the basis of her mystic, transcendental philosophy.

As Mary wandered the desert, she sensed that the land instructed. Teaching school to make money, Mary spent all her free time in the outlands, forming her desert ecology philosophy, which is the interrelations of land, water, crops, politics, and personal life. She became involved in the conservation issue of water usage, seeing and not supporting the desires of those who sought to own and control the water rights in the San Joaquin Valley.

She met and married Stafford Austin in 1891. They moved to the Owens Valley, which is east of the southern San Joaquin Valley and on the other side of the Sierra Nevada. She still continued her restless study of the desert and Native American lore. It was here that she began to write down her observations and interpretations. On a trip to San Francisco, Austin went to hear psychologist William James speak. He validated her belief in the need to experience wholeness and in spiritual connectedness, which she had learned earlier from the Paiutes. This validation solidified for Austin the interrelationship concept and cemented her environmental philosophy.

Finally, after at least twelve years in the desert, Austin felt the need to write her philosophy of the desert in a serious way. In 1900 she began a new life in the literary world. She was to spend the rest of her life in the midst of the most important writers and artists of the time, including John Muir, H. G. Wells, Jack London, Diego Rivera, Martha Graham, and Isadora Duncan. Her interest in Native American rhythm as the embodiment of the interrelationship of all natural elements became her passion and her expertise; she advocated that art and the land could not be separated. This belief was intensified by her association with David Starr Jordan, a writer, a scientist, and president of Stanford University. His interest in geographical distribution of animal and plant species and their relation to the environment was a catalyst for Austin's study and writing. She spent her middle and late years writing about connectedness, about personal freedom, and about native folklore. Known and appreciated both in the United States and in Europe, Austin continued to write and give lectures.

Two other individuals influenced Austin. One was scientist Daniel T. MacDougal, who convinced Austin to study the Pueblo Native American culture and the Arizona–New Mexico desert with the same intensity that she had studied the San Joaquin Valley and the Paiutes. The other influence in her later years was anthropologist Frank Applegate, who was working on a project to record the Spanish colonial arts in New Mexico. She used both influences to further her ecological-cultural philosophy and writings.

Because of her identification as an ecologist, Austin was chosen by the governor of New Mexico to be a delegate to the 1927 Seven States Conference discussing the controversy over water rights on the Colorado River. She condemned the Boulder Dam project being discussed as an unnatural ownership of water.

Austin's final years were spent in Santa Fe, leading the Indian Arts Fund project and the Spanish Colonial Arts Society; she hoped to preserve the spiritual connectedness to land and rhythm found in the two cultures. She delighted in working with Ansel Adams on a pictorial representation of the Southwest she loved.

A prolific writer of thirty-one books and many articles, Austin's best writing concerned nature: *The Land of Little Rain* (1903), about the San Joaquin Valley; *The Flock* (1906), about the sheep herding in the Sierra Nevada mountain range; and *The Land of Journeys' Ending* (1924), about the Arizona–New Mexico desert and people. Other notable Southwest books include *The Basket Woman* (1904), juvenile short stories; *Lost Borders* (1909); *California: The Land of the Sun* (1914); *The American Rhythm* (1923); *The Children Sing in the Far West* (1928); *Taos Pueblo* (photographs by Ansel Adams; 1930); and short stories *One-Smoke Stories* (1934) and *Mother of Felipe, and Other Early Stories* (1950). Austin produced one successful drama about the Southwest: *The Arrow Maker* (1911).

Often compared to John Muir and Henry David Thoreau, Austin also found direction and answers in the natural world. But she differed in that she did not want to separate from humankind; she believed that social consciousness could be enhanced by ecological connectedness. Austin is important because she studied the relationships among the natural world, the people, and their culture. Furthermore, she is important to the environmental movement because she gave a new perspective, a woman's perspective, which she identified as different from men's and yet just as valid. Austin's perspective sought to change society by deriving positive behavior from people by their connection to land and place.

KELLY A. FOTH

Burns, Lois. "Mary Hunter Austin." *American Women Writers: A Critical Reference Guide from Colonial Times to the Present*, ed. Lina Mainiero. New York, 1979.

Fink, Augusta. *I-Mary*. Tucson, AZ, 1983. A thorough, critical biography of Mary Austin.

"Mary Austin." *Twentieth-Century Literary Criticism*. Vol. 25. 1987.

Bailey, Florence Merriam (August 8, 1863–September 22, 1948). Born at "Homewood," her family's home in Locust Grove, Lewis County, New York, Bailey was the second daughter and fourth child of Clinton Levi Merriam, a successful banker, businessman, and future congressman (1871–1875), and Caroline Hart Merriam. Her only sister Ella Gertrude died the day before Florence Augusta was born. Soon afterward, Florence's father retired from business to pursue other interests and manage the family property.

Florence was tutored at home, then attended several private schools in the local area and in New York City, also spending some time at Miss Piatt's School in Utica, New York. As a child, she absorbed informal guidance in natural history—particularly in local birdlife—from her father and later from her much-loved brother Clinton Hart Merriam, known in the family as Hart to distinguish him from their father. Hart, eight years Florence's senior, taught her about birds and, later, when she briefly considered a medical career, shared some of his experiences as a medical student at the College of Physicians and Surgeons in New York. As a child and young woman, Florence also acquired an interest in writing. At nineteen, she entered Smith College in Northampton, Massachusetts. Lacking some of the credits in preparation for the regular course of study, she enrolled as a special student. She stayed four years and left without a degree in 1886. However, many years later, in 1921, she was awarded a B.A. "as of the Class of 1886."

In her final year at Smith, Florence began writing articles about birds, some of which appeared in *Audubon Magazine* and in various local newspapers. She then revised these and added some new pieces, all of which were published in her first book, *Birds through an Opera Glass*, released by Houghton Mifflin in 1889, when she was just twenty-six. The book was an impressive achievement for someone her age. It was soon reprinted for the Chautauqua Literary and Scientific Circle, a thriving year-round literary organization centered in Chautauqua, New York. Thanks in part to her brother Hart, one of the founders of the American Ornithologists' Union (AOU) in

1883, Florence became the first woman associate of that organization in 1885. Women were not permitted full membership for many years, but she became a full member in 1901 and was elected a fellow in 1929.

In 1891, Florence was briefly active in social work, volunteering one summer month to mentor young, employed girls at Jane Addams's Hull House in Chicago and offering her help that winter with a similar club formed by Grace Dodge in New York City. Perhaps as a result of exposure to these venues, she contracted tuberculosis and was obliged to spend time recovering at an uncle's ranch in California, which she had previously visited. She returned in 1894 to attend classes at Stanford University for six months, then stayed for a time in Utah, an experience that resulted in another book, *My Summer in a Mormon Village.* This publication, which appeared in 1894, focused on nature, especially the Great Salt Lake and its surroundings, but it also reflected Florence's ambivalence toward Mormon society. The spring of 1895 found the young author near San Diego, watching the local birds at "Twin Oaks," where she assembled notes and other materials that subsequently appeared in another book, *A Birding on a Bronco* (1896). On the recommendation of her brother Hart, a published authority on the biogeography of San Francisco Mountain in Arizona, she visited the mountain, finding the climate beneficial for her lungs.

Florence's fourth book, *Birds of Village and Field: A Bird Book for Beginners,* one of the first bird guides of its kind, was published in 1898. The volume included a few photographs and many black-and-white drawings and other illustrations by noted bird artists Louis Agassiz Fuertes and Ernest Thompson Seton. Her brother Hart may have been instrumental in getting these two well-known illustrators to work with her. Appendixes dealt with migration, winter birds and their variants, an "observation outline" (what to look for when watching birds), and a list of birds observed in New England towns.

During the 1890s and into the new century, Florence continued to publish articles on bird topics in the *Auk* (the journal of the AOU), *Bird Lore, Chatauquan,* and other publications. By 1900, as she approached her thirties, her books and articles had brought her recognition as one of the four leading women ornithologists in the United States, together with Olive Thorne Miller, Mabel Osgood Wright, and Neltje Blanchan. In an article written after her death, Paul Oehser characterized her as "one of the most literary ornithologists of her time, combining an intense love of birds and remarkable powers of observation with a fine talent for writing and a high reverence for science."

During her frequent visits to her brother Hart's home in Washington, DC, Florence became acquainted with Vernon Orlando Bailey, a young mammalogist employed by Hart in the Bureau of Biological Survey, where Hart was chief from 1885 to 1910. The unmarried Bailey was then living in Hart's house. In December 1899, Florence and Bailey were married. Although very fond of her nieces and nephews, the children of her brothers Collie (Charles Collins Merriam) and Hart, the Baileys themselves had no children. It is believed she may have miscarried on more than one occasion. The couple established their home on Kalorama Road in Washington, a much-loved objective for several generations of visiting naturalists and friends. Bailey grew accustomed to most of her husband's idiosyncrasies, although these occasionally conflicted with her Victorian sensibilities. Many of his habits resulted from long years working as a field naturalist in the western United States. She was, for instance, unable to squelch his habit of picking up his dinner plate and blowing nonexistent sand from it, even when dining with friends. They remained a devoted couple until Vernon's death in 1942.

In 1902, Bailey published her *Handbook of Birds of the Western United States,* a well-illustrated guide to species west of the Mississippi River. This 600-page book, containing almost as many illustrations, was specifically designed to complement the *Handbook of Birds of Eastern North America,* by Frank M. Chapman of the American Museum of Natural History. By now an expert in the field, she was careful to use the latest scientific sources, including successive editions of the *AOU Checklist of North American Birds.* In some cases, her husband drafted the scientific descriptions. This book subsequently went through some ten editions and was considered a standard work for over fifty years. Such was her reputation that in 1908 Dr. Joseph Grinnell, for many years director of the Museum of Vertebrate Zoology at the University of California, named a new subspecies of chickadee in her honor.

During her many years in Washington, Bailey gave unstintingly of her time to the educational activities of Washington. She was one of the local Audubon Society's founders in 1897. She wrote a foreword to Mrs. L. W. Maynard's *Birds of Washington and Vicinity and Adjacent Parts of Maryland and Virginia* (1898), which was used in the DC public schools. She helped plan bird classes taught by members of the Biological Survey, which were offered through the Audubon Society to teachers of nature study in the city's normal schools. These classes continued for a quarter century until the mid-1920s. Bailey was also active in the Cooper and Wilson Ornithological Clubs and the Biological Society of Washington.

For the first three decades of their marriage, the Baileys frequently traveled together to the western states—he, to complete fieldwork on mammals and other terrestrial vertebrates for the Biological Survey, and she, to observe and write about the birds. When Professor Wells W. Cooke of the Survey died in 1916, he left his work on the birds of New Mexico incomplete. Dr. Edward M. Nelson, then chief of the Biological Survey, asked Florence Bailey to finish the project, an undertaking that took a dozen years to complete. This made sense, because she knew the birds of the region intimately, and her husband had completed a biological and biogeographical survey of the state. Nelson asked her to check and update Cooke's conclusions, and she was given all necessary support by Nelson's successors as Survey chief. When her *Birds of New Mexico* was published in 1928, it was hailed as a model of its kind. Bailey was awarded the AOU's Brewster Medal, its highest honor, in 1929. Bailey died in Los Angeles at age eighty-five on the day after Christmas 1948.

KEIR STERLING

Horner, Elizabeth, and Keir Sterling. "Feathers and Feminism in the 'Eighties." *Smith College Alumnae Quarterly*, April 1975.

Kofalk, Harriet. *No Woman Tenderfoot: Florence Merriam Bailey, Pioneer Naturalist.* College Station, TX, 1989.

Oehser, Paul H. "In Memoriam: Florence Merriam Bailey." *Auk* 69 (January 1952).

Schafer, Elizabeth. "Florence Merriam Bailey." *American National Biography.* New York, 1999.

Bailey, Liberty Hyde (March 15, 1858–December 25, 1954). Born on a Michigan farm, Bailey was a teacher, conservationist, scientist, author, poet, and philosopher. In grade school he was an avid reader and learned to observe things firsthand from a teacher who instructed him in Latin, a subject he put to use later in taxonomy.

Young Bailey was fascinated by plant and animal life and especially by birdlife. He was deeply moved by the hardships that farmers faced during the Panic of 1873, and he noted how the railroad monopoly made things even more difficult for those who worked the soil. An early interest in politics led him to the Grange Movement, a grassroots rebellion against railroads in particular and monopolies in general.

Bailey entered Michigan Agricultural College (MAC) in 1877, mainly because the study of agriculture was free for residents of the state. He left his studies to teach grade school in a backwoods settlement in 1878–1879. On his return to MAC, he met Dr. William Beal, a famous nineteenth-century botanist whose method of teaching directly from plants made a profound influence on Bailey's later teaching. He was forced to leave his studies again because of a serious ear infection, which was fortunately treated by surgery. When he returned again to MAC he took an active part in student life and organized and edited an undergraduate quarterly, the *College Spectrum*; affiliation with this journal was the beginning of a writing career that was to span some seventy years. At MAC he won several honors. He was elected president of the Natural History Society and was chosen to head the Student Government Association. He was graduated from MAC on August 25, 1882.

Bailey's first job after graduation was as a cub reporter on the *Springfield (Illinois) Morning Monitor*. Because of his attention to detail, his hustle, and his excellent writing skills, he was promoted to city editor. At the time of his promotion he received a letter from Dr. Beal, who had recommended his former student to Dr. Asa Gray, director of the Harvard Herbarium. Although the position paid less than that available at the *Morning Monitor*, Bailey decided to go to Harvard, where his duties would require the sorting and classifying of specimens from Kew Gardens, England.

At the end of his stint at the Gray Herbarium, Bailey was offered a newly created position of professor and chair of horticulture and landscaping at MAC in 1885. In the same year, he published his first book, *Talks Afield about Plants and the Science of Plants*. The following year he published a paper on North American *Carices*, and shortly thereafter he was recognized as one of the leading authorities on sedges, a most difficult taxonomic group to work with. His fame and esteem in the scientific community were now such that he was offered a chair in botany at the University of Wisconsin, but he declined the position. In the winter of 1887 he first met the president of Cornell University, who offered him a newly created chair of Practical and Experimental Horticulture; after accepting this position, he moved to Ithaca, New York.

At Cornell, Bailey was treated rather contemptuously by some of his colleagues because of his agricultural background. His competence in the classroom and his scholarly research, however, endeared him to his students, and he was appreciated by scientists around the world. In addition to his teaching and research, he also presented many lectures at Grange meetings, at county and state fairs, and at scientific conferences. His reputation was now such that the president of Macmillan informed Bailey: "I will publish every book you write." In all, Bailey went on to write sixty-five books, some of which reached twenty editions and were in demand thirty years after their initial publication. Among his most significant titles are such works as *The Principles of Vegetable Gardening* (1901), *The Training of Farmers* (1909), *The Standard Cyclopedia of Horticulture* (1914–1917), and *How Plants Get Their Names* (1933).

Bailey was so prolific that he was often asked how he could accomplish so much. "There are two essential epochs in any enterprise," he quipped, "to begin and to get done." At times, when terms he required did not exist, he would coin them. He is especially known for such neologisms as "graftage" and "cuttage." He was as creative with the camera as he was with words. "Take your time," he once commented about photography. "At the end of a year tell me if you are not a nature lover." He also believed that children should have an opportunity to plant gardens. "Let them grow what they will," he held. "It matters less that they grow good plants than they try for themselves." For Bailey, "All nature was a garden."

In 1893, Bailey envisioned Cornell University as the leading institution of agriculture. He lobbied for state support, and in 1894 Cornell received a large sum for horticultural experimentation. During this period, he organized the nature study movement at the university. By providing leadership in the development of the movement, Bailey's name became well known among elementary school teachers. Under his leadership, teachers were provided with leaflets for their pupils on seeds, trees, insects, birds, cuttings, and more. By 1903, nearly 3,000 teachers were receiving nature study guidance by correspondence, and some 30,000 children were raising plants in school gardens.

Since Bailey viewed nature study as an affair of the heart, he cautioned teachers not to make it into a *course*, in the usual sense of the term. "A rigidly graded and systematic body of facts," he maintained, "kills nature study; examinations bury it." Never should the emphasis be shifted from the outdoors to books and pictures and mere collections: "Stuffed birds do not sing and empty eggs do not hatch. Let us go to the fields and watch the birds."

In 1903, Bailey was appointed dean and director of the School of Agriculture at Cornell. For ten years he served with distinction, stepping down in 1913. During his tenure the school grew from a department to a first-rate, state-financed agricultural institution of national reputation. The faculty grew from eleven to one hundred; students, from fewer than 100 to approximately 1,400. State funding increased greatly.

As dean, Bailey, who was a prolific writer and publisher himself, was forever encouraging his faculty to publish the results of their research. "I will restrict no man's investigation" was one of his credos. He also always found time for his students. Sunday evenings at the Bailey Ithaca residence were set aside for "at homes." At such informal gatherings, Bailey always gave encouragement to all those who had the good fortune of studying with him.

In 1906, he was elected president of the Association of Agricultural College Experimental Stations, an honor bestowed on him for his leadership ability, diligence as a researcher, and skills as an administrator. Teddy Roosevelt was one among many who admired his competence. When Roosevelt visited Cornell in 1911, he spent the day with Bailey and requested that Bailey chair the Commission of Country Life.

On his retirement from Cornell, Bailey had the opportunity to devote all his time to research and writing. He traveled to New Zealand in 1914, and the following year he published *The Holy Earth*, the theme of which is that we must make righteous use of the vast resources of the earth as founded on religious and ethical values. In one sense, he was ahead of himself; in another, he was reiterating the thoughts of great men such as Henry David Thoreau who had preceded him.

In 1919, Bailey traveled to Europe to examine the herbaria of notable collectors. Impressed with what he saw, he vowed to make his own herbarium one of the best in the world. By 1935 his herbarium contained over 125,000 specimens, which he willed to Cornell, along with his gardens of cultivated and native plants. But Bailey continued his travels in search of rare palms. Because of the lawlessness of some of the inhabitants of various regions, some of his trips proved dangerous; yet he always took such danger in stride. His work on the palms of the world became extensive and comprehensive. He increased the known number of palms from about 700 species to several thousand. He believed that there might be as many as 6,000 species of palms in the world.

Though he planned to visit Africa in 1950 to gather more data on palms, he broke his left thigh in a fall. Recovery from the injury was very slow. His leg never really healed, and at times he suffered pain. As a consequence, he never had an opportunity to visit Africa and collect its palms. He continued with his writing, however, and in 1953 he published one of his most interesting works, *The Garden of Bellflowers.* He died the following year on Christmas day at the age of ninety-six.

Bailey married Annette Smith on June 6, 1883; they had two children, Sarah Mary and Ethel Zoe.

RICHARD STALTER

Dorf, Philip. *Liberty Hyde Bailey: An Informal Biography.* Ithaca, NY, 1956.

Knudson, L., G. H. M. Lawrence, and W. I. Myers. "Liberty Hyde Bailey." *Necrology of the Faculty* [Cornell University, 1956]. Ithaca, NY, 1956.

Rogers, Andrew. *Liberty Hyde Bailey: A Story of American Plant Sciences.* Princeton, NJ, 1949.

Beebe, (Charles) William (July 29, 1877–June 4, 1962). A curator, naturalist, explorer, and oceanographer, Beebe was born in Brooklyn, New York, the son of Charles Beebe and Henrietta Marie (Younglove) Beebe. At an early age, he moved with his family to East Orange, New Jersey. It was here, in what was then a rural area, that he began to develop his great love of nature and birds.

As a young man he attended Columbia University as a special student in zoology. Although he attended Columbia from 1896 to 1899, he did not obtain a degree. Despite this, in November 1899 he became curator of ornithology at the New York Zoological Park (now the Bronx Zoo) on the recommendation of one of his professors, Henry Fairfield Osborn, who was the president of the Zoological Society.

Beebe's love of nature and outdoor life did not coexist well with the position of curator. In 1904, he undertook his first major expedition to Mexico with his first wife, Mary Blair Beebe. *Two Bird Lovers in Mexico* (1905) is an account of this trip. Throughout the rest of his life, Beebe made numerous trips abroad. In 1908 he visited Trinidad and Venezuela. In 1909 he visited British Guiana, and during 1910 and 1911, he traveled throughout Asia, collecting data for a four-volume work titled *A Monograph of the Pheasants* (1918–1922).

In 1916, Beebe established a research station at Kalacoon, British Guiana, where he collected a tremendous amount of data on the region's birds, mammals, reptiles, and insects. This research was subsequently published in his book *Tropical Wild Life in British Guiana* (1917). Beebe continued to travel to the Galapagos, the West Indies, and the Far East. In 1919, he became the director of the Zoological Society's Department of Tropical Research. His explorations shifted from the field of ornithology to oceanography. In 1929, he began extensive research on the life and ecology of the waters surrounding the island of Bermuda.

Beebe's most important exploration occurred during the 1930s, when he and Otis Barton made more than thirty descents off the coast of Bermuda in a bathysphere developed by Barton. (In this bathysphere, a spherical diving device, Beebe descended to a depth of 3,028 feet. This dive held the record until 1949.) Beebe made sixteen deep descents in the bathysphere. During these descents, he made detailed observations of the unknown or little-known deepwater-life species. He even broadcast one of his visits to a radio audience. A description of these visits is chronicled in his book *Half Mile Down* (1934).

In 1942 Beebe once again made jungle explorations in Venezuela. He established another research station there and studied the rain forest and the migration of birds. Despite his official retirement from the society in 1952, Beebe continued working on and off at a research station in the Arena Valley of Trinidad until his death in 1962.

Beebe's writings are viewed as both literary and scientific. He was a great popularizer of scientific and naturalist concerns. He published over 800 articles and twenty-four books.

Beebe had no children. After divorcing Mary Blair in 1913, he married author Elswyth Thane Ricker in 1927.

ARTHUR SHERMAN

Guberlet, Muriel Lewin. *Explorers of the Sea; Famous Oceanographic Expeditions.* New York, 1964.

Welker, Robert Henry. *Natural Man: The Life of William Beebe.* Bloomington, IN, 1975.

Bennett, Hugh Hammond (April 15, 1881–July 7, 1960). When his campaign for soil conservation resulted in his selection to be director of the Soil Erosion Service on September 19, 1933, Bennett had been a career soil scientist in the Department of Agriculture for thirty years. In April 1935 he became the first chief of the Soil Conservation Service. Shortly thereafter he became known as the "Father of Soil Conservation."

Bennett was born near Wadesboro in Anson County, North Carolina, the son of farmers William Osborne Bennett and Rosa May Hammond. He earned a degree in chemistry and geology from the University of North Carolina (UNC) in June 1903. At that time, the Bureau of Soils within the U.S. Department of Agriculture (USDA) had just begun to make county-based soil surveys, which would in time be regarded as an important American contribution to soil science. With USDA offering employment for soil scientists, Collier Cobb, professor of geology, had developed a course in soil surveying. Quite a number of UNC graduates had joined the Bureau of Soils. Bennett accepted a job in the bureau headquarters' laboratory in Washington, DC, but agreed first to assist on the soil survey of Davidson County, Tennessee, beginning on July 1, 1903. The acceptance of that task, in Bennett's words, "fixed my life's work in soils."

The outdoor work suited Bennett, and he mapped the soils and wrote a number of soil surveys. The 1905 survey of Louisa County, Virginia, in particular, profoundly affected Bennett. He had been directed to the county to investigate its reputation of declining crop yields. As he compared virgin, timbered sites to eroded fields, he became convinced that soil erosion was a problem not just for the individual farmer but also for rural economies. While this experience aroused his curiosity, it was, according to Bennett's recollection shortly before his death, Thomas C. Chamberlain's paper "Soil Wastage," presented in 1908 at the Governors' Conference in the White House (published in *Conference of Governors on Conservation of Natural Resources* [1909]), that "fixed my determination to pursue that subject to some possible point of counteraction."

In addition to supervising the soil surveys in the Atlantic Division, a position he assumed at the bureau in 1908, Bennett accepted numerous opportunities to study soils abroad and in U.S. territories. He made two surveys in Alaska, one a reconnaissance at the request of the Alaskan Railway Commission (1914) and the second at the request of the Forest Service for the purpose of eliminating agricultural lands from the Chugach Forest (1916). He made a survey of the agricultural possibilities of the Panama Canal Zone (1909); worked on the Guatemala-Honduras Boundary Commission (1919); and at the behest of the Department of Commerce, surveyed rubber-growing potential in Central America and northern South America (1923–1924). Most of these surveys appeared as USDA technical publications. The Tropical Plant Research Foundation published Bennett and Robert V. Allison's *The Soils of Cuba* (1928), which had been commissioned by the foundation.

Bennett wrote steadily and increasingly about soil erosion in the 1920s in an array of journals, from popular ones such as *North American Review* and *Country Gentleman* to scientific ones such as *Scientific Monthly* and the *Journal of Agricultural Research*. He was establishing himself as USDA's expert on soil erosion and was becoming recognized as such. His campaign received quite a boost when, effective New Year's day of 1928, Henry G. Knight, chief of the Bureau of Chemistry and Soils, placed Bennett in charge of a special study of the extent of soil erosion and methods of control. The travels and studies provided grist for Bennett's articles. Eventually he succeeded in arousing national attention where others had failed. Among his writings of the 1920s, probably none was more influential than a USDA bulletin coauthored with William Ridgely Chapline and titled *Soil Erosion: A National Menace* (1928). Bennett expressed the motivation for his later actions: "The writer, after 24 years spent in studying the soils of the United States, is of the opinion that soil erosion is the biggest problem confronting the farmers of the Nation over a tremendous part of its agricultural lands." The bulletin was not a manual on the methods of preventing soil erosion; rather, it was intended to draw attention "to the evils of this process of land wastage and to the need for increased practical information and research work relating to the problem."

Bennett followed up the attention from the bulletin and some well-placed magazine articles with a campaign for a national soil erosion program. He knew the few soil erosion researchers at the state agricultural experiment stations. Important as their investigations were, the experiments covered but a few spots on the vast agricultural landscape. In Bennett's mind, a national program of soil erosion was needed. Bennett's ally in cause, A. B. Connor of Texas Agricultural Experiment, enlisted the aid of Representative James Buchanan, who inserted a clause in the USDA appropriations bill for fiscal year 1929–1930 that authorized the soil experiment stations. (Eventually the stations would be renamed soil conservation experiment stations.) Bennett was disappointed that some of the funds were allotted to the Forest Service and the Bureau of Agricultural Engineering. Despite this disappointment, he sought out locations and cooperators for the stations that he would supervise from his new position in charge of the Bureau of Chemistry and Soils' soil erosion and moisture conservation investigations.

While Bennett set up the soil erosion experiment station, recruited the few known soil erosion specialists to staff them, and gave overall supervision to the stations, the country slid into an economic depression. In FDR's New Deal administration, Commissioner of Indian Affairs John Collier reversed national Indian policy from assimilation to one of cultural pluralism: Native cultures were to be respected and perpetuated. As part of the plan to preserve Navajo culture, Collier wanted to reverse erosion on the reservation. Collier's son John learned that Bennett was USDA's expert in soil erosion, and Bennett was appointed chairman of an Advisory Conservation Committee for the Navajo Reservation. At the conclusion of the committee's survey of the reservation, Collier and the Navajo Council accepted their recommendations. On July 17, Harold Ickes, acting in his capacity as federal emergency administrator of Public Works, allotted $5 million for soil erosion work. Collier had been favorably impressed by Bennett's interdisciplinary approach and heartily recommended him to Ickes to head up the soil erosion work. Collier assessed Bennett thus: "He sees the matter *steadily and whole,* and is not an engineering fanatic nor a reseeding and ecological fanatic nor an animal husbandry fanatic." Secretary Ickes heeded Collier's advice and both established the Soil Erosion Service and made Bennett its director on September 19, 1933.

Bennett charged the directors of the soil erosion experiment stations to select nearby areas where the conservation methods developed at the stations could be demonstrated to the farmers and field-tested on farms. The Soil Erosion Service furnished equipment, seed, seedlings, and assistance in planning the measures. Young men from nearby Civilian Conservation Corps (CCC) camps or workers hired by the Soil Erosion Service would do much of the work. A project staff might consist of an engineer, soil scientist, forester, economist, and biologist to work with farmers on rearranging the farm for conservation practices. Eventually the service developed the conservation farm plan, written cooperatively by the farmer and the service, to detail the needed work. The personal connection between the trained conservationist and the land user became the hallmark of agency activities.

The work found favor with farmers, and more requests came for demonstration projects from other areas of the country. As the Soil Erosion

Service (SES) became visible in the countryside, Secretary of Agriculture Henry A. Wallace argued that its work more properly belonged in USDA. President Roosevelt heard the arguments from Wallace and Ickes and even called Bennett for a consultation. He decided to transfer the SES to USDA—to be renamed the Soil Conservation Service (SCS)—and to support legislation to change its status from that of a temporary New Deal agency to one with continuing authorities. Having a national policy of soil conservation anchored in legislation had been one of Bennett's objectives. The congressional hearing provided one of the most memorable events in conservation history. In the mid-1930s it was not uncommon during the early spring for one of the dust storms from the Dust Bowl region to be swept up into the atmosphere and to be carried to the Atlantic seaboard. Bennett recounted the events five years after his testimony before the Senate Public Lands Committee on Public Law 46:

> The hearing was dragging a little. I think some of the Senators were sprinkling a few grains of salt on the tail of some of my astronomical figures relating to soil losses by erosion. At any rate, I recall wishing rather intensely, at the time, that the dust storm then reported on its way eastward would arrive. I had followed the progress of the big duster from its point of origin in northeastern New Mexico, on into the Ohio Valley, and had every reason to believe it would eventually reach Washington.
>
> It did—in sun-darkening proportions—and at about the right time—for the benefit of Public Forty-six.
>
> When it arrived, while the hearing was still on, we took a little time off the record, moved from the great mahogany table to the windows of the Senate Office Building for a look. Everything went nicely thereafter.

With passage of legislation giving the SCS a promising future, the work expanded rapidly. By mid-1936 there were 147 demonstration projects, forty-eight nurseries, twenty-three experiment stations, 454 CCC camps, and over 23,000 Works Progress Administration workers on the rolls.

As Bennett guided the young agency, his concepts proved wise. Various disciplines, not just one, would contribute to designing conservation methods and practices for the farm, and by the same token, no single effective conservation practice existed. Conservation farming meant rearranging the operations of the farm in the interest of conservation and productivity. Soil conservationists worked on the land, directly with farmers, to develop conservation farm plans for the benefit of the land and the farmer.

With help from specialists in SCS, Bennett published *Soil Conservation* (1939), a nearly 1,000-page treatise that encapsulated the state of soil conservation knowledge. Soil conservation assumed a place in the curricula of the land-grant college, aided, not inconsequently, by the prospect of employment in SCS. Bennett also published a college text, *Elements of Soil Conservation* (1947). Bennett pioneered in the use of soil survey data for soil conservation planning. In published soil surveys written early in his career, he identified soil types that should be left in trees or grass. In *The Soils and Agriculture of the Southern States*, he cited these soil types in recommending that some soils should be used for forest or pasture, less intensive uses than cropland. Bennett's belief—that susceptibility to erosion should be a guiding principle in farm planning—became systematized in the land capability classification developed by SCS staff. The first four classes in this classification were arable land; the limitations on their use and necessity of conservation measures and careful management increased from class I to IV. The criteria for placing a given area in a particular class involved the landscape location, the slope of the field, depth, texture, and reaction of the soil. The remaining four classes, V through VIII, were not to be used for cropland but might have had uses for pasture, range, woodland, grazing, wildlife, recreation, and aesthetic purposes. Within the broad classes were subclasses that signified special limitations such as erosion, excess wetness, problems in the rooting zone, and climatic limitations. SCS soil conservationists working with farmers used the system in developing conservation farm plans. The system greatly aided acceptance by farmers of soil conservation.

Bennett brought national attention to the soil erosion problem where others had failed. The works of Eugene Hilgard, Thomas C. Chamberlain, and Nathaniel Southgate Shaler had influenced the specialists but had not reached the general public as Bennett's magazine articles did. Bennett succeeded in seeing his soil conservation program institutionalized in a federal agency to carry on the work. Stephen Mather's work for a National Park Service and Gifford Pinchot's successful advocacy for a Forest Service come to mind as comparable in some regards. Unlike Mather and Pinchot, who were well connected socially and politically, Bennett was unique as the career civil servant. He had served a little over thirty years in USDA and in the army when he became director of Soil Erosion Service in 1933. Soil conservation became established as a profession and discipline, taught in the land-grant university system. During Bennett's time as chief of the Soil

Conservation Service, a professional organization, the Soil Conservation Society of America, was established. Bennett possessed the energy and single-mindedness of an evangelist in his promotion of soil conservation. His abilities as a writer and speaker were no small part of his success. He steadily wrote articles about soil conservation and was a welcome and inspiring speaker not only at farm-field demonstrations but also at scholarly gatherings. His crusading zeal brought many converts to soil conservation and made him the embodiment of the movement, the Father of Soil Conservation.

The recipient of several honorary degrees, Bennett was also president of the Association of American Geographers in 1943 and was awarded the Frances K. Hutchinson Award by the Garden Club of America in 1944; the Cullum Geographical Medal by the American Geographical Society in 1948; and the Distinguished Service Medal by the USDA and the Audubon Medal by the National Audubon Society, both in 1947. He was a fellow of the American Society of Agronomy, the American Geographical Society, the American Association for the Advancement of Science, and the Soil Conservation Society of America. On his retirement, the *Raleigh (NC) News and Observer* opined that Bennett might come to be "recognized as the most important North Carolinian of this generation."

J. DOUGLAS HELMS

Brink, Wellington. *Big Hugh: The Father of Soil Conservation.* New York, 1951.

Daniels, Jonathan. *Tar Heels: A Portrait of North Carolina.* New York, 1941.

Marbut, Curtis F., Hugh H. Bennett, J. E. Lapham, and M. H. Lapham. *Soils of the United States.* Washington, DC, 1913.

Morgan, Robert J. *Governing Soil Conservation: Thirty Years of the New Decentralization.* Baltimore, MD, 1965.

Berry, Thomas (November 9, 1914–). While maintaining an empirical approach to ecological problems that leads to a new definition of humankind's relationship to our planet, Berry vigorously rejects Cartesian dualism. At the same time, Berry is about reminding us that we live on a "magic planet," one that contains and controls our ultimate fate as much as we hold the planet's more immediate ecological fate in our hands. A Catholic priest, Berry redefines *spirituality* to mean an attitude that has at its core intimacy with the planet. A well-published historian of cultures, Berry sees humankind currently on the brink of ecological disaster—a disaster that can be salvaged only by a new approach to our understanding of how we interact with the natural world. He posits that we are in the "terminal phase" of the Cenozoic era and that the twenty-first century is the doorway to a new age, the age of the Ecozoic, which will be as dynamic and transformative of life on earth as when the dinosaurs were wiped out 65 million years ago, allowing for the ascendancy of mammals.

Berry was born in Greensboro, North Carolina, as William Nathan Berry, the third of thirteen children. Berry entered a monastery of the Passionist order in 1934 and was ordained a priest in 1942, at which time he chose the name "Thomas" after his spiritual and intellectual mentor, St. Thomas Aquinas. He wrote a doctoral dissertation on Giambattista Vico's philosophy, receiving his doctorate in history from Catholic University in Washington, DC, in 1948. He moved to China, studying the language and culture, but left after Mao Tse-tung rose to power. He served as a chaplain in the army in the early 1950s. He taught history at Seton Hall University (1957–1961) and St. John's University in New York (1961–1965), directed the graduate program in the history of religions at Fordham University (1966–1979), founded and directed the Riverdale Center of Religious Research in New York (1970–1995), and served as president of the American Teilhard de Chardin Association (1975–1987). He currently resides again in Greensboro, North Carolina, and continues to speak and write on important ecological and spiritual issues. He has written a number of books, including *The Historical Theory of Giambattista Vico* (1949), *Buddhism* (1968), and *The Religions of India* (1972). His most recent books, however, are what establish Berry as one of the most important nature mystics and ecologists now living and include *The Dream of the Earth* (1988), which won the Lannan Prize for nonfiction in 1995; *Befriending the Earth* (1991, with Thomas Clarke); *The Universe Story from the Primordial Flaring to the Ecozoic Era* (1992, with Brian Swimme); and *The Great Work: Our Way into the Future* (1999).

Berry writes that he had an epiphany when he was eleven, as he walked through a meadow near his house that he had never encountered before. At that moment, he apprehended "the natural world in its numinous presence," and the experience has remained with him as a lifelong touchstone: "This early experience . . . has become normative for me throughout the entire range of my thinking. Whatever preserves and enhances this meadow in the natural cycles of its transformation is good; whatever opposes this meadow or negates it is not good." The meadow is more than a metaphor; it is an actual physical place with a presence and a heft that convey significant and bolstering spirituality and friendship in Berry's schema. Such a presence requires protection, forethought, and compassion.

He has written poetry that encapsulates his perspectives. Many of his poems ask us to consider what the truly holy structure is, a building such as a church or our planet? In one poem, "Morningside Cathedral," Berry speaks of the sin of humankind as we continue to destroy our planet: "What sound, / What song, / What cry appropriate / What cry can bring a healing / When a million year rainfall / Can hardly wash away the life destroying stain?" His literary, religious, and mystical antecedents include Hildegard of Bingen, Richard of St. Victor, Meister Eckhart, St. John of the Cross, Black Elk, and Wang Yang-ming, whose "Questions on the Great Learning" (1527) speaks of the respectful and subtle approach ancient Chinese civilization took toward nature.

Intimacy with our planet is at the core of Berry's teachings. Just as we can be intimate with a loving human partner, and with a spiritual presence, so should our intimacy extend to our physical surroundings: "To appreciate our immediate situation we might also develop a new intimacy with the North American continent. For we need the guidance and support of this continent as we find our way into the future." Berry comes close to personifying the earth, but not quite. He recognizes that our current technological prowess distances us from the planet, but for our survival (and for the planet's), that must change: "We need only see that our human technologies are coherent with the ever-renewing technologies of the planet itself." The planet is not to be personified, but it must be viewed as a partner with humankind, and we must recognize that "[o]ur future destiny rests even more decisively on our capacity for intimacy in our human-Earth relations." Intimacy implies relationship and understanding (and presumes love), and it is an open channel of communication that goes both ways. Berry (along with most scientists in general) contends that our planet is speaking to us loudly and unequivocally and that if we continue along our path of exploitative and oil-based behavior, disaster looms.

Righteous moral indignation is evident throughout Berry's ecological works, a quality he shares along with another severe critic of human behavior—Jonathan Swift. Like Swift, Berry is a master of many rhetorical styles that carry his thought across multiple registers. Matthew Fox, the founder of Creation Spirituality, speaks about Berry's abilities and the almost Old Testament prophetic mode in which Berry often writes:

> I find in Thomas Berry and his passion for eco-justice and cosmic storytelling a true descendant (might we say reincarnation?) of the Celtic spiritual genius. His love of the Earth, his sense of humor, his gift of language, his poetic consciousness, his moral outrage, his primal appreciation of the aesthetic, his common sense, and his prophetic storytelling all point to the interconnectivity of nature and human nature, the sacred and daily life, the divine and the human and the more-than-human.

A large part of Berry's agenda is to reform the intellectual and cultural systems that make civilized life possible. His chief target of reform is the educational system, but he recognizes that the four "fundamental establishments that control the human realm[,] governments, corporations, universities, and religion" have all been structured so that what has been established is a "radical discontinuity between the human and other modes of being." The university receives the brunt of Berry's criticism. Instead of educating our youth to develop a sustained intimacy with the planet, we inculcate them in ways of perpetuating culture and civilization that lead inexorably toward ecological dysfunction: "In recent centuries the universities have supported an exploitation of the Earth by their teaching in the various professions, in the sciences, in engineering, law, education, and economics. Only in literature, poetry, music, art, and occasionally in religion and the biological sciences, has the natural world received the care it deserves."

Berry and his teachings have been the inspiration for a number of seminars, meetings, and movements. As part of this vast educational reform effort, in 1987 the first North American Conference on Christianity and Ecology was held in Indiana. In 2001, a seminar was held at Bellarmine University called EarthSpirit Rising, based on his teachings. Harvard has founded the World Religions Center that now sponsors a Forum on Religion and Ecology, directed by Mary Evelyn Tucker and John Grim. The rapper Drew Dellinger has founded an organization called the Center of the Universe to promulgate Berry's thought. The Society for Ecological Economics has been established by Herman Daly and Robert Costanza. It is clear that Berry's principles are percolating through society and may lead one day to a fuller reform of the educational system so that ecological literacy and planet intimacy become standard and valued aspects of the curriculum.

Related to reform is an acknowledgment of the social revolution that must occur if we are to reorient ourselves correctly toward the natural world. Currently, "we make everything referent to the human as the ultimate source of meaning and of value, although this way of thinking has led to catastrophe for ourselves as well as [to] a multitude of other beings." In a forceful statement reminiscent of Ralph Waldo Emerson, Berry states that we have taken all rights away from every presence on this planet and conferred them only to the human but that in reality "every being has rights to be recognized and revered. Trees have tree rights, insects have insect rights, rivers have river rights, mountains have mountain rights." Rights are not absolute licenses to exploit and desecrate but are reciprocal and relativistic; humans have rights only so far as they do not impinge on the rights of rivers.

In his latest book, Berry speaks of the Great Work that is before us. As a historian, he comments on the handful or so of sweeping cultural, spiritual, and economic transformations across all civilizations on this planet that have brought humankind to where we are today, the Great Works of the past 7,000 years. Now we have entered a precarious time, where we are tipping from one era into another. In spite of the great ecological damage the planet has suffered, Berry in the end remains optimistic. Human beings have it within their power to be part of the great change. In a truly religious sense whereby work can be redeeming, Berry says our "Great Work, now, as we move into a new millennium, is to carry out the transition from a period of human devastation of the Earth to a period when humans would be present to the planet in a mutually beneficial manner." The choice is ours to partner with our planet or to let our planet move in this direction without us. Recognizing spirituality and the power of the will in all things, Berry is confident that balance will be achieved in the end. His hope, prayer, and thought are that human beings journey with our earth and all the creatures on it as equal and intimate partners to regain that ecological balance together.

BRIAN U. ADLER

Center for Ecozoic Studies. Dir. Herman F. Greene. www.ecozoicstudies.org.

Collins, Daniel. "The Great Work: Recomposing Vocationalism and the Community College English Curriculum." *Beyond English Inc.: Curricular Reform in a Global Economy*, ed. David B. Downing et al. Portsmouth, NH, 2002.

Fox, Matthew. "A Profile of Thomas Berry, Scholar and Lover of the Earth." *EarthLight Magazine* 34 (1999): 23–28.

Jensen, Derrick. "The Universe Anew: An Interview with Thomas Berry." *Bloomsbury Review* 15 (1995): 5, 8.

O'Hara, Dennis P. "The Implications of Thomas Berry's Cosmology for an Understanding of the Spiritual Dimension of Human Health." Ph.D. diss., University of Toronto, 1998.

Thomas Berry: The Great Story. Prod. Nancy Stetson and Penny Morell. CD-ROM. Oley, PA: Bullfrog Films, 2002.

Berry, Wendell Erdman (August 5, 1934–). A self-declared "agrarian," Berry has been perhaps the strongest voice in twentieth-century America seeking the preservation of rural communities. He has lovingly depicted, and passionately defended, both the nonhumans and humans living in those communities. Berry is unusual in his concern for both of those groups—his concern not just for the wild or for agricultural production but also for domestic animals, traditions, soil, and rural people.

His poetry, fiction, and nonfiction are widely read in the environmental movement. He served as chair of his local Sierra Club chapter, has written for *Organic Gardening and Farming*, and has published many pieces in environmentalist periodicals. So even though he himself refuses to characterize himself as belonging to any movement, and is uneasy with the concept of an "environment," he has played an important role in modern American environmentalism. He is most important for his work as a farmer, as an advocate for farmers, and as someone committed to one *place.* As such, he has been the leading inspiration for the sustainable agriculture movement from the 1980s through the present.

Berry was born in Henry County, Kentucky, living in its county seat of New Castle for most of his childhood. He earned his bachelor's and master's degrees at the University of Kentucky. He met Tanya Amyx there, whom he married in 1957. In his stories written while in college, and in his first novel (*Nathan Coulter,* 1960), he depicted the rural life that he learned about from visiting family in the town of Port Royal, also in Henry County.

After graduating, Berry lived a life typical of an aspiring writer of that era—studying with Wallace Stegner for two years at Stanford, visiting Italy at length, and landing a job in the literary capital, New York City. During those years, he wrote poems on this variety of places, often influenced by the example of William Carlos Williams.

But then Berry made a decision that would change his life and define the meaning of his work. The path to literary success in New York was not the one he would follow. Instead, he moved back home. But not to New Castle—to Port Royal, home of farms that his family had worked for generations. He committed to that place, that land, and the people who lived there. Henry County, and his feelings about it, would dominate all of his future writings. In his 1960s works he explored that place and what it meant to him; in the 1970s he became an agricultural commentator, to defend the values of that county he believed were threatened.

His agrarian sympathies—love for soil and farm animals, for farmers and their skills—were evident as early as his college writings. It was in the mid-1960s when Berry first expressed concern in published writings about conservation issues. His 1966 essay "The Landscaping of Hell: Strip-Mine Morality" reflected his encounters with the damage done by strip mining in eastern Kentucky and presented the perspective through which he has viewed conservation matters ever since. He feared giving outsiders the power to manage the land in a locality, for they sought money, and believed in abstractions. Those outsiders did not know the place and would not live with (or learn from) the consequences of their actions.

Harry Caudill and the Appalachian Group to Save the Land and the People influenced Berry's views on mining and other matters as well: "to save the land and the people." Few influential Americans worked to protect *both* land and people. But Berry believed that local people, local ways of making a living, local animals and plants, and local soil were all threatened by the same forces. Only people committed to staying in a place, and developing local knowledge, could protect rural humans and nonhumans. By writing about his concern for one place, he suggested how others might attach to, and defend, their places as well.

His 1969 book of essays *The Long-Legged House* and his 1971 book *The Unforeseen Wilderness* read like fairly typical works of nature writing—an author viewing nearby nature and reflecting on it. But in his later work Berry less often discussed nondomesticated places, and the farming life became his main theme. In 1966 the Berrys planted their first garden on their Lanes Landing property in Port Royal. By the early 1970s, they were producing nearly everything they ate from their land. And Berry's written works reflected this immersion in farming, as he explored it in fiction, nonfiction, and poetry. Most directly, *Farming: A Hand Book* (1970) presented a narrator-farmer reflecting on his farm, manual labor, and the relationship of farm cycles to marriage and spirituality.

He wrote in *A Continuous Harmony* (1972, 1975) that "in the work is where my relation to this place comes alive. The real knowledge survives in the work, not in the memory. To love this place and hold out for its meanings and keep its memories, without undertaking any of its work, would be to falsify it." As that passage suggests, Berry grew very interested in manual labor in the early 1970s, for unlike most environmentalists, Berry worked to transform and harvest the land he sought to protect. He committed again to his place, this time not just to live in it but to *work* in it. And he suggested that knowledge is stored tacitly in farmwork; for decades Berry has championed farmer skills and knowledge against those who see farmers as "hicks" and those who believe that science—universalist and abstract—is a better guide for using land than the experience of those who live and work on it.

Most involved in agribusiness during the 1970s believed that new developments in science and technology had developed more effective uses of farmland and would continue to do so. In practice, that meant experienced farmers were forced off the land, in a wave of farm consolidations that left only one-third as many people farming in 1970 as were farming in 1940. New chemicals and new machines were at the forefront of these trends. Berry responded by exploring the possibilities of older methods and tools, particularly in pieces written for periodicals published by antichemical Rodale Press, drawing on the ideas of Sir Albert Howard. Berry's calls for appropriate technology came as a "small is beautiful" impulse, and this catchphrase and all it inferred became popular with environmentalists, particularly those who moved "back to the land."

Berry wrote his most famous book, *The Unsettling of America: Culture and Agriculture* (1977), as part of this exploration of farming issues. There he attacked scientists who envisioned a future agriculture that would rely on technology and the ability to control nature, and he found particular fault with the work of Secretary of Agriculture Earl Butz. He found hope in what he called the "margins," which included traditional farming, organic methods, draft horses, and the Amish. Perhaps no other text has done more to interest environmentalists in thinking about how to *work* in nature. (Indeed, in that Sierra Club–published book, he claimed that the club misguidedly tried to deal with preservation by isolating it from other land-use and cultural issues.)

In that sense, Berry's work might be seen as part of a conservationist tradition of wise use, rather than a movement that (he feared) believed in an "environment" separate from humans. But it makes yet more sense to consider him an *agrarian*, his favored term, as he often builds on the ideas of Thomas Jefferson, Liberty Hyde Bailey, and the Southern Agrarian authors of *I'll Take My Stand* (1980). Like them, even when he wrote about nature, it was often part of a discussion of issues involving human politics and human communities.

Community—or more specifically, rural community—became the chief focus of his work in the 1980s. He concluded that the best way to preserve the values, lifestyles, and species he cared most about was to preserve such communities and to celebrate the idea of a local economy. Only such communities could nurture the skills, systems of knowledge transmission, and patterns of landownership that he felt were necessary to have humans care properly for creation.

In his 1980s work and onward, Berry has been less concerned with new explorations of nature and agriculture and more interested to develop a fuller vision of the kind of culture that was, and could be, sustainable. In his "Sabbaths" poems, he presented a vision of spirituality, taken from days relaxing and reflecting in the fields of his home. His fiction continued to explore what a good rural life could be like, in one place, always exploring a fictionalized version of the town he lived in. This functioned as a rare model, in this era, of the positives of farming life. In his earlier work, Berry had at times noted the shortcomings of the rural society he grew up in, but in his later work, he tended to focus on celebrating its charms and menacingly depicting the forces that threatened it.

Wes Jackson became a good friend of, and the closest intellectual collaborator of, Berry's in the 1980s and onward. To Berry, Jackson's agricultural research at the Land Institute, particularly the perennial polyculture work, was a model of how appropriate science that used "nature as model" would function. The organization's periodical the *Land Report* became the closest thing to an activist community promoting Berry's ideas that he had ever been a part of, often promoting agrarianism.

The work of many authors who write about place and about sustainable agriculture today, including Brian Donahue, Eric Freyfogle, Frederick Kirschenmann, David Kline, Michael Pollan, and Scott Russell Sanders, often builds on Berry's work and sometimes was directly inspired by him. Collections of contemporary agrarian writing, such as *The New Agrarianism*

(2001) and *The Essential Agrarian Reader* (2003), pay tribute to Berry as the leading contemporary American agrarian.

Berry reached a new audience in the new century due to his writings on war. His essay "Thoughts in the Presence of Fear," written in response to the political atmosphere after the terrorist attacks of September 11, 2001, was widely reprinted. As he had since the 1960s, and as he would do during the American occupation of Iraq, Berry held to a strongly pacifist position. Wars, he suggested, allowed large organizations to make decisions (not local communities) and encouraged the dangers of abstract thought that damaged both human and nonhuman life.

Characteristically for Berry, he argued that developing an economic system based on local self-sufficiency was the best way to deal with all the country's dilemmas. Berry's thought has been marked by consistency and holism, for he believes that no problem can be dealt with in isolation. The fates of culture and agriculture, wilderness and farms, peace and soil fertility, the handing down of farming wisdom, and the health of humans are all inextricable. Similarly, his farming and his writing combine to produce his legacy: a model of a life devoted to sustaining ideas, nature, farmers, and one particular farm.

JEFF FILIPIAK

Angyal, Andrew J. *Wendell Berry.* NY, 1995.
Freyfogle, Eric T., ed. *The New Agrarianism: Land, Culture, and the Community of Life.* Washington, DC, 2001.
Merchant, Paul, ed. *Wendell Berry.* Lewiston, ID, 1991.
Peters, Jason, ed. *Wendell Berry: Life and Work.* Lexington, KY, 2007.
Smith, Kimberly K. *Wendell Berry and the Agrarian Tradition.* Lawrence, KS, 2003.
Wirzba, Norman, ed. *The Essential Agrarian Reader.* Lexington, KY, 2003.

Beston, Henry (June 1, 1888–April 15, 1968). Born Henry Beston Sheahan into an upper-middle-class Catholic family in Quincy, Massachusetts, Beston was a writer-naturalist (his name for his principal vocation), World War I ambulance driver and *Atlantic Monthly* war correspondent covering the submarine service (when America entered the war), editor, historian, children's story writer, and farmer of herbs and grapes. His father, of Irish descent, was a physician. His French mother made certain that he would become bilingual, a fact that may have helped him develop a distinctive poetic prose style in his later writings about nature. As a boy, he spent much of his time exploring the seacoast to the east and the granite hills to the west of his hometown, living, he wrote, "a New England boyhood of sea and shore." But "Nature" at Quincy Bay, he wrote to his wife of his summer vacations, "was in a suburban mood there; she went yachting in a catboat on Saturday afternoons and arranged her tides as if they had to take trains. There was no poetry." It was while staying in Sainte-Catherine-sous-Riviere in the Mont Lyonnais near Lyons, where he spent a post-Harvard year teaching English at the university, that he first "encountered and knew and loved the earth."

Beston's education fit a pattern of that of other well-off young men of his generation: the Adams Academy and then Harvard (B.A. in 1909 and M.A. in 1911). In 1916 he volunteered for the American Field Service in France and, in the same year, published *A Volunteer Poilu* under the name of Sheahan. By 1923, he had published under the name of Henry B. Beston a second book dealing with his war experiences and two books of fairy tales. Thereafter, beginning with the publication of a collection of historical adventures and a children's story of American Indians in 1925 and 1926, respectively, he dropped the meaningless middle initial.

It was not until 1928, when he was forty, that he published *The Outermost House: A Year of Life on the Great Beach of Cape Cod*, the book that has since been recognized as an American classic in the nature writing genre. His diversity of interests in naturalist writing is exhibited in the titles of later and lesser-known works, including *Herbs and the Earth* (1935); *The St. Lawrence* (1942), a volume in the widely read Rivers of America Series; and *Northern Farm* (1948), a seasonal chronicle, first published weekly in Robert La Follette's *Progressive*, detailing life at "Chimney Farm" near Nobleboro, Maine, where he and his wife, the noted children's story author Elizabeth Coatsworth Beston, made their family home. (*Chimney Farm Bedtime Stories* [1996] is an example of the collaborative talent Beston and his wife exercised to bring an appreciation of nature to the very young. She began as early as 1938 to record for publication in the *Christian Science Monitor* these imaginative tales that he had improvised for their little girls about events and animals around the farm.)

Especially Maine: The Natural World of Henry Beston from Cape Cod to the St. Lawrence (1970), collected and introduced by his wife, provides a useful way into his life and writings about nature through her memories of his personality and work habits and through a generous selection from his notes, letters, poems, books, and anthologies. These latter include *American Memory: Being a Mirror of the Stirring and Picturesque Past of Americans and the American Nation* (1937), accounts from "nonliterary" documents revealing "the American as a man of his own earth"; and *White Pine and Blue Water: A State of Maine Reader* (1950).

Following World War I, Beston edited *The Living Age* and wrote and lectured widely. Still, in spite of his earlier publications, *The Outermost House* "seems to have happened, full-grown," as Winfield Townley Scott has observed. At the age of thirty-eight, Beston bought over fifty acres of dunes near Eastham on the outermost beach of Cape Cod and had builders erect a twenty-by-sixteen-feet house replete with ten windows. He called it the Fo'castle because of its precarious perch on the edge of the ocean. In the fall of the next year, 1926, from within and without this shelter, Beston began to record nature's visible changes over four full seasons, noting such events as bird migrations in September; the awesome violence of winter storms; the welcome reoccurrences of the tides and stars and sun-filled days; the behavior of fish migrating and mating, feasting and being feasted upon; and the immense universe of summer insects. All the while he included his observations of the behavior of humans within this world that offered a continual challenge to human life. He included himself, a coast guard crew who were his neighbors, shipwrecked sailors, the citizens of Eastham, and in a remarkable passage, a solitary bather emerging naked from a late summer sea.

In his "Foreword to the 1949 Edition," Beston explained why he was careful to observe why human beings should learn to function in a natural environment that so often appears adverse in its wildness:

> Nature is a part of our humanity, and without some awareness and experience of that divine mystery man ceases to be man. When the Pleiades and the wind in the grass are no longer a part of the human spirit, a part of very flesh and bone, man becomes, as it were, a kind of cosmic outlaw, having neither the completeness and integrity of the animal nor the birthright of a true humanity.

When he left the Cape in the late summer of 1927, Beston had his notes but no book. The story is told that Elizabeth Coatsworth, then his fiancé, refused to set the date for their marriage until the book was finished. Although the story may be apocryphal, it suggests what she understood of his propensity to lose himself in the intensity both of observation and of crafting his prose. After his death, she remembered how he worked both mentally and physically in his garden: "At long intervals he might crumble a piece of earth between his fingers, or pull up a weed. But mostly he was just staring and staring. When he came in, he would say, 'I've been working in the herb garden all morning.'" He worked, like his predecessor Henry David Thoreau, "as deliberately as nature." Suffice it to say that he and Elizabeth married the year following publication of the book.

For the next forty years, Beston continued to concentrate on nature and writing, living much of that time at his Maine farm to which he moved in 1932. In 1960 he donated the Fo'castle and the fifty-six acres of dune property on which it stood to the Massachusetts Audubon Society, and in 1964 the cottage became a National Literary Landmark. In 1978, the Fo'castle, twice moved to protect it from the encroaching surf, was destroyed by a raging winter storm, much like one Beston described in *The Outermost House.*

Beston's influence on writers and naturalists is wide-ranging. Government officials have said that *The Outermost House* helped lead to the establishment of the Cape Cod National Seashore. Although Beston was not a trained scientist, his work continues to be cited by naturalists, primarily because of the accuracy and comprehensiveness of his observations of flora and fauna. But in the long run he may be best remembered for what he says about the relationship between man and nature in a lucid, aphoristic style that characterizes all of his writing. Although he is not known for a specific environmental philosophy (as Aldo Leopold is for the "land ethic," for example), his insistence on reconstituting the imbalance between the manfactured world and the natural world led Rachel Carson to think of him as her signal inspiration. He describes a world "sick to its thin blood for lack of elemental things, for fire before the hands, for water welling from the earth, for air, for the dear earth itself underfoot." But his particular vision is singularly positive, for he insists that given determination and willingness, humankind still has the capacity and the opportunity to recapture "elemental things." He writes, "Living in nature keeps the senses keen, and living alone stirs in them a certain watchfulness" and "To be able to see and study undisturbed the processes of nature . . . is an opportunity for which man might well feel reverent gratitude." Placing oneself in that position, he says, "One may stand at the breakers' edge and study a whole world in one's hand."

Beston received much recognition during his lifetime. He was made the honorary editor of *Audubon Magazine*, and in 1960, for distinguished achievement in literature, he was awarded the Emerson-Thoreau Medal of the American Academy of Arts, an honor previously accorded only to Robert Frost and T. S. Eliot. He became a fellow of the American Academy of Arts and Sciences, and he received honorary doctorates from the University of Maine and Bowdoin. He often lectured on nature writing, primarily at Dartmouth College, and he remained active in the Audubon Society, World War I veterans' groups, and the Grange. He was also a member of the Authors' Club of London. Beston and Elizabeth had two daughters, Catherine, who married Richard Barnes, and Margaret, who married Dorik Mechau.

JAMES BALLOWE

Beston, Elizabeth Coatsworth. *Especially Maine: The Natural World of Henry Beston.* Brattleboro, VT, 1970.

Finch, Robert. Introduction to *The Outermost House*, by Henry Beston. New York, 1988.

Paul, Sherman. "Coming Home to the World: Another Journal for Henry Beston." *For Love of the World: Essays on Nature Writers.* Iowa City, IA, 1992.

Scott, Winfield Townley. "A Journal for Henry Beston." *Exiles and Fabrications.* Garden City, NY, 1961.

Braun, E. Lucy (April 19, 1889–March 5, 1971). A pioneer ecologist of the twentieth century, Braun helped ecology become a recognized scientific discipline. A professor of botany who mentored many ecologists, she is best known for, and took the greatest pride in, her 1950 book *Deciduous Forests of Eastern North America*, a reference on forest ecology. She is less famed but equally notable for her work on the prairies of the unglaciated limestone region of southern Ohio. Braun's efforts led to the establishment of nature preserves in Ohio.

Braun, born in the Walnut Hills suburb of Cincinnati, was the youngest of two daughters of a school principal and schoolteacher. Annette Braun, older of the sisters by five years, later became a respected microlepidopterist and the first woman to earn a doctoral degree at the University of Cincinnati in 1911. The two women, who never married, lived together throughout their lives, with Annette assisting Lucy on field trips.

Braun grew up in a family that spent time examining wildlife. Her progression to botanist came as simply an extension of her family's interest in the natural world. Braun's interest in ecology likely had the same roots. As a child, she joined her parents on trips to the woods, where they identified wildflowers. Her mother, especially interested in botany, compiled a small herbarium. Braun began collecting and pressing plants during her high school years.

Braun was educated in the Cincinnati schools before receiving her undergraduate degree at the University of Cincinnati in 1910. She earned an M.A. in geology in 1912 and a Ph.D. in botany in 1914, both from the University of Cincinnati. Beginning in 1914, Braun spent three years as an assistant in botany. She served as an instructor from 1917 to 1923 and became an assistant professor in 1923. Braun taught world botany, the vegetation types in the United States, and plant succession, using the communities in the Cincinnati region as a laboratory. She illustrated her lectures with slides that she had made and encouraged her students to consider land conservation. Braun earned tenure in 1927 and became a full professor of plant ecology in 1946. She became emerita professor in 1948 and continued in this status until her death. Braun apparently did not enjoy teaching as much as she liked research. Early retirement gave her the time to conduct fieldwork and to publish three books and several major papers.

Braun's professional achievements are all the more notable for coming in an era when female scientists garnered little professional respect. Perhaps in an effort to compensate for the perceived softness of women, Braun became known for a blunt, tactless style. Nevertheless, Braun still had a wide circle of friends. Known and admired by botanists in her field, she carried on an active correspondence, often feuding with them if they opposed her views. Braun did not take challenges to her views especially well.

After leaving Walnut Hills in the 1940s, the sisters spent the rest of their lives on a two-acre lot at 5956 Salem Road in the Cincinnati suburb of Mount Washington. The Braun home and garden, surrounded by mostly undisturbed woods, contained a "scientific wing" that served as a laboratory. The garden contained rare and unusual plants that were grown for closer observation and study such as box huckleberry (*Gaylussacia brachycera*), Cumberland azalea (*Rhododendron cumberlandense*), nodding mandarin (*Disporum maculatum*), Allegheny-spurge (*Pachysandra procumbens*), and stoneroot (*Micheliella verticillata*). Braun had a herbarium on the second floor that housed 11,891 specimens shortly before her death, when she donated the collection to the Smithsonian Institution.

Starting in 1915, Braun made thirteen research trips to the western United States, five of them to the Pacific Coast. From 1934 to 1963, she drove her own car, often on roads that barely qualified for the word. She did this research at a time when few women traveled alone and, in the 1930s and 1940s, when few amenities existed for travelers. On all of these trips, Braun made observations on the vegetation and prepared herbarium specimens. These observations provided information for some of her popular articles in the magazine *Wild Flower.* However, most of Braun's fieldwork supporting her scientific publications was concentrated in the eastern deciduous forest, primarily in southwestern Ohio and the mountains of eastern Kentucky.

Braun was an original thinker in the fields of plant ecology, vascular plant taxonomy, plant geography, and conservation. She helped develop the field of plant ecology, particularly forest ecology. Among her 180 publications are works on the physiographic ecology of the Cincinnati region (1916), the vegetation of conglomerate rocks of the Cincinnati region (1917), the vegetation of the Mineral Springs region of Adams County (1928), the forests of the Illinoian drift plants of southwestern Ohio (1932), the undifferentiated deciduous forest climax and the association-segregate (1935), an ecological transect of Black Mountain, Kentucky (1940), and forests of the Cumberland Mountains (1942).

Braun's landmark work *Deciduous Forests of Eastern North America* stressed the importance of geologic processes in molding the development of the forests of the unglaciated regions. She contended that the forests remained exceedingly stable over millions of years. Critics argued that conditions of soil and climate determined the distribution of forest types, but Braun's work became the basic reference on the condition of the continent before the arrival of humans. She developed the theory that the southern Appalachians were the refuge during the Pleistocene era and the dispersal of the deciduous forest communities. To support this claim, Braun cited evidence from the pollen and macrofossil record of the eastern United States. Most of the plant species in this area could withstand temperature extremes far greater than those present during the twentieth century. She argued that the deciduous forest zone, although narrowed, maintained itself on the Appalachian Plateaus in southern Ohio and Kentucky, while glaciers extended southward in Ohio.

Braun also made major contributions to the study of the flora of Ohio. In 1926, she joined Lynds Jones in publishing *The Naturalist's Guide to the*

Americas. The book represented the first attempt at an inventory of Ohio's natural areas. In her paper "The Lea Herbarium and the Flora of Cincinnati," she compared the flora of 1920s and 1930s Cincinnati with the flora found in the city a hundred years earlier by Thomas G. Lea. In 1951, Braun organized an Ohio Flora Committee with the Ohio Academy of Science to prepare a comprehensive book on the vascular flora of the state. The project resulted in *The Woody Plants of Ohio* (1961) and *The Monocotyledoneae of Ohio* (1967). Throughout her career, Braun produced 180 publications, which appeared as four books and as articles in twenty scientific and popular journals.

Braun's ecological studies led her to advocate for the protection of the midwestern prairies and forests. She argued that no fire should ever touch the prairie patches in Adams County, Ohio, because she believed that the rocky soil was too shallow to withstand burning. Appointed in 1927 as the representative from the Ohio Academy of Science to the state of Ohio to formulate regulations to preserve the biological status of the public forests, she had the clout to make her views into law. Her interest and knowledge inspired the creation of the nature preserve system in Adams County, beginning in 1959 with 42 acres for Lynx Prairie. Buzzardroost Rock, a 152-acre unglaciated dolomite promontory in the county, was acquired in 1960 by the Ohio Chapter of Nature Conservancy at Braun's urging. Both Lynx Prairie and Buzzardroost Rock were designated national landmarks in 1967. The nature reserve system has subsequently grown to over 3,000 acres.

Braun received a number of professional honors. She served as the first female president of the Ecological Society of America. The E. Lucy Braun Award for Excellence in Ecology is presented annually by the Ecological Society of America in recognition of her achievements. In 1971, she became the first woman to be inducted into the Ohio Conservation Hall of Fame.

Braun died in March 1971 at eighty-two, following several months of illness. Her papers are housed at the Cincinnati Museum Center.

CARYN E. NEUMANN

Stuckey, Ronald L. *E. Lucy Braun (1889–1971): Ohio's Foremost Woman Botanist: Her Studies of Prairies and Their Phytogeographical Relationships: An Anthology of Papers.* Columbus, OH, 2001.

Brewster, William (July 5, 1851–July 11, 1919). Born in Wakefield, Massachusetts, Brewster took hunting trips with his father into the fields and woods that then still surrounded the town of Cambridge, kindling his love of nature. A copy of James Audubon's *Birds of America* was in the family library, and from an early age, Brewster amassed a collection of sample eggs and nests from the local species, augmenting them with stuffed specimens as well. Graduating from high school in 1869, a problem with his eyes prevented him from pursuing further study at Harvard. Accepting a clerkship in his father's bank in Boston, he continued to expand his knowledge of ornithology. His passion for birdlife grew into a calling that he attempted to institutionalize through the foundation of the Nuttall Ornithological Club in 1873 (the only forum for serious debate on the condition and study of American birds prior to the founding of the American Ornithological Union in 1883), which he served as president for over forty years. Following his marriage to Caroline Kettell of Boston on February 9, 1878, he left the business world to accept a position as assistant in charge of the collection of birds and mammals for the Boston Society of Natural History, where he remained until 1887, leaving for a similar post at the Cambridge Museum of Comparative Zoology. It was while he was with the Boston Society that his first research monograph, *Descriptions of the First Plumage of Various Species of North American Birds* (1879), appeared. His writings, eventually totaling more than 300 articles for scientific and popular periodicals, continued in 1902 with *Birds of the Cape Region of Lower California.*

Speaking at the second annual meeting of the union in 1883, he decried the damage being done to nongame species of American birds by the millinery industry, and he became the vital spirit behind the Committee on the Protection of North American Birds, the first formal effort made toward the conservation of domestic birdlife. In this capacity, he served as one of the organizers of the first Audubon Society and was a member of a committee of five appointed by the union to revise the system of nomenclature and classification used for North American birds. On the state level, he served as a president of the Massachusetts Audubon Society and on the board of the Fish and Game Commission, later chairing it from 1906 to 1908. His organizing talents were also applied to the creation of the American Game Protective and Propagation Association in 1911, which later became the North American Wildlife Foundation.

In 1891, he purchased a tract of land on the Concord River, beginning a series of acquisitions that would eventually total some 300 acres. Known as October Farm, this haven would remain his home until his death, as well as offering him a site for the study of regional birds and their habits. Diaries of birds observed on his property and in the Concord region were posthumously published in 1937 under the title *October Farm.* The data from these journals, combined with other observations, fueled some ten years of writing, culminating in the massive 1906 volume *The Birds of the Cambridge Region of Massachusetts.* Formal recognition of his place in American ornithology came with the awarding of honorary degrees from Amherst in 1880 and Harvard in 1899 and appointment as curator of the Cambridge Museum in 1900. His last years were spent at October Farm, writing and tending the extensive private museum of ornithology he had established. He died there in his sixthy-eighth year.

His significance for American ornithology lies in his leadership in creating the first lobbying group to consider protection of domestic birdlife a serious part of more general conservation issues and his promotion of the cause of ornithology in New England. Later students of changing regional ecology would find his collection of specimens invaluable. With a clear expository style of writing, his observations added much to the rising literature on American birds while making the topic accessible.

ROBERT B. MARKS RIDINGER

"Brewster, William." *Appleton's Cyclopedia.* New York, 1895.
"Brewster, William." *National Cyclopaedia of American Biography.* 1932.

Bromfield, Louis (December 27, 1896–March 18, 1956). An author and a farmer, Bromfield was born in the industrial town of Mansfield, Ohio. His father, Charles Bromfield, moved from the family farm to the town to earn sufficient income to support himself and his family. His mother, Annette Coulter Bromfield, disgusted with the values of the townspeople and disillusioned with life on the farm, encouraged her children to pursue artistic careers. The Bromfields were descendants of Jeffersonian agrarians who converted the wilderness of northern Ohio into farmland early in the nineteenth century.

In 1914 Bromfield enrolled at Cornell University to study agriculture following the return of the Bromfields to the Coulter farm near Mansfield but withdrew after one semester to help with the farm responsibilities when his grandfather injured himself in a fall. Unhappy with farm life, Bromfield entered Columbia University to study journalism in 1916 but again withdrew from school before completing a year's work, this time to serve as driver and interpreter in the American Ambulance Service attached to the French army during World War I. Following the war, he returned to New York to work for newspapers and publishing houses. In 1924 Bromfield published *The Green Bay Tree*, his first novel, and in 1925 he went to France for an extended visit of fourteen years. In 1939 he returned to the United States, bought three unproductive farms in the vicinity of Mansfield, and combined them into one farm, which he called Malabar. He devoted the rest of his life to restoring fertility to the soil at Malabar and making the farm productive and self-sufficient.

At his death, Bromfield was the author of thirty books and numerous shorter works. His novel *Early Autumn*, published in 1926, won the Pulitzer Prize. *The Rains Came*, published in 1937, was made into a Hollywood film. Bromfield's real commitment, however, was always to the land. A Jeffersonian agrarian democrat for most of his life, he was seldom out of touch with the land, even to the point of achieving gardening honors while he was living in France. When he returned to Ohio in 1939, he focused most of his writing on the relationship of people to the land. In *The Farm* (1933), a somewhat autobiographical story of farm life in Ohio from 1815 to 1914, he criticized both the system and the people responsible for the degradation of the land. In his 1945 book *Pleasant Valley*, he continued his criticism of American ways but presented his experiment at Malabar Farm as a viable alternative to exploitative uses of the land. In his 1948 book *Malabar Farm*, he reported on his accomplishments at Malabar at a more technical level, and in *Out of the Earth*, which appeared in 1950, he produced a work that was practically an agricultural textbook. *Animals and Other People*, published in 1955, is further testimony to the close, personal relationship that Bromfield had with the earth and its inhabitants.

Bromfield attributed his love for the land to his family, especially to Grandpa Coulter, who scarcely let a day pass without walking over his fields and through his orchards and woods and trying "some new if only small experiment." Bromfield felt that he had inherited his grandfather's "love of fields and cattle and forest and stream," that his grandfather was in his "flesh and bones." For Bromfield, education was important, but it was no substitute for a "passionate feeling for the soil" and a "sympathy for animals." This bonding with the earth Bromfield called being "teched." A teched farmer works not for profit but for pleasure. The success and profit that follow are "merely incidental."

Conservation of America's farmland was Bromfield's primary concern. Bromfield wanted to change the way Americans treat the land. He believed that Americans were unaware of the important role that land played in the national economy, that they often had no respect, no reverence for nature, that they were a wasteful and greedy people. He was particularly concerned about the loss of topsoil. Some of the changes in farming that he proposed were designed to check soil erosion. Cover crops and mulch, he believed, would help to anchor the soil, and he advocated the use of both. He also opposed the indiscriminate use of the moldboard plow. Turning the earth put the mulch below the surface of the ground and created a barrier between topsoil and subsoil. He deplored exclusive dependence on chemical fertilizer to make the land productive and the tendency to plant the same crop on the same land year after year. He promoted the planting of legumes and grass to restore humus to the soil. He believed that some land should not be cultivated but left in its natural state, that farmland should provide habitat for wildlife, and that good farming included good forest management. Above all, he believed that the motivation for good farming practices should be inspiration, not regulation, and that management decisions should be natural, practical, and economically feasible.

Many of the changes in agriculture that Bromfield advocated have taken place. To what extent he can be given credit for these changes is uncertain. He was not the only person in his day advocating changes. But he had an audience. His books were popular; his articles appeared in leading journals; he had a radio program. He also had Malabar Farm, an exhibit that people could see and touch. Today that farm is a state park. As such, it is a monument to a man who was as much a farmer as a novelist, one who will probably be remembered primarily as a pioneer in soil conservation.

LYNN DICKERSON

Anderson, David D. *Louis Bromfield.* New York, 1964.

Brown, Morrison. *Louis Bromfield and His Books: An Evaluation.* Fair Lawn, NJ, 1957.

Geld, Ellen Bromfield. *The Heritage: A Daughter's Memories of Louis Bromfield.* New York, 1962.

Brower, David (July 1, 1912–November 5, 2000). The *New York Times* once called Brower the most effective conservation activist in the world. The *Los Angeles Times* referred to him as the person who had accomplished more through the force of will than anyone since John Muir. In many other sources, it was noted that he was a writer, editor, and lobbyist for environmental causes. He held positions as the first executive director of the Sierra Club (1952–1969), founder of Friends of the Earth (FOE; 1966), founder of Earth Island Institute (1982), cofounder with Marion Edey of the League of Conservation Voters (1969), and originator or collaborator in a long list of conferences and forums dedicated to saving the earth.

Brower was nominated three times for a Nobel Prize, in 1978, 1979, and 1998 (with Paul Erlich), for his efforts on behalf of environmentalism. An environmental prophet who always seemed one step ahead of his opponents, of which he had many, he was often uncompromising in his views. He spared no individual or organization, no matter how prominent; yet he was as truthful about the failings of organizations he led as about those in the highest reaches of the federal government.

In an effort to save the wilderness and park systems, government lands, and private places from what he considered mindless commercial incursion, Brower often found himself in ongoing struggles to protect what remained of the wilderness in the United States. Eventually he extended his message around the world. Brower's obituary in the New York Times quoted him saying "We're not blindly opposed to progress, we're oppossed to blind progress." The same obituary also noted Russell Train's observation, "Thank God for David Brower. He makes it so easy for the rest of us to be reasonable."

The foundation for his environmental perspective began in the Berkeley Hills area of California, where he was born the son of Ross J. Brower and Mary Grace Barlow. The ancestral family (Brouwer) had arrived in America in the seventeenth century and lived in modest means. Brower's father often piled his family into a 1916 Maxwell and took them into Yosemite and Sierra Mountains on camping trips. The outings became a major influence on his eventual absorption with mountains and the wilderness. When he was twelve, he took up butterfly collecting, which he later observed served as an introduction to his understanding of ecosystems. At sixteen he graduated from Berkeley High School and planned to study entomology. He took courses in forestry, botany, and chemistry but abandoned an academic career. Later in life he was heard to say that he was "a graduate of the

University of the Colorado River." By the time his lengthy service in the cause of environmentalism ended, he was the recipient of nine honorary degrees.

Brower's formal entry into the arena that would eventually form his life and career began in the summer of 1930 when he was hired by Echo Lake Camp in Yosemite, first as a clerk and then as a leader of mountaineering trips, an activity at which he became expert. Captivated by mountaineering, his winters and summers were spent conquering new peaks and honing his skills—the latter contributed to by his membership in the Sierra Club in 1933. (Eventually he would be credited with seventy first ascents in the Sierras and the western United States, including leading the first successful ascent of New Mexico's Shiprock in 1939.)

His Sierra Club activities also led to writing for the club's bulletin and learning the mechanics of publishing. Later he credited the twin activities—mountaineering and publishing—as two of the most important influences of his life. By 1935 his new skills helped him obtain a job as public relations director for the Yosemite Park and Curry Company. He wrote articles and brochures, learned how to use a movie camera, and was eventually offered a position as an editor for the University of California Press. It was there that he met his future wife and fellow editor, Anne Hus. In 1939, Sierra Club friends found paid part-time work for him in the club office. He wrote, mapped, and edited Sierra publications, and in 1941 he became a member of the board.

With the advent of World War II Brower's civilian life was over for the duration. He enlisted in the army in 1942 with the U.S. Mountain Troops and, having recently edited the Sierra Club's *Manual of Ski Moutaineering*, became an instructor in the U.S. 10th Mountain Division, serving as a combat intelligence officer in Italy until the end of the war. Having entered as a private, he retired as a major, had been awarded a Bronze Star, and had had a hand in teaching 10,000 young men the mountaineering skills necessary to their training. He returned to both the editorial desk of the Berkeley Press and his work with the Sierra Club, "with wilderness at the top of my list." He had seen what war and industrialization had done to some of Europe's pristine areas and was determined to keep the U.S. wilderness from a similar fate.

With his hiring in 1952 to the Sierra Club's executive directorship, Brower's advocacy for the earth and its threatened natural treasures took on new power and new passion. Under his guidance, the membership of the club grew from 2,000 to 77,000 and became an important political force. (Its membership today is more than 750,000.) From 1953 to 1955, Brower testified before Congress numerous times to block building of two dams in Dinosaur National Monument. When the dam proposals failed, Brower and the club had their first big victory. John McPhee viewed the Dinosaur victory as "the birth of the modern conservation movement—the turning point at which conservation became something more than contour plowing."

The successful effort on behalf of Dinosaur National Monument was only the beginning of Brower's role as prime mover on behalf of America's wild lands. He successfully led campaigns that resulted in the establishment of Kings Canyon National Park (1953), Olympic National Park (1962), Point Reyes National Seashore (1963), Fire Island National Seashore (1964), Cape Cod National Seashore (1964), Alagash Wilderness Waterway in Maine (1967), Redwoods National Park (1968), Northern Cascades Wilderness (1968), Red River Gorge in Kentucky (1976), and Great Basin National Park in Nevada (1986). After years of lobbying for it, he also witnessed the signing of the National Wilderness Preservation Act in 1964. The National Environmental Policy Act of 1969 followed.

Whenever and wherever possible, Brower spoke, wrote, and initiated conferences. He was keynote speaker for hundreds of events and gave talks on every continent except Antarctica. "I never," he wrote in his autobiography, "missed an opportunity to preach. More and more I felt the sermons to be necessary to slow the loss of more and more wilderness." His was a voice congressional leaders heard often and came to respect. In his innovative work with the Sierra Club, he pioneered the use of journalism, films, and advertising to support his environmental arguments, and like few before him, he understood the impact of media in forwarding his causes. Perhaps the most noted and effective advertising campaign came in defense of Grand Canyon and the two dams that were planned for the Colorado River within its winding course. With hyperbolic charisma, large newspaper ads asked, "Should we also flood the Sistine Chapel so tourists can get nearer the Ceiling?" The dams were not built, but the success of the club's efforts led to the revocation of its tax-exempt status by the Internal Revenue Service (IRS). The measure, however, boomeranged, leading to greater rather than less public support. Brower cheerfully noted, "People who didn't know whether or not they loved the Grand Canyon sure knew whether they loved the IRS."

In the course of his lifetime, Brower edited over fifty publications and created the "exhibit format" books that became a hallmark of the Sierra Club. The first was *This Is the American Earth,* published in 1960. These were "coffee table" books of stunning photographs by prominent photographers such as Ansel Adams and Eliot Porter—*The Place No One Knew* (1963) (a eulogy to Glen Canyon, now Lake Powell), *Time and the River Flowing: Grand Canyon* (1964), *Earth and the Great Weather: The Brooks Range* (1971). There were nineteen in all published by Sierra and ten others by Friends of the Earth. The forewords to most were written by Brower. Like the High Country hiking and backpacking trips introduced by Sierra Club founder John Muir and renewed by Brower, the excursions and books were meant to introduce the public to the wonders of wilderness and natural places—reading about them, seeing the photographs, or experiencing them firsthand, people would join in the struggle to preserve.

In his own view, the most significant failure of his directorship of the Sierra Club came in 1963 when the club failed to stop construction of the Glen Canyon Dam—a failure to which he often referred with profound regret. To his more and more intractable perspective, Glen Canyon had been sacrificed to a club compromise: the two dams planned for Dinosaur were scrapped, but unique Glen Canyon was flooded, and the reservoir that is Lake Powell was born. The reservoir was not needed, Brower argued; it would both silt up and lose much of its volume to evaporation, conditions that the *National Geographic* reported on in April 2006. The lake had dropped 145 feet below full pool in 2005, and silting was, and would be, as one of the rangers observed, "inexorable. This is the land of erosion."

The loss of Glen Canyon became added impetus for Brower's subsequent militancy on behalf of the natural world. It also led to his ouster as director of the Sierra Club when in 1969 a slate of board candidates friendly to him were defeated, and he resigned. The reasons for his ouster seemed to include his opposition to compromise of any kind, the club's accommodation of corporate interests, and his free spending on behalf of environmental publications and issues. Although often described by colleagues, friends, and other writers as "a shy man," in his leave-taking of the club he had led for sixteen years, he summed up his steely perspective: "Nice Nelly won't get the job done." As he later wrote in a 1989 letter, "Compromise is often necessary, but it should not originate with the Sierra Club. We are to hold fast to what we believe is right, fight for it, and find allies and adduce all possible arguments for our cause. If we cannot find enough vigor in us or in them to win, then let someone else propose the compromise."

In spite of the rift within the club, in 1977 he received its John Muir Award, the highest honor the Sierra Club grants for service to the earth. By 1983 he was back on the board and was reelected to that position in 1986, 1995, and 1998.

Following his 1969 resignation, however, he founded Friends of the Earth the same year and added opposition to nuclear weapons to that organization's green issues, a cause the Sierra Club had been unwilling to embrace. FOE's well-known slogan became "Think globally; act locally." Though he resigned from the organization in 1986, he had led it to international prominence: it is now multinational, operating in sixty-eight countries, with headquarters in Washington, DC.

Perhaps no other record of Brower's tactics and persuasive power is more illustrative than John McPhee's 1971 work *Encounters with the Archdruid*, in which he pairs Brower with three zealous opponents—a mineral engineer, a resort developer (who considered all conservationists "druids, religious figures who sacrifice people and worship trees"), and a hydrologist/dam builder. The encounter trips and conversations that form the narrative clearly demonstrate the complex factors involved in deciding environmental issues and reveal in depth the unswerving core, quiet diplomacy and sense of humor in Brower's resolve when faced by his "natural enemies."

In spite of his dedication to protecting mountains and wild lands from further degradation, he was not insensitive to the strain this sometimes put on economic sectors, a fact later evidenced by his cofounding in 1999 the Alliance for Sustainable Jobs and the Environment.

In the second and third volumes of his autobiography, he had outlined his farsighted plans for the environmentalism of the future. In *Work in Progress* (1991), he offered thirty projects for a new "green century," among them worldwide Eco-Tours, a Peace Park in Tibet as proposed by the Dali Lama, a Council for Primeval Forests, and a World Heritage Library. In *Let the Mountains Talk, Let the Rivers Run* (1995), he envisioned a World Ecological Bank "as devoted to conserving, preserving, and restoring the earth's capital as the present World Bank is to spending it." He looked to recycled wood and alternative materials for building (plastics and steel), books printed on kenaf (as was *Let the Mountains Talk*—a tough, fibrous annual richer in cellulose than wood); natural "highways" for migrating animals as a way to restore ecosystems; and an Earth Corps to take up where the Peace Corps left off, "to be fully concerned with the endangered species and the endangered earth." His message to everyone, from farmer to architect, was to "put some eco-spin on your jobs, meaning you carry on with your work, but you carry on with the best interests of the Earth as you do so." Environmentalism, he later said, must be bold enough to continually ask, "What kinds of growth must we have, and what kinds of growth can we no longer afford" on our finite planet?

Not one to relax after any particular success, Brower often reminded his supporters that when they had won, it was only a temporary win. "All a conservation group can do is to defer something. There's no such thing as permanent victory. After we win a battle, the wilderness is still there, and still vulnerable. When a conservation group loses a battle, the wilderness is dead." In his passionate regard for mountains and wild places as "genetic reservoirs of bio-diversity" as well as sources of pleasure, serenity, inspiration, and renewal, Brower believed that interests as diverse as human health and aids to technology (as contemporary finds had proven) were waiting to be served by an endangered biodiversity thus far untapped.

In 2000, Brower died at his home in Berkeley at age eighty-eight.

SHIRLEY LOUI

Brower, David, with Steve Chapple. *Let the Mountains Talk, Let the Rivers Run.* New York, 1995.

Collins, Lawrence, with Martin Schweitzer and David Brower. *Only a Little Planet.* New York, 1972.

Glick, Daniel. "A Dry Red Season." *National Geographic*, April 2006, 64–81

McPhee, John. *Encounters with the Archdruid.* New York, 1971.

Reed, Christopher. "Obit of David Ross Brower, Environmental Campaigner." *Guardian*, November 8, 2000.

Schrepfer, S. R. "Environmental Activist, Publicist, and Prophet: David Brower." Regional Oral History Office, University of California, 1980.

Severo, Richard. "David Brower, an Aggressive Champion of U.S. Environmentalism, Is Dead at 88." *New York Times*, November 7, 2000.

Buell, Dorothy Richardson (1886/1887–May 17, 1977). "We are prepared to spend the rest of our lives, if necessary, to save the Dunes!" Buell famously proclaimed. This tenacious woman was responsible for the creation of Indiana Dunes National Lakeshore, located along the southern coast of Lake Michigan. Already well into her sixties when she joined the fight to preserve the dunes' ecologically rich land in 1949, Buell stands as a testament to the power of citizen activism.

Born in Neenah-Menasha, Wisconsin, Dorothy Richardson, along with her six siblings, spent her summers relaxing at a cottage near the dunes in what is now known as Ogden Dunes, Indiana. During this time, young Dorothy participated in dunes pageants, an experience that no doubt laid the groundwork for her later activism on the dunes' behalf. As a young adult, her dramatic side was further developed through training in elocution, first for two years at Milwaukee-Downer College, then at Lawrence College, from which she graduated in 1911 with a B.A. in oratory. After the completion of her studies, she became an English teacher, and in 1918 she married Major James H. Buell, with whom she moved all over the country, living in Gary, Indiana; Tulsa, Oklahoma; Flossmoor, Illinois; and her childhood haunt, Ogden Dunes, Indiana. During this time Buell gave birth to a son, Robert, and also kept up her interest in theater and literature, frequently hosting dramatic readings and book clubs and participating in community theater.

Although Buell's enjoyment of the dunes extended back to her childhood, her activism on their behalf was not kindled until 1949. While vacationing in the Southwest, Buell and her husband visited White Sands National Monument in New Mexico, where they came to the conclusion that the majesty of their hometown dunes—Ogden Dunes—was superior to that of White Sands'. Coincidentally, on the same trip, they stopped at a hotel in Gary, Indiana, and discovered that a local conservation group, Indiana Dunes Preservation Council, was holding a meeting to discuss threats to the dunes. Buell decided to attend, and in the process she realized that although her beloved dunes were in danger of development, it was not too late to work for their preservation. Thus, she soon found herself catapulted into a conservation struggle that ultimately received national attention.

The movement to save the Indiana Dunes already had a long and complicated history when Buell entered the picture. Beginning in 1916, a woman named Bess Sheehan—dubbed "Lady of the Dunes"—with the help of the newly formed National Park Service's first director, Stephen Mather, fought to set aside dunes land for national and state protection from industrialization. Their efforts were partially successful, resulting in the 1926 opening of

Indiana Dunes State Park. But by 1949, when Buell came into the picture, the dunes were again threatened by industry. In particular, the Central Dunes, which occupied the five miles between the subdivisions of Ogden Dunes and Dune Acres, were threatened by proposals to create a deepwater port at Burns Ditch and possibly a steel mill. Despite the impending threat, conservation efforts on behalf of the dunes were slow to get off the ground—that is, until Chicago Conservation Council president Reuben Strong urged Buell to take the lead in the struggle, a challenge she proved entirely able to take on.

With Strong's backing, Buell hosted about twenty women, among them the Lady of the Dunes herself, Bess Sheehan, at her house on June 20, 1952, to discuss the future of the dunes. After Sheehan gave a rousing speech about her own struggle to protect the dunes, Buell and her peers decided they would do all in their power to continue the fight and save the remaining lakeshore from further development. And so the Save the Dunes Council (SDC) was born, with Buell as its energetic president.

Under Buell's creative leadership, the group sprung into action, publishing a brochure about the dunes, assembling slides for educational programs, creating a short film, researching development plans, commissioning and writing newspaper articles and editorials, and even sponsoring a "Children's Crusade to Save the Dunes." To gain public support, Buell utilized the organizing experience she had gained from doing community theater and successfully reached out to her peers in key women's groups across the state. She also demonstrated publicity savvy by creating in 1954 a Save the Dunes Council advisory board composed of well-known citizens, including scientist Edwin Way Teale and artist Frank V. Dudley. By 1956, her outreach efforts were such a success that SDC had garnered 1,000 members. Only two years later the save the dunes petition that SDC had begun circulating included 500,000 names. In the end, the coalition Buell helped build included the League of Women Voters, teachers' unions, the Auto Workers Union, the AFL-CIO, the Sierra Club, the Izaak Walton League of America, National Audubon Society, Hoosier Environmental Council, and numerous others. Not wanting to threaten SDC's tax-exempt status, Buell originally strived to keep the organization strictly out of the political arena, focusing instead on educational outreach. Later, when it became clear that this approach was not enough to save the dunes, she and her fellow SDC members demonstrated an equal amount of political chutzpah, teaching themselves lobbying skills and making personal visits to all 435 House members and 100 senators to entreat the lawmakers to set aside the dunes for public enjoyment.

One of Buell's first crucial victories as Save the Dunes Council president was the purchase of Cowles Tamarack Bog. When, in 1953, Porter County put up for sale fifty-six acres of the bog, Buell led a drive to fund-raise the money necessary to buy and set aside the land. Notably, a portion of the funds used to make this acquisition came from the treasury of Sheehan's National Dunes Park Association, while the remaining funds were collected from generous donors, including Buell and her husband.

Despite this initial success, SDC and Buell found themselves up against formidable opposition: Bethlehem Steel, National Steel, Northern Indiana Public Service Company (which wanted to build a power plant), and others. Buell attempted to rally Indiana's then senators Homer Capehart and Albert Jenner to SDC's cause, but her pleas were to no avail; instead, she used her connections to reach out to Illinois Senator Paul Douglas. Initially Buell received only rejections from Douglas, but with the help of naturalist, writer, and SDC board member Donald Curloss Peattie and his wife Louise, she was able to entreat Douglas further, and on Easter 1958, Douglas phoned that he would be pleased to introduce a bill recommending the establishment of an Indiana Dunes National Monument. Although this phone call proved to be only a baby step on the long road to achieving legislation protecting the dunes, it marked an important tactical landmark in Buell's activism. Douglas had grown up loving the dunes and eventually embraced wholeheartedly SDC's mission to preserve them; his enlistment to the cause, owed to Buell's charm and determination, proved to be the critical edge that the council needed to win the fight to establish the dunes as a protected space.

Meanwhile, Buell found herself increasingly relying on her training in oratory for significant public effect. When the issue was framed as one of "picnics versus paychecks," Buell shot back in her 1959 comments to the Senate Committee on Interior and Insular Affairs, "Mr. Humphrey [president of National Steel] has said he prefers jobs to picnics. We ask, why is it not possible to have jobs *and* picnics? Surely this is the viewpoint of a humanitarian" (*Sacred Sands*). Guided by religion, and appealing to people's sense of civic responsibility, Buell managed to deepen the dunes debate. She gave it a spiritual and emotional component emphasizing the capability of nature to nurture humanity. The dunes are "one of the most beautiful natural shrines in all of America," she argued, adding that "though we share the pride of an advancing, industrial nation, we would save the inner core of its strength through preservation of its beauty and natural resources" (*Sacred Sands*). Later, in a *Chicago Daily Tribune* letter to the editor dated June 6, 1961, regarding the creation of Burns Ditch, she penned, "Conservationists feel that to snatch this remaining 4 miles of unique duneland and incomparable beach on the lake front from the people is a travesty on human justice. The dunes belong to the people." Needless to say, nature was for Buell a human right, and one of her key contributions as an activist was in making the public realize that nature generally—not just the dunes—is worth fighting for.

Despite certain setbacks involving the loss of a portion of the remaining unspoiled dunes, SDC, Douglas, and Buell could finally claim success in 1966 after Congress authorized the creation of Indiana Dunes National Lakeshore, permanently protecting a total of 8,330 acres of land and water. The ceremonial dedication of the park occurred on September 8, 1972, and Buell—a woman with little prior activism experience who managed to transform herself into a politically savvy, citizen environmental advocate—was an honored guest.

With her health in decline, Buell eventually set aside her work for the dunes in 1968 and retired with her husband to California to live with their son in 1970. However, her heart remained with the dunes, and she purportedly lamented, "Nobody in California wants to hear about the Dunes, and that's all I've got to talk about." In the end, she acted as SDC president for sixteen years. During her lengthy tenure as council president, Buell became known for her inspired and charismatic leadership, not to mention her dogged pursuit of dunes preservation. A staunch Republican, Buell was known for her eloquence, spiritedness, dignity, and integrity. Her work on behalf of the dunes also made her the recipient of numerous awards and honors, including being named one of Indiana's top female newsmakers. Without a doubt, when Buell passed away in 1977, the dunes lost an important advocate, and the conservation movement found itself lacking a powerful voice.

Today the Indiana Dunes National Lakeshore has been expanded to 15,000 acres, and the Save the Dunes Council continues to work for environmental protection, committed to preserving, protecting, and restoring Indiana's dunes and Lake Michigan watershed's natural resources. Meanwhile, visitors to the Indiana Dunes National Lakeshore will find the Dorothy

Buell Memorial Visitor Center—permanently honoring Buell's passionate advocacy on behalf of the dunes.

ERICA WETTER

Engel, J. Ronald. *Sacred Sands: The Struggle for Community in the Indiana Dunes.* Middletown, CT, 1983.

Franklin, Kay, and Norma Schaeffer, eds. *Duel for the Dunes: Land Use Conflict on the Shores of Lake Michigan.* Chicago, IL, 1983.

Greenberg, Joel. *A Natural History of the Chicago Region.* Chicago, IL, 2002. Shirley Heinze Environmental Fund. *The Indiana Dunes Story: How Nature and People Made a Park.* 2nd ed. Michigan City, IN, 1997.

Burnham, John B. (March 16, 1869–September 24, 1939). A writer, businessman, explorer, and conservationist, Burnham was born in New Castle, Delaware, in the historic Amstel House. After graduating from Trinity College, Connecticut, in 1891, he worked as business manager for *Forest and Stream.* Suffering from a severe depression at age twenty-seven, he quit his job and joined the first Klondike gold rush. In the Klondike his ingenuity was seen in his building and operating a timber railway for hauling freight over White Horse Pass.

Burnham had long been an ardent hunter—but a hunter with a concern for sportsmanship: "I love hunting; but I do not love it so much for the game in the game pocket as for the game in the fields and forests. I am ready always to give up shooting when the interest of wild life demands it. In this I think I represent the attitude of all true sportsmen."

This conviction, coupled with what seems to have been an aptitude for public service, led to Burnham's professional involvement in wildlife matters. From 1905 to 1911 he served New York State in three successive positions: chief game protector, deputy commissioner of fish and game, and acting commissioner of fish and game. In 1915 he was on the three-member committee that codified the New York Fish and Game Law.

In 1911 he was appointed the first president of the newly formed American Game Protective and Propagation Association, later called the American Game Protective Association (AGPA). The association was founded with the declared purpose of developing and promoting wildlife conservation projects. Among its supporters were some of the nation's leading gun manufacturers, a fact that troubled some environmentalists. Nevertheless, the AGPA was the "first sportsman-supported national organization" for the protection of game.

In his position as president of the AGPA, Burnham was a force to be reckoned with in environmental politics. He was one of those most responsible for the Weeks-McLean Act of 1913, insuring protection to migratory birds by severely limiting market hunting and outlawing spring hunting. In 1917, Burnham was also instrumental in the creation and ratification of the Migratory Bird Treaty between the United States, Great Britain, and Canada. In recognition of his effectiveness, he was named chairman of the Federal Advisory Committee to the Bureau of Biological Survey, a position he held for twenty years.

However, the fact that the AGPA and thus Burnham's salary were partly funded by gun-manufacturing companies led many fellow conservationists to perceive a conflict of interests. Beginning in the early 1920s, distrust widened into overt hostility with a game refuge bill drafted by Burnham, the passage of which was blocked by such powerful environmental activists as Dr. William Temple Hornaday and Will H. Dilg. Objections to the bill were based on its plan for establishing a federal hunting license and on its allowing hunting on the proposed refuges. The latter measure was seen as a profit-making move by the gun industry.

After New York Senator Fiorello LaGuardia read into the *Congressional Record* evidence of complicity between the gun companies and the AGPA, these companies withdrew their support from the organization. In 1928 Burnham resigned as AGPA president.

Further environmental service of Burnham was his chairing a U.S. Forest Service committee on Game in the National Forests. He was also a member of the Committee on Game and Fur-bearing Animals of the National Conference in Outdoor Recreation of 1924. He was perhaps the largest builder of log cabins in the United States, he was involved in the first experimental fur farm in the country, and he experimented with "scientific deer farming."

In 1921 Burnham was in charge of an expedition to Siberia to search for the rare Marco Polo mountain sheep. His book *The Rim of Mystery* (1929) relates his experiences on this mission.

KATHRYN HILT

Burnham, John B. *John Bird Burnham: Klondiker, Adirondacker, and Eminent Conservationist.* Ed. Maitland C. De Sormo. Saranac Lake, NY, 1978.

———. *The Rim of Mystery.* New York, 1929.

Fox, Stephen R. *John Muir and His Legacy: The American Conservation Movement.* Boston, MA, 1981.

"J. B. Burnham Dies; Explorer, Author." New York *Times*, September 26, 1939.

Trefethen, James B. *An American Crusade for Wildlife.* New York, 1975.

Burroughs, John (April 3, 1827–March 29, 1921). Born and raised on a rocky farm near Roxbury in the Catskill Mountains of New York, Burroughs was a teacher, writer, naturalist, bank examiner, and farmer. Burroughs's father, Chauncey Burroughs, was a hardscrabble farmer and a narrow-minded religious fundamentalist who never read anything other than the newspaper and the Bible. Chauncey had no interest whatsoever in nature except for what the soil could yield to his plow. According to Burroughs, his mother, Amy Kelly Burroughs, "never read a page of anything." However, she was probably the earliest inspiration for his sense of wonder of nature. She would take him and his nine brothers and sisters berrying and on nature walks in the hills above the small family farm. His mother was also responsible for his getting more than a primary education; his father put little stock in education beyond the basics.

Burroughs desperately wanted to be a writer from his early childhood on. He strived for as much education as possible and earned his own money to go to several private schools during his late teen years until, when he was nineteen, he became an itinerant teacher. At this point in his life, he was disconsolate about his future. Few bright, professional prospects were on the horizon, and his personal life was dominated by a woman who was never to appreciate his talents. On September 12, 1857, he began a stormy and rather heartless marriage with Ursula North that was to last until her death in 1917. Ursula never thought much of her husband's writing, and he, for his part, never thought much about being faithful to his overbearing wife. He used his writing and his love of nature to escape from his depressing domestic obligations.

Burroughs's early writing was pompous and overstuffed; he tried too hard to be literate and articulate. Only after he discovered Ralph Waldo Emerson's prose did he begin to form a style of writing that would take him to the heights of popular literature in the late nineteenth century. His first published work was an anonymous piece for the *Atlantic Monthly*; it resembled Emerson's style so thoroughly that many attributed the piece to the elder philosopher. Burroughs made a break from this hero worship and from his provincial life in New York when he moved to Washington, DC, in 1862, started a job as a clerk for the Treasury Department, and met poet Walt Whitman.

Whitman published his first edition of *Leaves of Grass* in 1855; it would become the inspiration that would influence Burroughs more than any other. The earthy "American" poet was to become Burroughs's friend and mentor until Whitman's death in 1892. Burroughs's first published book, *Notes on Walt Whitman as Poet and Person* (1867), was the first book written about the poet. The book was thoroughly edited and approved by Whitman, a practice that would continue at some level for years for most of Burroughs's writing. *Wake-Robin* (1871), Burroughs's first nature book, was both titled and edited by Whitman.

Burroughs's greatest contribution to conservation and the natural world was to inspire in readers a love of nature in their own backyards, nearby fields, and forests. He made nature something personal by writing about his own intimate encounters with the woods and mountains of his beloved Catskills. *Winter Sunshine* (1875), *Birds and Poets* (1877), *Locusts and Wild Honey* (1879), *Pepaction* (1881), and most of his earlier books were all personal accounts of his intimate encounters with nature or essays on other nature writers and their literature. Birds were a particular love of Burroughs's, and he used birding as the excuse to get into the woods as often as possible. Although he never gave up the nature essay form of writing, Burroughs's later work evolved more toward literary criticism and discourses on the evils of technology and his own personal religious beliefs. These books include a very mediocre book of poetry, *Bird and Bough* (1906), *Time and Change* (1912), and *Accepting the Universe* (1920).

During his lifetime, he became one of the most beloved people in the United States. Schoolchildren and young women especially were drawn to his personal, descriptive nature writings. As he grew more famous, he began to move in very impressive social circles. He became a personal friend of some of the most powerful men of his generation. He traveled with E. H. Harriman and John Muir to Alaska and California; he visited Yellowstone National Park with Teddy Roosevelt; he took camping trips with Henry Ford; and he grew up with Jay Gould. People who were encountered by Burroughs and his larger-than-life friends responded most heartily to Burroughs, not to his travel companions. He was feted by the common people who read his moving prose about the birds and the wildflowers.

Today the Burroughs Medal is still awarded for exceptional literary quality in nature writing. Despite this acknowledgment of Burroughs's contributions to nature literature, he is much less well known today than many of his contemporaries. His current stature is minimal considering the adoration heaped on him during his lifetime. The reasons for this are complex. Probably the most telling is that Burroughs was almost never controversial, nor did he actively promote conservation as a social responsibility. He loved what he called the "perfectly cultivated landscape." Wilderness in all its forms was antithetical to his comfortable Catskill retreat "Slabsides." Unlike John Muir, he did little to affect the future of the natural world. And unlike Henry David Thoreau, he did not challenge the philosophical underpinnings of commercial America. His legacy is that of a kindly and articulate grandfather who brought nature to the minds and souls of several generations; he did not bring conservation to the conscience of an industrializing nation.

Burroughs was survived by his only child, Julian Burroughs.

THOMAS P. HUBER

Barrus, Clara. *John Burroughs, Boy and Man.* Boston, MA, 1921.

———. *Our Friend John Burroughs.* Boston, MA, 1914.

Brooks, Paul. *Speaking for Nature.* Boston, MA, 1980. This book on the history of nature writers has a large section on John Burroughs.

Kelley, Elizabeth Burroughs. *John Burroughs, Naturalist.* New York, 1959.

———. *John Burroughs's Slabsides.* West Park, NY, 1987.

Renehan, Edward J., Jr. *John Burroughs: An American Naturalist.* Post Mills, VT, 1992.

———, ed. *A River View and Other Hudson Valley Essays by John Burroughs.* Croton-on-Hudson, NY, 1981.

Westbook, Perry. *John Burroughs.* New York, 1974.

Cain, Stanley Adair (June 19, 1902–April 1, 1995). A man who once noted that "innumerable people cannot enjoy wilderness together," Cain was born in Jefferson County, Indiana. After receiving a B.S. in botany at Butler University in 1924 and a Ph.D. in plant ecology at the University of Chicago in 1930, he taught botany at many universities around the country, including Butler, Indiana University, the University of Tennessee at Knoxville, the University of Michigan, and the University of California at Santa Cruz. Everywhere he taught, he left a legacy of environmental education and innovation. Even while still in graduate school in 1927, he became the first person to use aerial photography in biology when he went up in a biplane to do some ecological mapping for a research paper.

During his tenure at the University of Tennessee, Knoxville, from 1935 to 1946, Cain vigorously supported the enjoyment of the nearby Smoky Mountains. He not only worked with Arthur Stupka, the Smoky's first chief naturalist; he also wrote the first walking guide in the Smokies, the Greenbrier-Brushy Mountain Nature Trail guide, in 1937. In this guide, he recognized the need for quiet, leisurely attention to wilderness: "This trail can offer little to the sightseeing tourist in a hurry. Those who run cannot 'see,' and those for whom hiking is too great an exertion cannot stop to contemplate nature, only to pant."

At Michigan he created the first Department of Conservation in the country and helped develop the principles used by the Michigan Natural Areas Council to evaluate and create management plans for natural areas. The council, which began in 1946 as part of the Southeastern Chapter of the Michigan Botanical Club, has been instrumental in the dedication of more than 100,000 acres of natural areas in Michigan.

Besides teaching, Cain served the environment through various organizations, including the Ecological Society of America, where he was president in 1958. From 1963 to 1964 he chaired the Michigan Conservation Commission. In 1966 President Lyndon B. Johnson appointed Cain as the Assistant Secretary of the Interior for Fish, Wildlife, and Parks, a post he held until 1968. He then returned to Michigan to head the Institute for Environmental Quality until 1972, when he retired from Michigan.

However, his university work was not completed. When the University of California at Santa Cruz (UCSC) was being created, Cain helped Chancellor Dean McHenry design the campus with the idea of preserving its natural beauty. According to Cain, a natural environment was crucial to sound education. In 1972 he helped plan the development of College 8 at UCSC and taught there for the remainder of the 1970s.

Cain wrote many books and pamphlets and more than one hundred articles on both botanical and environmental concerns. While at the University of Tennessee, he wrote *Foundations of Plant Geography* (1944). Working as a botanist for the Cranbrook Institute of Science from 1946 to 1950, he contributed to its Bulletin No. 34, *Farwelliana: An Account of the Life and Botanical Work of Oliver Atkins Farwell, 1867–1944* (1953), along with Rogers McVaugh and Dale J. Hagenah. From 1955 to 1956, Cain was employed by UNESCO (United Nations Educational, Scientific and Cultural Organization) as a member of the United Nations Technical Assistance Mission to Brazil as an expert in ecology. There he worked with Dr. G. M. de Oliveira Castro, a specialist in tropical medicine, and coauthored with him a *Manual of Vegetation Analysis* (1959). In 1971, he chaired the Advisory Committee on Predator Control and helped prepare the report to the President's Council on Environmental Quality and the Department of the Interior, under a contract with the Institute for Environmental Quality at the University of Michigan. This report suggested a new approach to predator control. Rather than exterminating predators for the sake of livestock concerns, it indicated a need to balance those concerns with the value of wildlife, both for the natural environment and for the enjoyment of humans' wilderness experience.

Cain died at the age of ninety-two in a Santa Cruz, California, nursing home. He is remembered as a conservationist and a founder of the science of ecology, to which he brought his considerable botanical knowledge.

JANICE L. EDENS

Thomas, Robert McG. "S. A. Cain, 92, Conservationist and Ecology Pioneer, Is Dead." *New York Times*, April 3, 1995.

Caras, Roger (May 24, 1928–February 18, 2001). A man who donned many hats, Caras was an animal rights activist, Hollywood executive, radio and television personality, and naturalist. He was also an environmental activist who after joining ABC News in 1974 was assigned exclusively to cover animals and the environment. He covered, for instance, the calamitous 1988 *Exxon Valdez* oil spill in Prince William Sound, Alaska.

Caras was born in Methuen, a small rural town thirty-five miles northwest of Boston, the son of Jacob, an insurance salesman, and Bessie Caras, an accountant. His affection for animals developed at an early age. At home he was exposed to cats, dogs, and canaries, as well as "pet" snakes, frogs, and turtles. Then there was the wildlife. As Caras explained in his autobiography *A World Full of Animals* (1994):

> Methuen had open fields and meadows, streams and ponds, the lazy Spickett River, and dark, mysterious woodlands full of places to hide. With so much wild habitat there were, of course, wild animals. Round fat muskrats lived in the river, and rabbits and meadow mice scurried across the fields. In the woods there were opossums, squirrels, skunks, deer, and raccoons.
>
> There were also many kinds of birds: bluebirds, blue jays, indigo buntings, goldfinches, cardinals, horned larks, hawks, owls, geese, ducks, and gulls. In the spring, their songs were a nonstop musical performance.
>
> Butterflies of all kinds also flitted in our gardens and in the meadows. There were other insects too, like dragonflies, beetles, moths, flies, and of course, mosquitoes, especially near the river and ponds.

"Methuen was a wonderful place in which to learn and to explore," he recalled in his autobiography. "Little wonder that growing up there gave me the idea that the world was full of animals." At the age of ten Caras took a job as a kennel aide with the Massachusetts Society for the Prevention of Cruelty to Animals (a job he soon relinquished because he could not abide the painful job of putting unwanted animals to sleep).

As a boy, Caras wrote about animals. He also read a great deal, and his favorite book, which he read several times, was Charles Darwin's *The Voyage of the Beagle*. Young Caras resolved that he too "was going to travel around the world to see animals." After graduating from Boston's Huntington Preparatory School, Caras enlisted in the U.S. Army in 1946. In 1948, at the end of his tour of duty, he entered Northeastern College in Boston to study zoology; then in 1950 he transferred to Western Reserve University in Ohio. With the outbreak of the Korean War, Caras reenlisted in the army and served from 1950 to 1952. During his tour of duty he traveled to Germany, Japan, the Philippines, and Guam and studied filmmaking, writing, and natural history. Both in the army and subsequently at the University of Southern California, from which he graduated in 1954 with a major in cinema, Caras made notes, kept a diary, and eventually sold articles about animals to newspapers and magazines. He was laying the groundwork for his career as a naturalist.

Originally Caras intended to be an actor, but he grew doubtful that he had the talent to succeed in that profession. Still, he determined to enter the movie business. Beginning as a press agent in 1955, he advanced steadily up the corporate ladder as national director of merchandising at Columbia Pictures (1955–1965), vice president of British Hawk Films (1965–1969), and an executive of American Polaris Productions. In 1968 he crisscrossed the country while promoting Stanley Kubrick's epic *2001: A Space Odyssey*. During his free time, however, Caras shoved aside the feverish world of show business to read and write about Kodiak bears, wolves, and whales.

In 1961, while living in New York City and working at Columbia Pictures, Caras produced a record album of the nuclear submarine *Nautilus*'s trip to the North Pole. Because of this and his army experience, he was invited to join an American scientific expedition to the South Pole with the task of writing and broadcasting a story on the navy's role in the expedition. Caras, who leaped at the opportunity, soon recognized that the South Pole has no land animals but does have several species of sea animals, including whales, seals, and especially penguins, flightless birds that, he wrote, "capture my attention." Also during his trip Caras vowed never to write about an animal before he had observed it in its habitat.

The difficult journey provided Caras with material for his first book, *Antarctica: Land of Frozen Time* (1962), which had chapters on the exploration of Antarctica and on the frozen continent's fauna and flora. Moreover, following the book's publication, he began to appear regularly on radio and television programs, including the *Today Show*, where he was the "house" naturalist for eight years. His "adventurous traveling" had launched a new career for Caras, and in 1964 he decided to jettison the expense account and movie-star world to pursue a career as a full-time naturalist. Now he appeared regularly on radio and television. From 1975 to 1992, he was a regularly featured reporter on *ABC World News Tonight with Peter Jennings*, and he appeared on *Nightline*, *20/20*, and *Good Morning America*. He also hosted several radio programs, such as *Pets and Wildlife* (CBS), *Report from the World of Animals* (NBC), and *The Living World* (ABC); and he was the announcer for the Westminster Kennel Club Dog Show. In addition, he was a consultant to the Walt Disney Corporation on its Florida-based Animal Kingdom.

Caras traveled widely—most of the globe and every state in the United States—to gather material for his television and radio programs as well as for his books. Among the places that held a special fascination for him was Africa, so when Columbia Pictures invited him in 1971 to go to Africa, he accepted eagerly. His assignment was to write and direct a documentary film about the making of George Adamson and Joy Adamson's sequel to *Born Free* (to be called *Living Free*). Caras's journey took him to northern Kenya, where he saw giraffes, hippos, elephants, crocodiles, and lions in their natural environments. During the trip, Caras began the research for *Mara Simba: The African Lion* (1985), a fictionalized animal biography. (He visited Africa over twenty times, during the next fourteen years, before he felt informed enough to write *Mara Simba.*) Given his sympathies and his many sojourns abroad and throughout the United States, Caras was especially attentive to the state of wildlife. What he found was dispiriting. His book, in which he surveyed the state of about forty different animals, was a "clarion, a horn, an alarm," he declared, putting humankind on notice that due chiefly to overhunting and habitat destruction whole species were on the verge of extinction.

As he wrote:

> Without the possibility of a major species evolving anew anywhere save perhaps in the depths of the sea, and there only for a little longer, we must attempt to sustain those species that have evolved in the past. The decision as to what is to die and what is to survive is ours to make in this generation and the next. . . . We are about to commit ourselves once and for all time either to a planet rich in wonderment and beauty or to a planet that is a mockery of itself, drenched in poisons, littered with metal junk heaps, and stripped of timber, an ugly planet that will soon enough strangle itself on its own reeking gases and gag itself on its own self-spawned contaminated juices. This is mankind's last chance on earth. From here on, the world will be a heaven or hell of our own choosing.

And Caras added:

> Man at this point in history is deciding whether or not he shall share this planet or shall inhabit it alone. His is deciding whether there shall be concrete and steel and nothing else. The difference between the ways things are going now and the complete and utter destruction of everything natural on this planet is not a difference in kind, but only in degree. Stepping up the present pace will result in utter desolation.

Caras considered a tragedy in the making. "Wildlife is our connection with the real world," he declared. The Endangered Species Act (1923) suggests that the warning by Caras and other naturalists and environmentalists was heeded by Congress and the president. As early as 1964, in *Dangerous to Man*, Caras warned that the "wildlife of the world is disappearing at an alarming rate." Two years later in *Last Chance on Earth*, he asserted, "Man's record so far, as the custodian of [animal] life on the planet is bad, very bad indeed."

Coursing through Caras's nature books is his focus on the complex connections between living creatures and their environment. Perhaps nowhere in his many books is the relationship so clearly etched as in *The Forest* (1979). This volume is the tale of life for over eighty species of fauna and flora found in northwestern U.S. coniferous forests. The book's view of the balance of forest life covers a great deal of territory, from insect wars within hemlocks and Douglas fir through the struggles to survive of bears, bobcats, mountain lions, and snakes. In sum, *The Forest* is an informed portrait of the full range of the network of woodland life.

But the book's ending adds a further dimension to *The Forest*. As Caras concludes:

> A forest is a place. For any moment it is a total system, just like the one described in this book.
>
> But no one can contemplate a forest today, the one in this book or any other, without a sense of foreboding or even dread. It has been calculated that the tropical rain forest is disappearing at the rate of fifty acres a minute, night and day, every day of the year. If the assault on all forests were to be calculated, it is quite possible that two to three times as many acres per minute may be falling—perhaps as many as 150 acres every sixty seconds.
>
> Surely that is not progress. We do not need forest products, but the harvest should be orderly, the yield calculated, and the replacement planned before the harvest begins. But we need not dwell on this horror.

> Having read this book, you should be numbed by the prospect of all free, wild forests vanishing within the next fifty years—and this could happen.

After numerous trips abroad to study and write about animals, Caras elected to "help them more directly," and in 1991 he was named president of the American Society for the Prevention of Cruelty to Animals (ASPCA). In pursuit of the organization's main goal—to prevent cruelty to animals—Caras expanded its animal protection program, stressing population control rather than destruction of unwanted animals. Under his leadership the ASPCA established the nation's first animal poison center. "Perhaps more importantly," the *New York Times* (February 20, 2001) asserted, "Caras initiated a spay and neuter program that served as a model for other humane organizations."

Caras's interest in animals and his concern for the environment brought him a great deal of national attention and multiple honors, national and international. In 1977, he was awarded the Joseph Wood Krutch Medal for "outstanding contributions to the betterment of our planet." In 1984, he became the first recipient of Israel's Oryx Award for wildlife conservation. The following year he was named the first recipient of the Humane Award of the Year of the American Veterinary Medical Association. In 1988, he was singled out for the James Herriot Award; and in 1990, the Emmy Award. He was also awarded honorary doctoral degrees by Rio Grande College in 1979, the University of Pennsylvania, School of Veterinary Medicine, in 1984, and the State University of New York in 1987.

Having an affinity for the entire animal kingdom, Caras maintained that there was no such thing as a "dangerous" animal. Every creature, whether a water moccasin—whose bite is often fatal—or a lion gnawing on the carcass of a giraffe, has a place in the natural order. And for four decades, through hundreds of magazine and newspaper articles and television appearances—and almost one hundred books—he was in touch with millions of his fellow citizens; as a result, he helped shape American attitudes toward animals and the environment. As president of the ASPCA, moreover, he launched reforms that were subsequently adopted by other humane societies. Though he worked in a popular milieu friendly to environmental concerns, he should certainly be accorded credit for helping to forge that milieu.

RICHARD P. HARMOND

Day, Sherri. "Roger Caras: Animal Welfare Advocate." *New York Times*, February 20, 2001.

Harmond, Richard. "Roger Caras." *American National Biography*. 2000.

"Roger Caras." *Current Biography*. 1998.

"Roger Caras" [obit.]. *Washington Post*, February 20, 2001.

Carhart, Arthur Hawthorne (September 18, 1892–November 30, 1978). Because his ideas eventually resulted in the creation of the National Wilderness Preservation System, Carhart is known as the "Father of the Wilderness Concept" and the "Dean of American Conservation." He was born in Mapleton, Iowa, in 1892, to George W. and Ella Louise (Hawthorne) Carhart and became interested in conservation through his grandfather, David T. Hawthorne, a homesteader under the Tree Claim Act. In 1916 he received the first Bachelor of Science degree in landscape architecture and city planning given by Iowa State College. He married Vera Amelia Van-Sickle in 1918. During World War I he was a first lieutenant in the Sanitary Corps of the U.S. Army, where he gained experience working in greenhouses and nurseries as well as on various landscape projects.

Carhart joined the U.S. Forest Service in 1919 as its first permanent landscape architect, though he was initially designated a recreational engineer. In that role, he visited Trappers Lake in the White River National Forest of Colorado during the summer of 1919. Although the purpose of his trip was to create a plan for developing vacation homes, the incredible beauty of the region convinced him that the land should not be developed any further but should instead be preserved in its pristine wild state. On December 6 of that same year, he discussed his vision with Aldo Leopold, who endorsed his views. Carhart prepared a memorandum of their meeting, which became the first written document recommending wilderness preservation: "There are a number of places with scenic values of such great worth that they are rightfully the property of all people. They should be preserved for all time for the people of the nation and the world" ("Memorandum for Mr. Leopold, District 3," dated December 10, 1919). He finally convinced his supervisor Carl. J. Stahl to keep Trappers Lake roadless and to deny all of the many requests for vacation home sites then pending, although it was not until November 1922 that his suggestions received final approval. This unprecedented action initiated the Forest Service's role in wilderness planning and preservation. In 1920 he applied some of the same principles to Colorado's San Isabel National Forest, preparing its first Recreation Plan. He was aided by the San Isabel Public Recreation Association, which helped raise money to create the recreation areas Carhart had conceived. In 1921 he submitted a revised Recreation Plan for the Superior National Forest, which resulted in the preservation of what is now the Boundary Waters Canoe Area of Minnesota. This plan was approved by District Forester Peck on November 8, 1922. Because of Carhart's work, thousands of acres were restricted from roads, mining, logging, grazing, and other forms of human interference; and during the next seven years, the Forest Service set aside 5 million acres of roadless wilderness.

Although Carhart's vision eventually resulted in some dramatic steps in preserving wilderness, he resigned from the Forest Service in December 1922 because he had become discouraged by the lack of progress in implementing his plans. He did not, however, abandon the cause but went into private practice with Irvin McCrary and Frank Culley as a professional landscape architect and city planner; their firm became the largest of its kind in the West. In 1928 he served for a year as the first executive secretary of the Denver City Planning Commission. As such, he suggested placing city buildings around open malls and public parks. In 1938 Colorado Governor Teller Ammons appointed him as the director of the program for Federal Aid in Wildlife Restoration. In 1960, along with John T. Eastlick, he initiated the establishment of the Conservation Center Library in Denver to gather all environmental research into a single location. Although the center no longer operates, its materials still reside in the Denver Public Library.

Carhart was a prolific writer throughout his life. He wrote many articles for magazines such as *American Forestry, Parks and Recreation*, and *Field and Stream*, promoting his ideas to the public. In addition to his more than 4,000 short stories, novelettes, serials, articles, and essays, Carhart wrote a number of outdoor books including *How to Plan the Home Landscape* (1935), *Trees and Shrubs for the Small Home* (1936), *The Outdoorsman's Cookbook* (1944), *Hunting North American Deer* (1946). *Fresh Water Fishing: Bait and Fly Casting* (1949), *Fishing in the West* (1950). *Fishing Is Fun* (1950), and *Trees and Game—Twin Crops* (1958). Some of his books have become classics in their arguments for conservation of various natural resources. *Water—or Your Life* (1951) considers the need of a comprehensive plan for using and protecting the country's water supply. *Timber in Your Life* (1955) explores the attempts of the "Land Grab Gang," as he dubbed them, to exploit resources in the national forests for personal and corporate gain. In *The National Forests* (1959), Carhart addresses the importance of the national forests as watersheds and encourages a multiple-use ethic in forest management. He also wrote several works of fiction, such as *The Ordeal of Brad Ogden* (1929), *The Last Stand of the Pack* (1929), *Drum Up the Dawn* (1937), *The Wrong Body* (under the pseudonym V. A. Van Sickle) (1937), and *Bronc Buster* (1937) and *Saddle Men of the C-Bit Brand* (1937), both under the pseudonym Hart Thorne. For his conservation writings and other contributions, he won the Founder's Award of the Izaak Walton League (1956), the Jade of Chiefs award given by the Outdoor Writers Association of America (1958), and the American Forest Products Industries citation for Conservation (1966).

Today Trappers Lake is known as the "Cradle of Wilderness" because of Carhart's defense of it against development and his influence in the subsequent establishment of the wilderness movement in the United States, culminating in the Wilderness Act of 1964. As a result of his efforts, Trappers Lake and its shore retain the untouched beauty that Carhart first witnessed in 1919. Furthermore, by the end of 1989 he was indirectly responsible for the more than 90 million acres of the United States that had been set aside as wilderness.

Carhart's contribution to Trappers Lake was commemorated by the Forest Service on July 21, 1985, when it officially dedicated a hiking trail around the lake to him. Perhaps even more significant was the creation of the Arthur Carhart National Wilderness Training Center, established in 1993 at Lolo National Forest in Montana, which preserves Carhart's memory as the father of the wilderness system.

JANICE L. EDENS

Baldwin, Donald N. *The Quiet Revolution: Grass Roots of Today's Wilderness Preservation Movement.* Boulder, CO, 1972. This book covers Carhart's contributions extensively.

Martin, Erik J. "A Voice for the Wilderness: Arthur H. Carhart." *Landscape Architect* 76 (July–August 1986): 70–75.

Nash, Roderick. *Wilderness and the American Mind.* 3rd ed. New Haven, CT, 1982.

Watkins, T. H. "Untrammeled by Man: The Making of the Wilderness Act of 1964." *Audubon*, 1989, 74–91.

Carr, Archie (June 16, 1909–May 21, 1987). Author and educator, taxonomist and evolutionary biologist, Carr was deservedly recognized as one of the foremost experts on sea turtles. With his books on wildlife in general and reptiles in particular, he educated countless readers on the imminent dangers facing endangered species and the world ecosystem. Internationally, he was acclaimed a zealous advocate of conservation, and today he is widely acknowledged as an important forerunner of contemporary environmentalism.

Carr was born in Mobile, Alabama, the son of Archibald Fairly Carr, a Presbyterian minister, and Louise Gordon (Deaderick) Carr, a teacher of piano. His father instilled in his son a love and respect for animate and inanimate nature. Archie collected snakes, frogs, lizards, and turtles. To satisfy his boyhood curiosity, he avidly studied each of his unusual pets. When he grew older, his family moved to Fort Worth, Texas, and then to the state of Georgia. In 1927, he entered Davidson College as an English major. Encouraged by one of his professors, he decided to switch his major to biology, and he transferred from Davidson to the University of Florida at Gainsville. He received his bachelor's degree in 1932 and began graduate study. In 1934, he was awarded a master's degree. In 1937, he was the recipient of the University of Florida's first doctorate in biology.

After serving as an assistant biologist at the University of Florida (1933–1937), he was offered a fellowship at Harvard's Museum of Comparative Zoology (1937–1943). After his postdoctoral work at Harvard, he was appointed an assistant professor at the University of Florida. In 1945, he was promoted to associate professor, and four years later, he was elevated to full professor. In 1959, on the basis of outstanding research, he was advanced to graduate research professor.

Overlapping his teaching and research at the University of Florida, he also served as a professor of biology at Escuela Agrícola Panamericana, Hondurus (1945–1949). His reputation now was such that he was named a biologist for the United Fruit Company in 1949. In 1953, he was appointed a research associate at the American Museum of Natural History, with which he was affiliated for more than thirty years. In 1956, he was named technical adviser to the faculty of the University of Costa Rica. Three years later he was appointed technical director of the Caribbean Conservation Corporation, and its executive vice president in 1968.

In 1935, Carr published his monumental *Handbook of Turtles*, in which he not only classified some seventy-nine species but also dispelled folklore and myths about the horny plated reptiles. When he began his *Handbook*, Carr realized that practically nothing of a scientific nature was known about sea turtles. All species proved of interest, but the green turtle benefitted most from his insatiable curiosity. In 1954 he initiated a tagging project at Tortuguero, Costa Rica. Results of this and ancillary projects established that some sea turtles migrated hundreds, often thousands, of miles between their feeding grounds and nesting beaches. His work also yielded basic information on reproductive cycles, embryology and development, growth, feeding habits, and behavior and orientation of both young and mature sea turtles.

Research funds and the cooperation of the U.S. Navy helped Carr reestablish populations of the giant turtles along Caribbean shores where human predators who used them for food had virtually wiped them out. In 1964, for example, the navy airlifted 18,500 baby turtles from the species' last breeding sites in Costa Rica to a dozen beaches in Mexico, Colombia, Florida, and the Caribbean Islands. Even after elucidating migration patterns for decades, the "Turtle Man," as Carr was dubbed, remained baffled by what happens to the offspring of sea turtles after they slip into the ocean as hatchlings weighing but a few ounces, not to be seen again for a year or more, when they are discovered hundreds of miles from their birthplace and well on their way to becoming 400-pound adults.

In particular, his research concerned the whereabouts of sea turtles during the interval between their hatching and their reappearance in waters of the continental shelf. His laborious tagging techniques allowed him to document the remarkable navigational abilities of green turtles, which often journey over 1,000 miles each way from the beaches of Ascension Island, where they hatch, to the east coast of South America, where they mature, and then back to Ascension in the southern Atlantic Ocean to lay their eggs five or six years after their birth.

In 1964, after puzzling over the matter for several decades, Carr reached several valid conclusions. His documented records demonstrated that most baby turtles would hitch rides on rafts of sargassum, where they would feed on this floating seaweed that carried small shrimp, crabs, and jelly fish. Too many baby turtles, he also discovered, would chomp on almost anything they could get into their hungry mouths. All too often the items included fragments of Styrofoam, droplets of heavy oil, and globs of tar that would glue their mouths together and induce death. Obviously, pollution of the waters and beach habitats posed a serious threat to the turtles' survival.

Carr's research and consequent conservation efforts saved the sea turtles from near extinction. His work did not go without notice. Over the years, he was the recipient of dozens of awards. *Handbook of Turtles*, for example, won the Daniel Giraud Elliot Medal from the National Academy of Sciences in 1952. His *High Jungles and Low*, published in 1953, won the 1955 John Burroughs Medal for exemplary nature writing. One of his most widely acclaimed books, *The Windward Road* (1956; reprint 1979) inspired the formation in 1959 of the Caribbean Conservation Corporation.

In addition to many awards for his books, Carr was honored with a gold medal from the World Wildlife Fund and special recognition by the New York Zoological Society. He was declared an officer in the Netherlands Order of the Golden Ark and awarded the Edward H. Browing citation for outstanding achievements in conservation. In 1979, the Florida State Museum established the Archie F. Carr Medal in his memory, and the University of Florida established the Archie F. Carr postdoctoral fellowship in 1983.

The subtitle of *The Windward Road* is descriptive of its subject matter: *Adventures of a Naturalist on Remote Caribbean Shores.* Carr covers his relationships with the populace and his fascination with exotic Caribbean animals and its sea life. Of special interest is his search for sea turtles and the beaches on which they laid their eggs, as well as his exposition of the life cycle of giant turtles and what steps should be taken to save them from extinction.

Other naturalists and critics adulated the volume for a variety of reasons. Far from being a dry scientific tract, one of them observed that *The Wideward Road* "provides much meat . . . for the natural history fan, and much food for thought for anyone concerned about wildlife conservation." One chapter in the book was singled out for the 1956 O. Henry prize for the best nonfiction short story of the year.

Beautifully written, today *The Windward Road* is considered a conservation classic. Carr's fame, Marston Bates noted, should rest as much on his ability as a writer as on his knowledge of wildlife biology. With *The Windward Road*, Bates wrote, Carr "has established his place in that select company of first-class naturalists who write extremely well." Fully in agreement with Bates, another critic commented that *The Windward Road*, "while not a book of high adventure but a series of events naturally incidental to Mr. Carr's roaming of the Caribbean[, is a] . . . satisfying . . . read."

One success often leads to another, and the reception of *The Windward Road* led to the publication of one of Carr's best-known and most far-reaching works, *The Reptiles* (1963). The volume, an important addition to a Time-Life series on the animal kingdom, offers a wide overview of the subject illustrated with numerous photographs and sections on evolution, adaptation, and the dangers of the extinction of various species. Typical of many, one critic writing in the April 1964 issue of *Natural History* commented that *The Reptiles* "is a scientifically impeccable account of reptiles, their relationships and history, and their way of life. Dr. Carr is to be complimented on his clear, well-written and well-organized text."

Over the years Carr's reputation continued to grow. His status as a zealous conservationist and leading environmentalist was such that after his death in 1987 Congress honored his accomplishments with the naming of the Archie Carr National Wildlife Refuge. Formally established in 1991, its 900 acres were placed under the control of the U.S. Fish and Wildlife Service, State of Florida, Brevard County. The refuge, a twenty-mile section of coastline from Melbourne Beach to Wabassa Beach in Florida, is one of the most important nesting areas for loggerhead sea turtles in the Western Hemisphere. Some 25 percent of all loggerheads and 35 percent of all green turtles nests in the United States are found in this twenty-mile zone. Nesting densities of 1,000 nests per mile have been recorded.

Major refuge activities focus on protecting critical nesting sites from human activity and community developments. Coordination with local and state authorities regarding joint management of beaches, conducting index nesting surveys, and public education programs are also priorities. Protection of these beaches, it is said, is essential to the survival of loggerheads and green turtles in North America. Even the endangered mammoth leatherback turtle, it has been reported, climbs up on these beaches to deposit its eggs in the sand.

Without a doubt, Carr did more than any other individual to save sea turtles from extinction; and so much of what he initiated is being carried on today. The Archie Carr Center for Sea Turtles, for example, has for its central missions to conduct studies in all aspects of the biology of sea turtles, to train graduate students at the University of Florida, and to further sea turtle conservation through the communication of results to the scientific community, management agencies, and environmental organizations throughout the world.

Granted, Carr's efforts were narrowly circumscribed and somewhat emotionally limited, but he is still widely acknowledged for his fifty-five-year dedication to research and his consequent publications. His work is responsible for the stabilization and worldwide eventual increase in the total turtle population. After his death, his wife Majorie Harris Carr, a biologist and distinguished conservationist in her own right, did what she could to see that his work continued. In 1994, she culled the best parts of his ten books and over one hundred articles and papers on natural history that he had authored to be anthologized in *A Naturalist in Florida: A Celebration of Eden.*

GEORGE A. CEVASCO

"Archie Carr." *Who's Who in America with World Notables.* Vol. 9. 1985–1989.

"Archie F. Carr, Jr." [obit.] *Marine Turtle Newsletter* 40 (1987).

Bowen, J. David. "To Save the Green Turtle." *Americas* (1960).

Brody, Jane E. "Archie Carr Zoologist Dies: Devoted Career to Sea Turtle." *New York Times*, May 23, 1987.

Graham, Frank, Jr. "What Matters Most: The Many Works of Archie and Majorie Carr." *Audubon* (1982).

Carrighar, Sally (February 10, 1898–October 9, 1985). As a naturalist and a writer, Carrighar was a participant in the revolution that occurred in the writing of wild animal stories in the latter half of the twentieth century in the United States and Canada. As Ralph Lutts points out in his book *The Wild Animal Story*: "Earlier forms of animal stories tended to be fictional accounts in which the animals were little more than humans in furry or feathery coats, whose narrative role was to instruct and morally elevate the reader." Carrighar was instrumental in surmounting that and in debunking many myths concerning animals and their behavior. She succeeded in popularizing the animal world through a combination of cutting-edge science and a truthful approach to, and depiction of, animal behavior based on careful observation. Pointing out the affinities between animals and the human race, she at the same time tried to avoid the stigma of anthropomorphism, a charge more deserving of earlier writers of animal stories who tended to sentimentalism.

Few women wrote wild animal stories during the first part of the twentieth century, Lutts points out, and none of their works have endured. Two American women changed this, however, Rachel Carson and Carrighar. In her book *Wild Heritage* (1965), Carrighar explained her philosophy of writing about animals: "[T]here are many kinds of activity, nevertheless, which can be studied without drawing conclusions about the state of any animal's consciousness." Activities and interactions with the terrain, the climate, and other animals were her areas of investigation. By carefully examining animal life, a process that involved extensive fieldwork, she felt humans could see their own lives more clearly.

Carrighar was one of the first to write about the natural world utilizing the latest discoveries of eminent ethnologists. According to Lutts, Carrighar's animals became "ecological actors." To quote again from his book *The Wild Animal Story*:

> Both [Rachel] Carson and Carrighar allowed their readers to experience animal lives through the animals' own eyes and other senses. They were, however, very careful in their language to avoid humanizing their animal characters. Theirs was a more behavioristic representation of wild animals, in contrast to earlier writers who presented animals as personalities, although Carrighar was later criticized for anthropomorphism.

Carrighar was born in Cleveland, Ohio, to Perle Avis Harden and George Thomas Beard Wagner. Her birth name was Dorothy Wagner. She attended the public schools there and then went on to Wellesley College. Although she was at Wellesley for only two years, leaving because of fragile health, the Massachusetts school had a great impact on her and her future development. An influential geology course, the great emphasis placed there on writing, and the beautiful campus would all be important influences in her choice of career.

She became involved with, and interested in, nature at an early age. This was due in part to an extremely unhappy childhood spent at the mercy of a mentally unbalanced mother (many believe she was paranoid). Carrighar escaped her mother's cruelties by spending time at her maternal grandparents' home in a more rural part of Ohio (Painesville). At age fifteen she began spending summers at a resort in the Canadian woods. She became acquainted with a very knowledgeable guide connected to the resort who taught her a great deal. She, on her part, was an eager student, who eventually worked as a guide at a fishing lodge for six months while still a teenager. All this experience was laying the groundwork for a life devoted to nature and nature writing.

Carrighar's real involvement with nature would not commence until she had tried and abandoned many careers related to writing, motion pictures, advertising, and radio. Her 1973 autobiography *Home to the Wilderness* (1973) does not even begin to discuss her nature writing until the last one hundred pages. It was in midlife, after times of incredible hardship at the hands of a deeply disturbed mother and after some serious health problems, that Carrighar finally found her true life's work. Thanks to the time spent at Wellesley College, having a campus she described as "400 acres of sweeping, tree-shaded lawns, slopes and dells, a brook and a pond as well as serene Lake Wabam," as well as to a series of what she called "proxy parents" (adults who took the place of her mother and her absentee father), she was prepared to embrace a career as a nature writer. The incident that finally decided her fate is described by her as the "mouse in the radio" story. In an interview with Lewis Nichols for the *New York Times Book Review* of July 19, 1953, Carrighar put it this way:

> "Some years ago I was writing radio shows for a San Francisco advertising agency," she said. "I hated it. I hated the pressure and the dishonesty and was increasingly unhappy. Then, one night—this sounds fantastic, I know—a mouse got into my radio set. It was a singing mouse, and I didn't know there were such things. It sounded a little like a canary, and was building a nest there. After hearing it, I decided right away I'd write about animals."

Carrighar's major contributions to environmentalism are the popularization and, at the same time, truthful representation of animal species that inhabited specific sites in the United States and the Arctic. In an effort to find her way back to nature's principles, she felt it would be helpful to rediscover how animals live. In some ways her life until middle age was a preparation for this very undertaking. She learned valuable research skills at Wellesley. Her musical training (in childhood she took lessons on and played the piano) proved very helpful in accurately recording animal sounds. Her time spent as a professional writer helped her develop what she regarded as a new way to represent animal life. "A new writing technique: truthful as nothing I ever had written had been. At the start I made this test for myself: would what I am saying about this animal seem true to him?" To ensure this truthfulness, Carrighar based her "fictional nature narratives" on the best scientific research of the time, research being conducted by such prestigious institutions as the Museum of Vertebrate Zoology at the University of California at Berkeley, the California Academy of Sciences, and the U.S. Forest Service. As a result of mastering much of this research, she subsequently received a Guggenheim fellowship to study the natural environment of the Arctic, the subject of two of her books.

As "an eloquent chronicler of wildlife," Carrighar wrote many books on animals and their environments during her lifetime. She began with *One Day on Beetle Rock* (1944), written after many years of direct observation. The book presents the stories of nine animals in the High Sierras of California. Those included are a weasel, a Sierra grouse, a chickaree, a black bear, a lizard, a coyote, a deer mule, a steller jay, and a mule deer. Together with the lizard, these birds and mammals lead lives of constant struggle against hunger, other animals, and the weather of the California mountains they inhabit. They all strive to compete for places to live, for food, and among their own species, for life partners. All these nine animals interact with one another as they go about their daily existence in the mountains.

She went on to write *One Day at Teton Marsh* in 1947. This book has as its setting Jackson Hole, Wyoming, in Grand Teton National Park. It includes, because it is about a marsh, a more varied cast of characters than *One Day at Beetle Rock*. Here we have mammals, fish, a crustacean, an insect, and more. The activities of the creatures of Teton Marsh are divided into three locales. In the marsh in general, we have an otter, a cutthroat trout, an osprey, a mink, a varying hare, an American merganser, a moose, a trumpeter swan, and a beaver. In the willow cave are a mosquito and a scud. Finally, on a water lily we find a *Clepsine* leech, a leopard frog, and a *Physa* snail. Again, all these creatures interact with each other as they struggle to survive.

The next book, *Icebound Summer*, published in 1953, finds Carrighar moving beyond the boundaries of the continental United States to explore the Arctic, "the wilderness of the North." She looks at "the birds and animals, fish, seals and whales" that gather on the northwest coast of Alaska during the short summer season, the only season hospitable to animal life. She presents studies of lemmings, seals, a fox, a loon, a tern, a beluga whale, a walrus, a humpback whale, and a golden plover. She continues her exploration of northern areas with her next book *Moonlight at Midday* (1958). In celebration of Alaska's statehood in 1958, she studies its climate, terrain, animals, and most important, its indigenous people, the Eskimos. Her third book on the Arctic, *Wild Voice of the North*, was published in 1959.

In 1975, Carrighar published *The Twilight Seas: A Blue Whale's Journey*, a book about a single species. Concentrating on "the birth, life, travels and adventures, and untimely death, of an immense young Blue," she made accessible to a large audience the incredible story of the blue whale.

She also wrote *Wild Heritage* (1965), her theoretical musings on animals and animal behavior, as well as her autobiography *Home to the Wilderness.* (1973) She stands as an outstanding example of scientific self-education. As a layman she wrote for laymen, but what she wrote contained the latest scientific discoveries about animals and their behavior. Her fictional natural history is entertaining as well as scientifically accurate. Such television programs as *Mutual of Omaha's Wild Kingdom*, with Marlin Perkins, and PBS's *Nature*, with George Page, are a direct result of her work. In fact, the NBC TV network broadcast an episode of *Walt Disney's Wonderful World of Color* featuring *One Day at Teton Marsh*, a film based on Carrighar's book of the same name.

In addition to nature writings, Carrighar tried her hand at other genres including the novel and playwriting. *The Glass Dove*, published in 1962, is a novel about an Ohio family who run an Underground Railroad station at the time of the Civil War. Her play *As Far as They Could Go* (1956) takes place in the Eskimo village of Wales.

Carrighar's lasting contribution to the world is the change she effected in the way we view animals. In her book *Wild Heritage*, Carrighar points to her most important discovery: "In this book, which is a survey of some of the habits, 'ways,' tendencies of animals in the four fields, especially, that we share with them—parenthood, sex, aggressiveness, and play—the similarity between our human and their subhuman natures will not often be emphasized or even mentioned. I believe that in many cases it will nevertheless be evident."

What Carrighar did was to direct our attention to the evident. It was sorely needed.

ARTHUR SHERMAN

Blain, Virginia, Patricia Clements, and Isobel Grundy. *The Feminist Companion to Literature in English*. New Haven, CT, 1990.

Lutts, Ralph H., ed. *The Wild Animal Story*. Philadelphia, PA, 1998.

Mainiero, Lina, ed. *American Women Writers: A Critical Guide from Colonial Times to the Present*. New York, 1979.

Carson, Rachel (May 27, 1907–April 14, 1964). *Silent Spring* (1962), Carson's controversial, award-winning nonfiction work on the use and abuse of chemical pesticides, herbicides, and fungicides, provoked changes in government environmental policy and established the environmental movement as an important issue. Paul Brooks wrote in his 1987 foreword to *Silent Spring*:

> Twenty-five years after original publication, [the work] has more than a historical interest. Such a book bridges the gulf between what C. P. Snow called "the two cultures." Rachel Carson was a realistic, well-trained scientist who possessed the insight and sensitivity of a poet. She had an emotional response to nature, for which she did not apologize. The more she learned, the greater grew what she termed "the sense of wonder." So she succeeded in making a book about death a celebration of life.

He points out that rereading *Silent Spring* today makes one aware of implications that reach far beyond the immediate crisis with which it first dealt. Carson's awakening of Americans to the danger of poisoning the earth with chemicals helped us to recognize many other ways (some little known in her time) in which humankind is degrading the quality of life on our planet.

Before the publication of *Silent Spring*, Carson was already a highly acclaimed writer. She received perhaps more awards for her writing than any other environmentalist. For example, *The Sea around Us*, published in 1951, won the National Book Award for best nonfiction book of 1951 and the John Burroughs Medal for natural history book of outstanding literary quality. The book was later made into a film, which won an Oscar for the best full-length documentary of 1953.

As a writer, Carson enjoyed many other honors. She received honorary doctorates from Pennsylvania College for Women, Oberlin College, Smith College, and the Drexel Institute for Technology. She was made a Fellow of the Royal Society of Literature in England and was elected to the National Institute of Arts and Letters in the United States. Following her death, the Rachel Carson National Wildlife Refuge in Maine was dedicated in 1970; and in 1980, President Jimmy Carter awarded Carson a posthumous Presidential Medal of Freedom, the highest civilian award in the United States. In May 1981, the Rachel Carson stamp was issued by the U.S. Post Office in

Springdale, Pennsylvania, the day following her birthday and in the city of her birth.

Carson was born to Maria McLean and Robert Warden Carson. She had one sister and one brother, each older than she. She began her publishing life early, at age eleven, with contributions of short stories to *St. Nicholas* magazine. Carson graduated from Parnassus High School and attended Pennsylvania College for Women in Pittsburgh (now Chatham College) on scholarship. There she received a B.A. in science magna cum laude. In college she contributed to the student newspaper and literary supplement.

Following graduation, Carson was awarded a fellowship for summer study at Woods Hole Marine Biological Laboratory and a one-year scholarship for graduate study in zoology at Johns Hopkins University, from which she received an M.A. in 1932. The ocean was to become the focus of her writing life, but it was not until 1929, during her summer fellowship at Woods Hole, that she first saw and fell in love with the sea.

Following her father's death in 1935, Carson wrote radio scripts for the U.S. Bureau of Fisheries to help support her mother and herself. She scored first on her civil service examination and was awarded a permanent appointment as junior aquatic biologist at the U.S. Bureau of Fisheries. At this time, she also took on the responsibility of raising her two nieces, following the death of the girls' mother.

During the period from 1936 to 1939, Carson contributed articles to the *Baltimore Sunday Sun* and the *Atlantic Monthly.* Carson also began contributing to *Nature Magazine*, *Collier's*, *Coronet*, *Transatlantic*, and *Field and Stream*, along with other magazines.

In 1941, Carson's first book, *Under the Sea-Wind*, was published. She became an aquatic biologist and assistant to the chief of the Office of Information of the U.S. Fish and Wildlife Service, where she wrote and edited government booklets promoting the consumption of fish to conserve resources during World War II.

From 1947 to 1950, Carson prepared a series titled "Conservation in Action" for the Fish and Wildlife Service. She was also appointed the agency's editor in chief. It was at this time that she experienced undersea diving and a deep-sea voyage to Georges Bank on a government research ship. She also received the Eugene F. Saxon Fellowship to complete *The Sea around Us.*

In 1950, Carson won the George Westinghouse Science Writing Award for "The Birth of an Island," a chapter of *The Sea around Us*, which was published in *Yale Review.* In 1951, she was awarded a Guggenheim Fellowship for work on her next book.

Following the publication of *The Sea around Us*, Carson's life is a catalog of numerous honors and awards, some of which have already been mentioned. This success enabled Carson to quit her job with the Fish and Wildlife Service and devote herself to writing and caring for her great-nephew, following the death of his mother. She built a new home in 1957, and in 1958 her mother died, following a long illness. In 1960, Carson herself had a cancerous tumor removed from one of her breasts. The remainder of her life was consumed with writing and presenting *Silent Spring.* Her last literary contribution was in 1963. She wrote the foreword for Ruth Harrison's *Animal Machines*, a book depicting the cruelty to animals caused by intensified methods of raising livestock. Carson died the following year of cancer and heart disease.

Silent Spring was, according to Justice William O. Douglas, "the most important chronicle of this century for the human race." By the end of 1962, the year of its publication, there were over forty bills in state legislatures to regulate pesticides. For this work alone, Carson received many awards and citations. From the Animal Welfare Institute's Schweitzer Medal and the Women's National Book Association Constance Lindsay Skinner Achievement Award for merit in the realm of books to the Albert Einstein College of Medicine 1963 Achievement Award, Carson received recognition. She was named Conservationist of the Year by the National Wildlife Federation and received the Isaak Walton League's Annual Founders Award, not to mention a citation from the International and U.S. Councils of Women.

However, along with the citations and recognition, Carson also received a great deal of criticism, leveled at her by the chemical industry, which mounted both personal and scientific attacks. For her allegations in *Silent Spring*, Carson received vindication from the President's Science Advisory Committee. Her work also gained additional credibility when she received medals from both the National Audubon Society and the American Geographical Society. Ultimately, she was elected to the American Academy of Arts and Letters.

Carson, in her acknowledgments, explains what prompted her to write her landmark work. She had received a letter from a woman who recounted the bitter experiences she had "of a small world made lifeless" by pesticides. The woman "brought my attention sharply back to a problem with which I had long been concerned," she acknowledged. "I then realized I must write this book." *Silent Spring* opens with "A Fable for Tomorrow," which describes a town made silent by its abuse of chemicals. From first to last, *Silent Spring* stands as the premier book of its kind to use science persuasively to support environmental concerns.

SHELLEY ALEY

Brooks, Paul. *The House of Life: Rachel Carson at Work*. Boston, MA, 1972.

Gartner, Carol B. *Rachel Carson*. New York, 1983.

Graham, Frank, Jr. "Rachel Carson." *EPA Journal* 4 (1978): 5–7, 38.

——. *Since Silent Spring*. Boston, MA, 1970.

Hynes, H. Patricia. *The Recurring Silent Spring*. New York, 1989.

Lear, Linda. *Rachel Carson: Witness for Nature*. New York, 1997.

Sterling, Philip. *Sea and Earth: The Life of Rachel Carson*. New York, 1970.

Carter, Jimmy (October 1, 1924–). In the field of environmentalism Carter will long be remembered for his formulation of a national energy policy and in particular for the Alaska National Interest Lands Conservation Act. The Alaska Act set aside an unprecedented 103 million acres as a national park.

Carter was born in Plains, Georgia, where his family has lived since the early twentieth century. When he was four years old, he and his family moved less than three miles away to a farm in Archery. Named after his father, James Earl Jr. eventually took the more familiar Jimmy as his official designation. Carter was the son of James Earl Sr., a cotton and, later, peanut farmer and small-time merchant. His mother was Bessie Lillian Gordy, popularly known as Miss Lillian, a registered nurse.

When Carter was a child, his father was the central figure in his life and the object of his highest admiration. An expert fisherman, hunter, and agriculturalist, James Sr., by his example, instilled in his young son a love for, and fascination with, the outdoors. From early childhood Carter was a devoted fisherman and hunter.

Recognizing the boy's natural propensity for exploring nearby swamps and woods, the elder Carter taught him survival techniques and thereby encouraged and supported the youngster's self-education in wildlife appreciation. While plowing fields on his father's farm, he never wore shoes, and this quaint habit helped to instill in him a sense of intimacy with the earth. Many years later, upon reflection of his early influences, Carter wrote: "I've often said that, in another life, I'd like to be a forester or a game and fish ranger. There is an almost indescribable sense of peace and pleasure when I'm alone, especially quite early in the day, and wandering over the same remote areas that I first saw as a little boy following my father."

Throughout his childhood, it was Carter's most fervent wish to attend the U.S. Naval Academy. A gifted student, he received his earliest education in Plains Public School. In 1941 he left Archery for Americus to attend Georgia Southwestern College and focused his energies on courses that were recommended in the Annapolis catalog. In 1942 he transferred to Georgia Institute of Technology in Atlanta, where he studied engineering and served in the Navy Reserve Officers Training Corps. Finally, he was accepted at the Naval Academy in Annapolis, Maryland, where he graduated in 1947 with a baccalaureate degree, finishing an impressive fifty-ninth in a class of 820. He went on to study nuclear physics and reactor technology at Union College in Schenectady, New York.

In 1946 Carter married Rosalynn Smith, who lived next door in Plains and who was a friend of his younger sister Ruth. He later served under Admiral Hyman G. Rickover in the navy's nuclear submarine program. In 1953, following his father's death, Carter resigned from the navy, returned to Georgia, and took charge of the family-owned peanut business and farm.

His success in the family enterprise led to his becoming influential in the community, and Carter began to participate in local politics. In 1962 he ran for and won a seat in the state senate, serving two terms. In 1966 he ran for the Democratic gubernatorial nomination but finished third. However, within a brief period of time, he was laying out a strategy for another run in 1970.

Campaigning as the common man, Carter was able to capture the imagination of the Georgia voters and win the nomination. He went on to handily defeat his Republication opponent in the general election. At his inauguration, he proclaimed an end to the era of segregation.

As governor, Carter boldly opposed public works projects that needlessly harmed the environment, such as the construction of large dams simply for the sake of creating work. Carter also pressed state legislators to pass laws to help preserve and protect the environment.

Georgia's constitution did not permit consecutive terms for the governorship. In 1975 Carter left the governor's mansion and immediately began to run for president. As a national unknown coming into the race, he was a welcome addition to the list of Democratic candidates. Americans were disillusioned with Washington insiders due to the Watergate scandal, and Carter appeared to be a breath of fresh air.

After securing his party's nomination, with a strong lead he began his run against incumbent Gerald Ford. Ford had lost favor with the electorate due in large part to his pardon of former President Richard Nixon. Although Ford narrowed the gap considerably, Carter ended up winning, having received 50.1 percent of the popular vote.

On assuming office in 1977, Carter once again took on the persona of the common man. For instance, instead of riding in a limousine, he walked the length of the inaugural parade. This simple and endearing act proved to be a harbinger of his philosophy toward energy consumption. It soon became evident to the new president that part of America's energy problem was rooted in waste.

When he entered the White House, America's dependence on foreign oil had been steadily rising and was then at nearly 50 percent. Energy consumption had hit record highs. The rising cost of imported oil contributed greatly to the inflationary woes that beleaguered the economy. Carter recognized the need to manage natural resources more efficiently. He saw the energy challenge as the "moral equivalent of war." He created a new cabinet-level position, the Department of Energy, naming Dr. James R. Schlesinger its first secretary.

He encouraged Americans to reduce their energy consumption in many small ways, for instance, by turning down their thermostats. He set an example by having solar panels installed on the White House roof and a wood stove placed in his living quarters. In February 1977 he addressed the nation for the first of five times on the energy problem: "The amount of energy being wasted which could be saved is greater than the total energy that we are importing from foreign countries. . . . [W]e will emphasize research on solar energy and other renewable energy sources."

Carter successfully introduced legislation that, among other things, encouraged car pooling and gasohol production through tax incentives; strongly encouraged solar-power development as well as tax incentives for the installation of solar units in homes and other buildings; offered the first incentives to produce corn-based ethanol; and penalized the manufacturers of gas-guzzling automobiles.

Carter signed into law legislation that established a "Superfund"—that is, "a system of insurance premiums collected from the chemical industry to clean up toxic wastes." It was his hope that this fund would significantly reduce the risk that toxic waste sites presented to the well-being of millions of Americans.

As an environmentalist, Carter's crowning achievement was his Alaska Interest Lands Conservation Act, which he signed into law in December 1980. The secretary of the interior during the Carter administration, Cecil Andrus, wrote that "Carter managed to leave behind a legacy of volcanic craters, alpine lakes, ancient forests, tundras needed by grizzly bears, and federal land managers who weren't devoted only to drilling, digging up, and cutting down the great resources of America's forty-ninth state."

The 103 million acres of Alaska, more than 25 percent of the state, set aside as national park land was the largest land conservation initiative in the history of the United States. Andrus noted that "Jimmy Carter was, with Theodore Roosevelt, one of the two most committed conservationists ever to occupy the Oval Office." Although Carter's administration was not without its great successes, negative economic conditions and the hostage crisis in Iran combined to cost him reelection.

On leaving the presidency in 1981, he, working with Rosalynn, established the Carter Center in Atlanta. The Center concentrates on a number of social issues such as human rights, monitoring the electoral process around the world, charitable causes, global health care, and environmental concerns. In 2002 Carter was awarded the Nobel Peace Prize. His political enemies and public detractors, of which there were more than a few, questioned the judgment of the Nobel Foundation. There was no debating his humanitarian achievements, however, and he is widely renowned for his work with Habitat for Humanity.

Carter has likewise enjoyed considerable success as an author, having written more than twenty books that, for the most part, have been well received. In his first book *Why Not the Best*? (1975), he covered his early political career; more important, he stated the motives that led him to seek the presidency of the United States, which he "always looked on . . . with reverence and awe." Idealistically he wrote of an America with a government "as decent, as competent, as compassionate, as good as its people." Missing from this autobiography are Carter's philosophical and political convictions, adverse critics pointed out. One went so far as to carp that Carter had "an Everest complex": he wanted to be president, "to climb the highest mountain because it was there."

When he was governor of Georgia, Carter wrote, he would awaken every morning overjoyed with the challenge of solving the day's problems. As president, he expected his days to be the same, but when he won the office, he soon discovered that the problems were far more convoluted than those he had to deal with when he was governor. Most of those problems he covered in the books he wrote after leaving the White House.

Chief among them are *Keeping Faith: Memories of a President* (1982), *The Blood of Abraham* (1985), and *Turning Point: A Candidate, a State, and a Nation Come of Age* (1992). *Turning Point* has little to say about contemporary affairs, and the volume was attacked for overlooking domestic and foreign difficulties. Captious critics complained that even two-thirds of the more than fifty articles he had written for newspapers and magazines during the early 1990s slighted race relations, economic disparities, personal morality, and public malaise.

As though to demonstrate his verbal creativity, in 1995 he published *Always a Reckoning, and Other Poems*. He then went on to write *The Virtues of Aging* (1998), a volume that stirred little controversy. His latest book, on the contrary, *Palestine: Peace Not Apartheid* (2006), has been the recipient of both bouquets and brickbats. Though he attemped to be objective in his call for a revitalization of the peace process between Palestine and Israel, many concluded that he favored the former at the expense of the latter.

Be that as it may, it is still regrettable that among all that he wrote, he touched on environmentalism only tangentially. Nonetheless, for his work as president and for his contribution to the preservation of the environment, even after leaving office Carter has been recognized repeatedly. In 1979, Carter received the Conservationist of the Year Award, and in 1993 he was awared the Conservationist Year Medal from the National Wildlife Federation. In 1982 he was honored with the Ansel Adams Conservation Award from the Wilderness Society. In 2000 he received the William Penn Mott Jr. Park Leadership Award from the National Parks Association, and the Zayed International Prize for the Environment in 2001.

At a 2007 conference, former Carter adviser Stuart Eizenstat said, "The U.S. today is 50% more energy efficient . . . than we were in 1977, in significant part because of the foundation that Jimmy Carter laid."

MICHAEL HAYES

Borne, Peter G. *Jimmy Carter: A Comprehensive Biography from Plains to Post-Presidency*. New York, 1997.

Domin, Paul. *Jimmy Carter, Public Opinion, and the Search for Values*. New York, 2003.

Glad, Betty. *Jimmy Carter: In Search of the White House*. New York, 1980.

Hargrove, Edwin C. *Jimmy Carter as Peacemaker: A Post-Presidential Biography.* Baton Rouge, LA, 1988.

Lasky, Victor. *Jimmy Carter, the Man and the Myth.* New York, 1979.

Carver, George Washington (1864?–January 5, 1943). A naturalist, artist, agricultural reformer, and educator, Carver was a prophet of sustainable development for poor agricultural communities and one of the most farsighted environmental voices of his time. An advocate of small-scale, organic agriculture, Carver saw clearly the links between land use and poverty in the rural South, and he set about to rectify the situation by endorsing solutions that anticipated the rise of ecological agriculture. Ironically, however, this is not what he is most famous for; indeed, the environmental movement has paid scant attention to him. Instead, Carver is venerated as "the Peanut Man," an iconic African American scientist whose scientific accomplishments and influence on southern agriculture were inflated over time, obscuring his work along conservation lines. Even so, Carver's lasting significance lies in his agricultural and environmental thought.

Carver was born to slaves in the small agricultural community of Diamond Grove in southwestern Missouri, but the precise date—for that matter, the precise year—of his birth is not known. Orphaned as an infant, he was adopted by his former owners, Moses and Susan Carver, who raised him as if he were their own. His chores around the farm were comparatively light (in part because he was often sick), and he spent much of his time wandering the woods and fields of the farm where he first developed an affinity for the natural world. A curious child, he wanted "to know every strange stone, flower, insect, bird, or beast," as he recalled in an autobiographical sketch penned in 1922, though plants, in particular, captivated his attention.

By all indications, his home environment was a loving and enjoyable one, but by the mid-1870s, segregated schooling had effectively become state policy in Missouri, circumscribing Carver's educational opportunities. Consequently, after a brief stint with a private tutor and a few abortive months at a black school in a nearby town, Carver headed to neighboring Kansas in 1877, seeking an education. He spent a few years bouncing from town to town in the eastern portion of the state before settling in the central Kansas village of Minneapolis, where he graduated from high school.

He made plans to attend a small religious college in the northeastern corner of the state that had accepted him by post, but he was turned away at the door when the school discovered he was black.

Discouraged by this turn of events, Carver—like many young men with limited prospects at the time—turned his eyes toward the west, moving to Ness County on Kansas's western plains in 1886. There he built a sod house and began life as a homesteader. He impressed his neighbors, virtually all of whom were white; indeed, the *Ness County News* found "his knowledge of geology, botany and kindred sciences . . . remarkable" and concluded that he was "a pleasant and intelligent man to talk with, and were it not for his dusky skin—no fault of his—he might occupy a different sphere to which his abilities would otherwise entitle him." It is not entirely clear why he abandoned his homestead shortly after purchasing the title to it in 1888, though the fact that the late 1880s were rough years on the high plains marked by drought and blizzards certainly played a role.

Carver made his way to Iowa, where he was befriended by a white family who persuaded him to give college another try. Simpson College, a Methodist school in Indianola, admitted Carver, and he spent a year studying art. His teacher was impressed by his work—especially his botanical paintings, one of which would later win honorable mention at the Chicago World's Fair in 1893—but she grew concerned that few opportunities existed for African American artists. Consequently, she encouraged him to pursue his passion for plants in agriculture rather than art and prevailed upon her father, a well-respected horticulturist at the Iowa Agricultural College (IAC) in Ames, to facilitate Carver's admittance to the school.

Carver proved open to her suggestion. In fact, as an ardent Social Gospeler, he became convinced that God had chosen him to help "his people." Thus, viewing it as an act of martyrdom, he laid down his brushes to study scientific agriculture. The first African American to enroll at the IAC, Carver sat under the tutelage of some of the leading figures in the emerging field of agricultural science, including two future U.S. secretaries of agriculture. His most influential instructor, however, was the less well known Louis Hermann Pammel, a young botanist who introduced Carver to the nascent science of ecology.

Carver distinguished himself first as an undergraduate, then as a graduate student, and won not only the respect of his white professors and peers but their friendship as well. Shortly before Carver became the first African American to earn an advanced degree in agricultural science, Booker T.

Washington, the nationally renowned president of Tuskegee Institute in Alabama, heard of Carver and offered him a position as the head of the Agricultural Department at Tuskegee—a position for which he had almost despaired of finding a qualified black candidate. Carver accepted the offer, which he interpreted as a confirmation of his conviction that God had set him apart to be of special service to black farmers.

Carver's arrival at Tuskegee in October 1896 marked the first time he had ever been to the Deep South, and though Tuskegee would be his home for the remainder of his life, he was utterly unprepared for what he encountered there. The cotton belt of Alabama, as he wrote Pammel shortly after his arrival, was "indeed a new world to Iowa." Its denuded fields tended by impoverished tenant farmers and sown almost exclusively with cotton were a far cry from corn fields and more diversified farms he had known in the Midwest. And while he had certainly encountered racism in Missouri, Kansas, and Iowa, the racial animosities he had experienced there had, for the most part, lacked the violent edge they entailed in the Deep South. Nevertheless, the devoutly religious Carver was confident that with God's help he would be up to the task of restoring the hope of financial independence to tenants and vitality to the region's soils.

Initially Carver approached the task in much the same way as any formally trained agronomist of the era, but over time he developed a unique agricultural vision that diverged in some significant ways from the main currents of agricultural science. For Carver, the progressive farmer was not the one who produced the most cotton but the one who recognized the truth of the adage that waste did not exist except in ignorance. This abhorrence of waste along with an appreciation of the foundational principles of ecology—expressed by Carver as "the mutual dependence of the animal, mineral, and vegetable kingdoms, and how utterly impossible it is for one to exist in a highly organized state without the others"—served as the underpinning for his campaign on behalf of black farmers.

For a number of reasons, Carver's campaign ultimately failed, but it played an enormous role in shaping his environmental and agricultural visions. As the predominant trends of conventional agricultural science—which progressively emphasized increased production, technological innovation, and the application of chemical fertilizers—made little sense for the impoverished tenants of the region, Carver grew to eschew them. Instead, Carver encouraged black farmers to turn to the natural world for a measure of solace from the political and economic uncertainties that plagued them. By thinking "ecologically" and abhorring waste, they could meet many of their needs with things that they had previously overlooked or neglected. There was no need to buy commercial fertilizers when they could produce a better fertilizer by composting swamp muck, decaying leaves, and other organic matter that was free for the taking. Likewise, there was no need to go into debt for food at the local merchant's store when the Great Creator had provided plenty of food in wild fruits and vegetables that many disparaged as "weeds." For that matter, even "rare beauty and fragrance" were available to even the poorest farmers at a cost of only labor if they were willing to transplant wild flowers and shrubs. As an agriculturist, then, Carver advocated a way of thinking rather than a particular regimen.

By the end of his first decade at Tuskegee, Carver's considerable gifts as a public speaker had contributed to a growing reputation, and with the onset of World War I, much of what he had been advocating for tenant farmers attracted a wider, national audience as a result of wartime conservation measures. It was his research on promising southern crops other than cotton, however, that catapulted him to fame. More specifically, his work with peanuts—research he had undertaken as part of his campaign in an effort to find crops that, in contrast to cotton, poor farmers could both market and consume—caught the eye of the nation's peanut lobby, which saw in him a potential spokesman.

Though Carver had researched other plants as well, the pairing was a good one, and he appeared before Congress in January 1921 as an expert witness on behalf of the nation's peanut farmers who wanted a protective tariff. In winning over the skeptical committee (and a tariff for the peanut lobby), he earned for himself national renown. In time, he became an icon: the Peanut Man. Myriad groups, often with antithetical aims, endorsed him. African Americans, for example, held him up as a hero who scuttled the logic of Jim Crow segregation; southern whites embraced him as the exception that proved the rule, an example of what blacks could accomplish within the confines of Jim Crow. Indeed, in 1923 he was improbably honored by both the National Association for the Advancement of Colored People (which bestowed on him its highest award, the Spingarn Medal) and the United Daughters of the Confederacy (who passed a resolution extolling him for his work with peanuts). Religious groups, white and black, exalted him as a scientist who saw no conflict between faith and science. Some wondered how many other "Carvers" whose poverty, race, or educational opportunities had prevented them from fulfilling their potential might be out there, while others held him up as proof that America was the best poor man's country, where a man like Carver could rise from slavery to a position of national prominence.

With so many divergent groups endorsing him, it is hardly surprising that a mythology of sorts came to surround Carver. By the end of the 1920s, his scientific accomplishments attracted disproportionate attention. In fact, Carver was routinely referred to as the "world's greatest chemist," and though his humility played a significant role in the Carver mythology, he did little to refute the claims made about him.

Distracted by his growing fame, Carver essentially abandoned his campaign during World War I, though he returned to it later in life. He shifted his focus from improving the lives of poor, black farmers to convincing southerners that the region's economic salvation rested in developing new uses for overlooked resources and alternative ones for underutilized crops. His emphasis on the latter, especially peanuts, sweet potatoes, cow peas, and soybeans, made him a pioneer in the chemurgy movement—a movement that essentially collapsed in the early 1960s with the proliferation of inexpensive petrochemical products, though its successors today (including the New Uses Council) lead the charge for bio-diesel and ethanol fuels.

By the mid-1930s, Carver had grown disillusioned with the demands his fame entailed, and his health began a steady decline. His concerns about the direction of American agriculture, however, seem to have increased. While there was no way for him to anticipate the kind of widespread application of petrochemical pesticides and fertilizers that became routine in post–World War II America, he began to express some serious misgivings about the direction American agriculture was heading. "To our amazement," Carver wrote in 1936, "we are learning that a tomato may not be a tomato nutritionally speaking, but only a hull or shadow of the savory, nutritious, palatable vegetable it should be." Although it might look "in every way just like an ordinary tomato," he added in 1942, favorably citing the work of another scientist, it could have comparatively few of the nutritional "qualities of a well-grown unfertilized (artificially) tomato." Likewise, he anticipated the kind of argument Rachel Carson would make in *Silent*

Spring (1962) when he pointed out that chemicals put on fields made their way into the body. Those "who eat watermelons," Carver wrote, for example, "know that if they are not exceedingly careful they remain sick as long as the watermelon season lasts, because of the improper use of nitrate of soda."

Carver's deep religious sensibilities made him something of an environmental mystic. Though his beliefs were decidedly more Christian than those of many of his peers in the early conservation movement, his was a religion of nature as much as of the Bible, and he took daily nature walks to commune with the "Great Creator." Accordingly, many of his writings and speeches, especially those from his later years, contained paeans to nature and warnings about violating its limits that would do justice to many of the environmental movement's more celebrated heroes. It was a "fundamental" truth, he maintained, for instance, in 1940, that "nature will drive away those that commit sins against it."

When Carver died, the nation mourned. The mythology that surrounded him, however, overwhelmed discussions of him in subsequent years, obfuscating his conservationist legacy. Indeed, when the modern environmental movement came into its own in the 1960s and 1970s, few among it heralded Carver as a hero. The irony of his fame as the Peanut Man, then, lay not in the fact that his scientific accomplishments were exaggerated but rather inasmuch as it prevented his recognition by groups for which he legitimately served as a forebear.

Men like Theodore Roosevelt, John Muir, Gifford Pinchot, and Liberty Hyde Bailey are some of the more familiar names in the Progressive conservation movement that laid the foundation for modern environmentalism. Even if his concern for black farmers muted his influence to some degree at the time, Carver merits a place beside them. Perhaps no Progressive reformer gave primacy to ecological ideas to the extent Carver did. He was among the first agriculturists to endorse what would later be called organic farming, and his rejection of technological solutions anticipated the emergence of the "appropriate technology" movement, which drew its inspiration from Ernst Schumacher's *Small Is Beautiful* (1973) and advocated the application of technology appropriate not only to a particular environment but to a particular culture, given its circumstances at a particular time. Indeed, Carver's farsighted campaign on behalf of impoverished southern farmers, along with his intuitive and religious appreciation for the natural world, mark him not only as an important rural adviser and agricultural reformer but as a visionary conservationist.

MARK HERSEY

Burchard, Peter Duncan. *George Washington Carver: For His Time and Ours: Special History Study—Natural History Related to George Washington Carver National Monument, Diamond, Missouri.* National Park Service, 2005.

Carver, George Washington. *George Washington Carver Papers at Tuskegee Institute* (microfilm). Ed. John W. Kitchens and Lynne B. Kitchens. Produced with the assistance of and under the sponsorship of the National Historical Publications and Records Commission, General Services Administration, issued by the Division of Behavioral Science Research of the Carver Research Foundation of Tuskegee Institute. Tuskegee, AL, 1975: 67 reels.

Ferrell, John S. *Fruits of Creation: A Look at Global Sustainability as Seen through the Eyes of George Washington Carver.* Wynnewood, PA., 1995.

Hersey, Mark. "Hints and Suggestions to Farmers: George Washington Carver and Rural Conservation in the South." *Environmental History* 11 2 (April 2006): 239–268.

Kremer, Gary. *George Washington Carver: In His Own Words.* Columbia, MO, 1987.

Chapman, Frank Michler (June 12, 1864–November 15, 1945). A field and research ornithologist, museum curator, popular and scientific writer, editor, and public speaker, Chapman devoted his professional life to increasing the general knowledge, appreciation, and protection of birds. Born in the New Jersey countryside (now Englewood) to Lebbeus Chapman Jr., a Wall Street attorney, and Mary Augusta (Parkhurst) Chapman, a gifted musician and gardener, Chapman responded early and instinctively to birdsong. By his early twenties, he determined to pursue his interest through intensive study, leading to a lifelong career in ornithology.

He called himself a mediocre student in his youth, having always been more inclined to pursue out-of-doors activities than his schoolwork. When he graduated from the Englewood Academy in 1880, Chapman took a position at the American Exchange National Bank in New York City, where his father had been a legal council. The younger Chapman felt no particular calling for this line of work but remained with the bank for six years, commuting from his home in New Jersey.

His future as an ornithologist became clear to him when, in response to an American Ornithologists' Union (AOU) call in 1884 for volunteers for a study of bird migrations, he enthusiastically undertook to conduct his observations and prepare his reports in the hours before and after his work at the bank. Far from flagging, his passion for birds increased, as did his concern for their protection. His first efforts were directed at the use of birds in the millinery trade. During two afternoon walks along New York's busy 14th Street, Chapman "sighted" over forty species of birds—feathers as well as entire stuffed birds—decorating the hats of fashion-conscious women. He published his report of these "sightings" in 1886. He later delivered the opening address of the Washington, DC, Audubon Society, "Woman as Bird Enemy." In his *Autobiography of a Bird Lover* (1933), he stated his belief that widespread ignorance of birds led to such unthinking destruction. Convinced that they "must be made real before we could expect to appeal effectively in their behalf," Chapman initiated, through such practical services as public speaking and popular writing, educational efforts in the fundamentals of birdlore that continued all of his life.

At the age of twenty-two, Chapman left the bank to pursue a career as an ornithologist, embarking on an independent study and collecting trip to Florida. In 1888, he was invited by Dr. J. A. Allen, then in charge of the Department of Mammals and Ornithology at the American Museum of Natural History in New York, to join the museum staff as his assistant for a monthly salary of $50, considerably less than he made at the bank. Chapman was promoted to associate curator in 1901 and to curator of birds in 1908. When a separate Department of Birds was formed in 1920, he became its first chairman. Chapman remained with the museum without interruption until his retirement on June 30, 1942.

In his autobiography, Chapman himself divides his forty years with the museum into four periods, each nearly a decade long. In the first, from 1888 to 1897, he concerned himself with the growth of the museum as well as his own self-education as an ornithologist, fieldwork, research, and efforts to promote a popular interest in birds. From 1898 to 1910, he worked on improving the museum's methods of exhibiting birds and mammals, developing habitat groupings stressing the importance of understanding individual species within the context of their native environment. Chapman during this time began to use the camera in bird study. In the third period from 1911 to 1919, he was involved with ornithological surveys of the Andes sponsored by the museum and the ongoing study of the collections that resulted from these lengthy trips into the field. He was interested in issues of the geographical distribution of bird species, particularly their altitudinal distribution and the evolutionary relationships between birds of different altitudinal zones. During this period, from May 1917 to June 1919, he served as well with the Red Cross. Finally, from 1925 to 1932, Chapman focused his field study and research on the birds of the tropical American forest on his beloved Barro Colorado Island in the Panama Canal Zone, concentrating his studies on individual tropical species and their habitats.

Chapman is best remembered for his energetic efforts to deepen popular knowledge of birds and strengthen a public commitment to their preservation. In his autobiography, he stated that "I had . . . a growing belief, which in time became a religion, in the re-creational and spiritual value of close contact with nature, and birds, I was convinced, are nature's most eloquent expressions." He first published the *Handbook of Eastern North American Birds* in 1895. Widely influential and reissued twenty-three times, it was the standard reference on over 500 species of birds for amateur birders, serving both their home study and field guide needs, and remains today a worthy introduction to bird study. From 1894 to 1911, he served as associate editor of the *Auk*, the organ of the AOU. In February 1899 Chapman founded *Bird-Lore*, remaining its sole owner, publisher, and editor for thirty-five years. The precursor of the present-day *Audubon* and associated informally with the Audubon Societies from the beginning, *Bird-Lore* was intended to serve as a popular magazine encouraging bird study and reporting on bird protection as well as other activities of the state societies.

Chapman also saw his work as a museum curator as essentially educational in nature. For the public in general, he was interested in providing carefully and accurately designed habitat displays, employing artists such as Louis Agassiz Fuertes and Frances Lee Jacques in this endeavor. For professional ornithologists, he greatly enlarged the museum's collection of bird skins as a result of the many collecting trips the museum sponsored. His most important professional publications include *The Distribution of Bird Life of Colombia* (1917), *The Distribution of Bird Life of the Urubamba Valley, Peru* (1921), and *The Distribution of Bird Life in Ecuador* (1926). In his autobiography, he lists a selected bibliography of 189 popular and scientific publications.

Chapman has been criticized for his enthusiasm as a collector. He and his colleagues shot thousands of birds in the course of their field research.

It was Chapman, however, who in 1900 first proposed the Christmas Bird Count, which continues to this day, as an alternative to the many local hunts conducted after Christmas dinner. In later life, he gave up hunting.

Chapman was the recipient of a number of honors and held memberships as well as offices in a variety of learned societies. In 1912 he became president of the AOU. In the 1930s he served as chair of the board of directors of the National Association of Audubon Societies. He was given the honorary degree of Doctor of Science by Brown University in 1913. In 1921 he was elected both to the American Philosophical Society and the National Academy of Sciences. He was awarded medals honoring his achievements by the Linnaean Society of New York, the National Academy of Sciences, the AOU, and the memorial associations of both John Burroughs and Theodore Roosevelt. He was an honorary member as well of a number of international ornithological associations, among them the British Ornithologists' Union and the Sociedad Ornitológica del Plata.

Chapman married Fanny Bates Embury on February 24, 1898. She accompanied him and actively assisted him on many of his field expeditions. They had one child, Frank Jr.

SUZANNE ROSS

Brooks, Paul. *Speaking for Nature.* Boston, MA, 1980.

"Frank Chapman" [obit.]. *New York Times,* November 17, 1945.

Gibbons, Felton, and Debora Strom. *Neighbors to the Birds: A History of Birdwatching in America.* New York, 1988.

Murphy, Robert Cushman. "Frank Michler Chapman, 1864–1945." *Auk* 67 (1950): 307–315.

Commoner, Barry (May 28, 1917–). A biologist, educator, conservationist, ecosocialist, and political activist, Commoner is a prime mover in the modern environmental movement. *Time* featured him in its February 2, 1970, issue and dubbed him "the Paul Revere of Ecology." In its cover story, the magazine cautioned "the price of pollution could be the death of man" and adulated Commoner for his concern about the welfare and future of humanity.

In his lectures, articles, and books, he imparts to the general public the considerable dangers to the environment posed by scientific advances and everyday practices. Over the decades, he has warned about the delicate condition of the planet. "Environmental pollution is an incurable disease," he iterates. "It can only be prevented." He continually advocates concern. The main thing to realize, he lays stress on, is that everything is related. His professed hope is to provide an ecological balance to the biosphere.

Commoner was born in Brooklyn to Russian immigrant parents. Though a city boy, in his adolescence, fascinated by nature, he spent much of his free time roaming through Prospect Park. In high school he had a special aptitude for biology, and he began to collect specimens during his hours in the park that he would take home to examine under his microscope. An outstanding student, he was admitted to Columbia University, where he majored in zoology. In 1937, he was awared his bachelor's degree, conferred with honors. He qualified for a university fellowship at Harvard, from which he received a master's degree the following year. In 1941, he was awarded his doctorate in cellular physiology.

His first teaching position as a biology instructor at Queens College, New York, was interrupted in 1942 when he enlisted in the Naval Air Force. One highlight while in uniform was his part in the spraying of pesticides on various Pacific islands against insect-borne diseases. At the time he questioned the use of toxic chemicals and expressed a fear that such activity could bring on ecological disasters. After his discharge, he served as an associate editor at *Science Illustrated* (1946–1947).

In 1947, he was appointed an associate professor of plant physiology at Washington University in St. Louis. Over the years he was named chairman of the Botany Department and university professor of environmental science (1976–1981). In his research, he concentrated on viruses and their effects on the genetic material basic to all living matter; it developed into a long-range interest into fundamental problems inherent in the functioning of cells. Generous grants from the Rockefeller Foundation, the American Cancer Society, and the Lederle Laboratories supported his research.

In the late 1950s he first rose to prominence when he protested the testing of nuclear weapons. Troubled by the apparent indifference of governmental authorities to the dangers of radioactive fallout and the presence of strontium-90 in the atmosphere following bomb tests, he helped organize the St. Louis Committee for Nuclear Information (later known as the Committee for Environmental Information). He served as the committee's vice president from 1958 until 1965; as president, 1965 to 1966. Combined efforts resulted in the adoption of a nuclear test-ban treaty in 1963.

In 1966, he published his first book. *Science and Survival.* As its title indicates, the work is a study of humanity in a modern technological society. In it he asserts "the age of innocent faith in science and technology may be over." Science was advancing out of control, and scientists should no longer remain somewhat passive about their discoveries. Overspecialization had become such that individual scientists were often so engrossed in their individual projects that they often failed to see the whole picture. "Science can reveal the depth of the crisis," his book concludes, "but only social action can resolve it."

Commoner's remonstrations were widely being heard in the 1970s. His warnings began to generate political action. Congress passed legislation dealing with clean air, pure water, conservation, and other environmental concerns; still, he was disappointed in what was actually being accomplished. Too many laws, he complained, were being designed not to solve crises but rather to delay them. In his second book, *The Closing Circle: Nature, Man, and Technology* (1971), he amplified such topics as smog, fertilizer-poisoned water, and detergent foam. Technology was chiefly responsible for environmental degradation. "We must learn to restore to nature the wealth that we borrow from it," he pleaded.

In *The Poverty of Power: Energy and the Economic Crises* (1976) and *The Politics of Energy* (1979), he advocated "a national policy for the transition from the present, non-renewable energy system to a renewable one." Though not popular with the habits and preferences of the American public, he called for the use of solar power, trains in preference to private automobiles, and methane or gasahol rather than gasoline.

Always the activist, in 1980 Commoner founded the Citizens Party to serve as the political conveyor of his ecological principles. Since none of the presidential candidates gave voice to environmental issues, or did so at best in superficial fashions, he decided to run for president as the candidate of the Citizens Party. The peak of his campaign was reached in Albuquerque when a local reporter asked him: "Dr. Commoner, are you a serious contender or are you just running on the issues?" Despite his emphasis on the issues—or because of his emphasis on the issues—he finished fifth and received only a quarter of 1 percent of the vote.

On reflection, he admitted that it was probably a mistake to run a presidential campaign. It would have been more sensible "to run in the primaries and make a good showing in a few states and make a point there." A lesser person might have become depressed by such a political defeat—not Commoner. He consoled himself with the realization that his message, as radical as it seemed at the time, had still moved thousands to support him at the ballot box. "I'm an eternal optimist," he resolved, and added, "I think eventually people will come around."

In 1981, he accepted an appointment to Queens College as professor of earth and environmental sciences; here he established a Center for the Biology of Natural Sciences. He was also invited to serve as a visiting professor at the Albert Einstein College of Medicine. His research at Queens resulted in several important discoveries. Foremost, perhaps, was the detection of the origin of dioxin in trash-burning incinerators and the development of a computer model to track the long-range transport of dioxin and other pollutants from their initial sources through the food chain into the human diet. The model proved invaluable in evaluating dioxin contamination of milk on dairy farms from Vermont to Wisconsin.

Commoner's ability to write in nontechnical language made him one of the best-known environmentalists during the 1990s. His *Making Peace with the Planet* (1990) was virtually a bestseller. In it he covers the origins and significance of ecology and stresses what must be done to avoid further deterioration of the environment. As for global warming, he argues that "the only rational solution is to change the way . . . we do transportation, energy production, agriculture, and a good deal of manufacturing." Our problems originate in human activities. Most of the "greening" that we see now fails to realize that action has to be taken on what is produced and how it is produced. "No one . . . would deny that we're getting warmer," he points out. Is this due to things people have chosen? He responds, "All of the things we have chosen to do include the release of materials like carbon dioxide, which affects the retention of heat by the planet, and we are at a high point in the heating-cooling cycle." It seems obvious to Commoner, accordingly, that the argument that there are natural ways in which the temperature fluctuates is a spurious one. "If we accept that we're in a cycle," he contends, "it's idiocy to increase the high point."

At one time, he strongly advocated the practice of recycling. Clearly, he maintains, this is something we should all do, but it has to be remembered that many individuals live in cramped quarters. They cannot fit containers in the limited areas in which they live. Poverty has to be taken into account, since it conditions what can actually be accomplished.

Many individuals do not have the inclination to recycle simply because they live a day-by-day existence. Commoner concedes that he has grown more sensitive to recycling. "I have never been an eco-freak," he protests. His aim is to promote activities that result in sensible behavior.

He is inflexible, however, when it comes to using nuclear power instead of fossil fuels. Use of the former, he contends, is a shortsighted solution. "It may be a way of producing energy without carbon dioxide, but every activity that increases the amount to which we may be exposed is idiotic." There is not a life-and-death reason to do it. "We still have used fuel sitting all over the place. I think the fact that some people who have established reputations as environmentalists have adopted this is appalling." In short, he demands nothing less than the complete outlawing of all nuclear testing.

Commoner's reputation is such that in 1995 he was invited to be a featured speaker at an Earth Day Conference held at Darmouth College to commemorate the twenty-fifth anniversary of that most important day. In his presentation he called for the government to formulate an industrial policy that would promote such things as organic farming and the improvement of electric motors to serve as sources of clean energy. Most important, he encouraged a preventive strategy that increases production without pollution.

In the same year, on October 21 he was cited by the *Earth Times* as the scientist "who made a difference world wide," and he was labeled "The dean of the environmental movement." Such adulation is a consequence of his life's work supported by his five books and over one hundred articles and papers in lay and professional publications. He is a fellow of the American Association for the Advancement of Science and a board member of various scientific organizations from the American Society of Naturalists to the American Institute for Biological Sciences. Additionally, he is the recipient of several honorary degrees, and he has a star on the St. Louis Walk of Fame.

Now in his nineties, he is still affiliated with Queens College, though he has stepped down as director of its Center for the Biology of Natural Systems. He is currently working on another book, which focuses on whether DNA alone is responsible for an organism's traits. Whatever else he may accomplish, his ecological philosophy can be found in his four essential laws:

1. Everything is connected to everything else. There is one ecosphere for all living organisms, and what affects one affects all.
2. Everything must go somewhere. There is no "waste" in nature, and there is no "away" to which things can be thrown.
3. Nature knows best. Humankind has fashioned technology to improve on nature, but such change in a natural system is likely to be detrimental to that system.
4. There is no such thing as a "free lunch." In nature, both sides of the equation must balance.

GEORGE A. CEVASCO

"Barry Commoner." *Contemporary Authors.* Detroit, MI, 2000.
"Barry Commoner." *Who's Who in America.* Chicago, IL, 2004.
Chisholm, Anne. *Philosophers of the Earth.* New York, 1972.
Egan, Michael. *Barry Commoner and the Science of Survival.* Boston, MA, 2007.
Vinciguerra, Thomas. "A Conversation with Barry Commoner." *New York Times,* June 19, 2007.

Cottam, Clarence (January 1, 1899–March 30, 1974). A biologist, conservationist, and educator, Cottam is best known for authoring several books on wildlife and insects and for his research and testimony on environmental pollution, especially on the effects of pesticides on wildlife. Cottam was born in St. George, Utah, to Thomas P. and Emmaline (Jarvis) Cottam. He graduated with his B.S. (1926) and M.S. (1927) degrees in biology from Brigham Young University and his Ph. D. (1936) in biology and ornithology from George Washington University. A member of the Church of Jesus Christ of Latter-Day Saints, he served missions in Texas and Kansas for his church. His religious beliefs influenced his professional as well as personal life, giving deeper meaning to his views that both humans and wildlife could occupy the same territory.

Following his graduation from Brigham Young, Cottam was hired by the university to instruct chemistry and biology classes. In 1929, Cottam began a career in government service, joining the food habits research section of the Biological Survey of the U.S. Department of Agriculture as a biologist. In 1934, he was advanced to chief of the section, and when the Fish and Wildlife Service was formed in 1939 by consolidation of the Bureaus of Biological Survey and Fisheries of the U. S. Department of Interior,

Cottam was named senior biologist. From 1946 until his resignation in 1954, Cottam was assistant director of the service.

During his years with the Fish and Wildlife Service, Cottam promoted the development and implementation of sound national wildlife and game management policies by directing and participating in careful studies, testifying at public hearings on legislation, and lecturing before groups of state wildlife managers and citizens. Of particular note was Cottam's early leadership in pointing to the detrimental effect of environmental pollution by insecticides and pesticides. While he was an early proponent of DDT, the weed killer 2, 4-D, as a means of controlling the rodents, he soon became an active opponent. In the early 1940s, the U. S. Fish and Wildlife Service directed much of its research on creating an effective new rodent poison. After a long search, rats seemed to take to a highly poisonous insecticide, sodium fluoroacetate. Preliminary experiments showed the insecticide to be so poisonous that one pound was sufficient to kill 1.8 million rodents. Even though the toxicity levels of DDT were considered biologically dangerous, the desire to control rats and other unwanted rodents took precidence, and initially the benefits appeared to outweigh the deficits. As part of his endorsement, Cottam coauthored with Robert S. Zim an article titled "Bad News for Brother Rat," which appeared in the *Saturday Evening Post*, on November 10, 1945. Yet Cottam did not accept the initial "success" of DDT to be the final report. When the true extent of this pesticide began to indicate a long-term, widespread pollution of the environment, Cottam began pointing out the dangers of indiscriminate use of insecticides and pesticides. The following year he wrote a government circular titled "DDT—Its Effects on Fish and Wildlife," which represented an about-face from his previous endorsement.

Cottam was instrumental in defeating proposals for a number of large-scale engineering and pest-control projects that were sponsored by federal and state governments, presenting evidence that these programs were unnecessary, extravagent, and potentially damaging to the ecological balance and, in some cases, to human health. Specifically, he opposed a government-sponsored program to eradicate the fire ant, and in that connection he conducted meticulous studies to prove that there were no instances of economic loss, livestock depredation, or human injury attributable to the fire ant that were sufficient to justify the poisoning of land on a mass scale and at great expense. Cottam attacked the federal plan as "the most dangerous application of pesticides I've heard of" and said it could "sterilize the bays all along the Gulf Coast."

Cottam left government service in 1954 in response to a request by Brigham Young University that he set up its new College of Biology and Agriculture. He held the title of dean of the college to 1958, but after his organizational work was completed in 1955, he left the university to accept an appointment as the first director of the Rob and Bessie Welder Wildlife Foundation in Sinton, Texas, a post he held until the close of his life.

Welder, a cattleman with large oil holdings, bequeathed funds for the foundation and specified in his will that its purpose was to further public education in conservation, wildlife appreciation, and the relation of wildlife to domesticated animals. The foundation was also to perform, support, and encourage study and scientific research related to wildlife propagation, growth, and development. The foundation funded research in conservation by graduate students, a significant number of whom went on to achieve prominence in the field of ecology. During Cottam's tenure, 145 fellowships, involving a total expenditure of $575,000, were granted to college and university students. The foundation was credited with making a substantial impact on the growth of public interest in environmental problems, and for that it received the American Motors Corporation Conservation Award for 1971.

Cottam detected a lack of objectivity in government studies on wildlife-pesticide relations, and this led him to undertake further research on the matter. As a result, he became an ardent campaigner against environmental pollution and decided to join with Rachel Carson as a consultant on biological principles for her highly influential antipesticide book *Silent Spring* (1962). Cottam's efforts contributed significantly to bringing about the passage in 1970 of legislation banning or restricting the use of many pesticides on all lands controlled by the U. S. Department of the Interior and in programs managed by that department. Additionally, a law was passed providing that this department share with the U.S. Department of Agriculture all future decision making on pesticide use.

Complementing his work in the area of wildlife-pesticide relations, Cottam was active in the cause of conservation. During 1967–1968, he helped campaign successfully for the creation of the Padre Island National Seashore on the Gulf of Mexico and for a major expansion of the Arkansas Wildlife Refuge in southeastern Texas, an expansion that helped save the whooping crane from extinction. In both campaigns, he was able to win support by demonstrating the economic advantages of conservation, although his personal motives were scientific and aesthetic.

Throughout his career, Cottam engaged in extensive research. He was especially interested in populations of bird species in various areas, which led to several publications, notably *Food Habits of North American Diving Ducks* (1939), *Ecological Studies of Birds in Utah* (coauthored with Angus M. Woodbury; 1962), and *Whitewings: The Life History, Status and Management of the White-Winged Dove* (coedited with James B. Trefethen; 1968). Cottam researched the effects of mosquito-control ditching on marshes and of large-scale government engineering projects on wildlife and ecology. His earlier research and interest in insects led to the publication *Insects: A Guide to Familiar American Insects,* coauthored with Herbert S. Zim (1951). This book has become a standard reader as an introduction to American insects and is still in print. Additionally, Cottam contributed chapters or sections to several books and published some 250 articles for journals such as *Audubon Magazine, National Parks Magazine, Living Wilderness, Sierra Club Bulletin*, and *Biological Science.*

For his contributions to conservation and environmental quality, Cottam received many awards, among others the Aldo Leopold Award from the Wildlife Society, 1955, for recognition of outstanding achievement in the wildlife and natural resources conservation field; the Audubon Medal from the National Audubon Society, 1961, for distinguished individual service to conservation; and the Poage Humanitarian Award from the Society for Animal Protection, 1962. He was a member of a number of environmental, outdoor, and biological organizations and served on the boards of the Wildlife Society of America (president, 1949–1950), the Texas Ornithological Society (president, 1957), and the National Parks Association (president or chairman of the board, 1960–1974).

Cottam relied on meticulous research for his motivations, and this in turn allowed him to win over many individuals to his point of view. A lasting legacy was his attacks on the indiscriminate use of pesticides and its detrimental effects on wildlife, a legacy that profoundly influenced the management of the country's wildlife. He challenged public assumptions about insecticides and pesticides at a time when they were being used widely and in effect averted grave blunders in the handling of nature.

Following Cottam's death, the Zoological Society of Houston, Texas, named him the recipient of the Alban-Heiser Award, and the annual Cottam lectures on natural history were established at the University of Utah in honor of him and his brother, Walter P. Cottam, who was also a conservationist. He was married to Margery (Brown) Cottam, and together they had four daughters. A final legacy was that he helped inspire the next generation of conservationists.

BYRON ANDERSON

"Biologist Cottam Receives Audubon Medal." *Audubon*, January–February 1962, 10–11.

"Clarence Cottam: Nationally Acclaimed Conservationist." *National Parks and Conservation Magazine* 48 (June 1974): 26.

"Cottam, Clarence." *Contemporary Authors*. 1981.

"Cottam, Clarence." *National Cyclopaedia of American Biography*. Vol. 58. 1979.

Thorup, H. Christian. "Clarence Cottam: Conservatonist, the Welder Years." Ph.D. diss., Brigham Young University, 1983.

Curwood, James Oliver (June 12, 1878–August 13, 1927). Described by his biographer H. D. Swiggett as the man who knew the beautiful Canadian Northwest better than any other, Curwood was a journalist, adventurer, conservationist, and popular fiction writer of the out-of-doors. He was born the last of three children to James Moran and Abigail (Griffin) Curwood in Owosso, Michigan. His father was related to Captain Marryat, a writer of sea stories, and his mother, according to the legend, was remotely descended from an Indian princess. Much of Curwood's youth was spent in Vermillion, Ohio, near Lake Erie. According to his autobiography, he dreamed of being an author from early childhood. Though he lacked the writing skills, he improvised a 200,000-word novel by the time he was nine years of age. On November 23, 1894, his first published story appeared in the local newspaper, the *Argus*.

Curwood's education was lackluster and sporadic. He was expelled from school, and at sixteen years of age, he embarked on a bicycle tour of many of the southern states. At seventeen he traveled in a carriage, selling proprietary medicines. Curwood returned to Michigan and for two years studied at the University of Michigan. During the next eight years, he was employed as a reporter and later editor of the Detroit *News-Tribune*. He wrote stories he felt people would love to read by combining tales of nature and adventure. His stories reached beyond the newspaper and began to be accepted regularly by popular magazines such as *Gray Goose Magazine*, *Munsey's Magazine*, *Outing*, and *Good Housekeeping*. In addition, he contributed more than one hundred nature sketches to *Leslie's Weekly*. While employed by the *News-Tribune*, he was introduced to Canada by M. V. "Mac" MacInness, the Canadian Immigration Department representative in Detroit. The introduction proved vastly significant in Curwood's life and writing.

Beginning in 1906, Curwood started to write novels, and a year later, he resigned his position as editor to devote a full-time effort to writing. His first novel, *The Courage of Captain Plum*, appeared in 1908, followed in the same year by *The Wolf Hunters*. He then went on a long vacation to Hudson's Bay, an area that became the ruling passion of his life. He began to write about the vast wilderness of the Peace River country, the great reaches of the Athabasca and Mackenzie River districts, the solitary Arctic plains, and the uninhabited forests. The Canadian government recognized the rising popularity of his stories and hired him to explore the prairie provinces of the West and further up into the North. Curwood gathered material for articles and stories, and these, in turn, were intended to induce settlers into that country. He was the only American ever to be employed by the Canadian government as an exploratory and descriptive writer.

All of Curwood's stories dealt with the outdoors, and usually the setting was the northwest territory of the United States and Canada. Curwood lived many of the stories about which he wrote. He and his second wife, Ethel Greenwood Curwood, would for months bury themselves in the wilderness hundreds of miles from civilization. With wilderness as the background, his novels were infused with adventure and romance. A flavoring of the novels can be sensed, in part, by the titles, for example, *Philip Steel of the Royal Mounted Police* (1911), *The Honor of the Big Snows* (1911), *Isobel: A Romance of the Northern Trail* (1913), and *The Valley of Silent Men* (1920). Of great success was his describing the northern wilderness areas as "God's country." This phrase ended up in numerous articles as well as in the titles of several novels including *God's Country and the Woman* (1915). *Back to God's Country* (1920), and *God's Country—Trail to Happiness* (1921). The latter was composed of four essays that summarized the author's pantheistic views and feelings about nature. Curwood defined God's country as "green forests and waters splattered with golden sun."

Curwood's success as a writer did not happen over night. He wrote for ten years before he sold his first story and twenty-five years before he made a comfortable living at it. His advice to authors was the same that he took for himself: "hard and steady work for years, with a fixed purpose." Curwood had great success in lining up publishing contracts first with Bobbs-Merrill, later bought out by Harper and Brothers, followed by contracts with Doubleday, Page and Company, and finally, beginning in 1919, the Cosmopolitan Book Corporation.

Curwood's novels and stories depicted well the immensity of the Far North—the forest isolation, untrodden fields of snow, the silent places where no trail had yet been blazed, the glory of the Northern Lights, and the quiet, loneliness, and peace of the wilderness. Set against this background, Curwood's stories of romance and adventure slowly gained in popularity. Sales of Curwood's novels increased steadily. His contract with the Cosmopolitan Book Corporation was enhanced with the use of modern advertising and sales methods and proved unexpectedly lucrative. Prior to the contract, Curwood's novels sold an average of 10,000 copies. *The River's End: A New Story of God's Country* (1919), the first Cosmopolitan title, sold more than 100,000 copies in the initial edition, no doubt facilitated by the addition of illustrations by Dean Cornwell. *The River's End* is considered by Swiggett as his greatest and finest work. This was followed by *The Valley of Silent Men* (1920), which had an initial print run of 105,000 copies. The novel with the greatest sales overall was *Kazan* (1914), a story about a wolf dog, which sold over 500,000 copies. The success of this novel was followed by a sequel, *Barbee, Son of Kazan* (1923). The Kazan novels proved popular in both England and France and competed for a place held by Jack London's *White Fang* (1906).

The success of Curwood's later novels kept his earlier works in print. As a result, most of his books ran to five editions or more, and a number were translated into foreign languages. Curwood's novels proved popular in Europe, especially France. French readers, it was surmised, saw America as wild and untamed, and Curwood's novels met their expectations and stereotypes. Curwood's success later found its way to new mediums. Eight of his books were eventually made into full-length feature films, though these appeared after his death, mostly during the latter 1940s and early 1950s. Exceptions were *The River's End*, which was released by Warner Brothers in 1930 and again in 1940, and *The Trail Beyond*, from Lone Star Productions in 1934, starring John Wayne and Noah Beery and based on his novel *The Wolf Hunters* (1908). His novel *Nomads of the North* (1919) was released by the Associated British and Pathe Film Distributors as *Northern Patrol* in 1954 and by Disney as *Nikki, Wild Dog of the North* in 1961. The movie *The Bear*, released by RCA Columbia Pictures in 1990, was based on Curwood's *The Grizzly King* (1916). Several of the movies have since been released on videocassette and DVD, giving Curwood a modern-day presence.

Curwood's writing was not without its critics. He accredited the popularity of his writing to the fact that it was clean and wholesome. His critics charged that his heros and heroines were flawless in virtue, unbelievably strong in character, and too good to be true. The gray areas of life were missing—the bad were always punished, the good rewarded, and justice served. Curwood weaved these general themes through all his novels, causing later critics to complain that his works had become standardized and nothing of consequence occurred that had not been previously written. His novels were criticized for having well-worn plots with saintly and uninteresting characters. Curwood, on the other hand, believed that happiness did come to those who deserved it, and he wrote of things to which all individuals aspire.

While writing was central to Curwood' s life, he was also active in committees and associations working for conservation of wildlife and the forests. He asserted that the country's animals and forests were being destroyed by humans and politics. Curwood was particularly appalled by the situation in his own state of Michigan. Here he urged that politics be eliminated and that individuals properly prepared by study and experience be delegated to take charge of the state's natural resources. In 1927, he was made chairman of the Game, Fish, and Wildlife Committee of the Conservation Department of the state of Michigan. In this capacity, he was able to limit the capture of certain species of birds, set a lower bag limit on others, and eliminate spears as a hunting weapon. Later, he accepted an offer to become the head of the Izaak Walton League.

He furthered his involvement in conservation by personally changing from being an enthusiastic hunter to one who embraced wildlife and nature. He wrote, "It is not wild life that is at war with man, but man that is at war with wild life." He admitted to being a "killer" of animals and foreswore the activities of a hunter. Where at one time he took pride in the diversity of his wildlife killings, as a writer he took pride in getting others to love wildlife and nature. It appears that he was particularly stirred in his role as a conservationist by the magnanimous conduct of the bear, an animal he came to know and love. The bear shows up in a number of Curwood's writings.

Curwood's primary contribution to conservation, however, lay in his stories and writing. He knew it was impossible for the great multitudes to go out and find nature as he had experienced it, so he brought what he could to millions of readers through his writings. In his books, readers could take, for example, an 1,800-mile trip along the Arctic coast and live with the Eskimo as presented in *The Alaskan* (1923). In all, Curwood wrote thirty-two novels; one historical work, *The Great Lakes* (1909); and two autobiographical works, *The Glory of Living: The Autobiography of an Adventurous Boy Who Grew into a Writer and Lover of Life* (1928) and *Son of the Forests* (completed by Dorothea A. Bryant; 1930).

Other novels by Curwood include *The Gold Hunters* (1909), *The Danger Trail* (1910), *Flower of the North* (1912), *The Hunted Woman* (1916), *The Courage of Marge O'Doone* (1918), *The Flaming Forest* (1921), *The Golden Snare* (1921), *The Country Beyond* (1922), *A Gentleman of Courage* (1924), *The Ancient Highway* (1925), *The Black Hunter* (1926), *Swift Lightning* (1926), *The Plains of Abraham* (1928), and *The Crippled Lady of Peribonka* (1929). One additional work, *Green Timber*, was completed by Dorothea Bryant and published in 1913.

In all, it has been estimated that Curwood's novels reached some 7 million readers and his articles and short stories many more. There is no question that he loved nature and all living things and that he placed this love into a narrative of color and adventure that popular fiction readers enjoyed. Publication rights to many of Curwood's books were later purchased and reprinted by other publishers, some for years beyond his death, including many in foreign languages. As late as 1990, Newmarket Press brought out *Barbee, the Story of a Wolf Dog.* The popularity of the movie *The Bear* caused Newmarket to publish Curwood's original *The Grizzly King* under a new title, *The Bear: A Novel* (1990), which was reprinted by Curley Publishers in 1992.

Overall, it can be said that while some of the themes in Curwood's novels have maintained their popularity, Curwood as a novelist has fallen into obscurity. Whereas he was once popular and mentioned along with Jack London and Upton Sinclair, few today would recognize his name. Given that Curwood never rewrote or redrafted any of his writing except to correct grammar and spelling suggests some limitations in his style. Curwood's literary reputation did not hold up in the long run. Then why was he so popular? The answer is, in part, that he was a writer for his time. His novels and stories were simple and straightforward, and he possessed all the elements of a good storyteller.

Curwood was a vivid, forceful, and immensely popular writer who brought the message of nature, wildlife, and conservation to millions of readers. He wrote, "It is my ambition to take readers with me into the heart of nature." The out-of-doors life was for him a universal panacea. He encouraged all to partake and experience nature and enjoy all its benefits. He even believed that he would live to be a hundred years old, but his life was cut short by blood poisoning possibly caused by a spider bite. He died in Owosso, Michigan, his birthplace.

Curwood was twice married. On his death he left behind his second wife, Ethel, who had experienced firsthand so much of what Curwood had written, and their son. His legacy lies in his writings, which acted as a liaison between the masses and nature. What Curwood did—perhaps not by design but inadvertently—was to lead millions of readers a step closer to an appreciation for nature and wildlife.

BYRON ANDERSON

"Curwood, James Oliver." *Dictionary of American Biography.* Vol. 2. 1958.

"Curwood, James Oliver." *National Cyclopaedia of American Biography.* Vol. 21. 1931.

Swiggett, H. D. *James Oliver Curwood: Disciple of the Wilds.* New York, 1943.

Darling, Jay Norwood (October 21, 1876–February 12, 1962). According to his biographer David Lendt, Darling is remembered today for his double career—the first as a Pulitzer Prize–winning cartoonist with the *Des Moines Register,* the second as an influential conservationist. While only a hundred of his more than 15,000 political cartoons carried an environmental theme (more popular topics included the plight of the farmer and the failings of FDR's New Deal), Darling used his notoriety as an illustrator to become one of America's most effective Depression-era advocates for wildlife.

Darling was born in Norwood, Michigan. His parents, Marcellus and Clara, left the Michigan town soon after his birth and relocated the family several times before Marcellus assumed the position as minister of the First Congregational Church of Sioux City, Iowa. Northwestern Iowa, in 1886, was still a fairly wild place, and at the fork of the Missouri and Big Sioux Rivers, Jay's hunting and fishing excursions were his lessons in nature. The progressive deterioration of this pristine countryside throughout Darling's lifetime also was the root of his environmental ethic. Reminiscing in a letter in 1944, he wrote, "If I could put together all the virgin landscapes which I knew in my youth and show what has happened to them in one generation, it would be the best object lesson in conservation that could be printed."

Darling often drew pictures as an adolescent, but his first published sketches were in his college yearbook at Wisconsin's Beloit College. These works were the first to carry the signature "Ding," the name that would accompany his political cartoons for the next fifty years. The term *Ding* was a contraction of his family name and had been identified previously with both his father and his older brother Frank. Unfortunately for Jay, the yearbook drawings were unflattering caricatures of Beloit College faculty, and they may have contributed, along with poor grades, to his suspension from school when he was a junior. He was readmitted a year later and graduated in 1900.

Darling's career as a cartoonist began as a reporter for the *Sioux City Journal.* When a local attorney refused to be photographed for a trial story, Darling ran a caricature instead. The drawing was popular, and Darling soon was providing cartoons on a regular basis. In 1906, Darling married Genevieve Pendleton, and while on their honeymoon, he was offered a position as political cartoonist for the *Des Moines Register and Leader.* With daily front-page cartoons for the *Register and Leader,* fame quickly reached beyond the boundaries of Iowa, and twice during the 1910s, Darling took positions with newspapers in New York. He realized, however, that he was happier and more productive in Iowa, and both times returned to Des Moines. In 1916, Darling realized the ideal situation, simultaneously working as staff on the *Des Moines Register and Leader* and as syndicated cartoonist with the *New York Herald Tribune* syndicate.

Ding Darling twice won the Pulitzer Prize. The first was in 1924 for a drawing titled "In Good Old USA," which proclaimed the value of hard work by noting the simple beginnings of Herbert Hoover, Warren G. Harding, and Frederick Peterson, a well-known New York neurologist. The second Pulitzer was in 1943, for a drawing that depicted Washington, DC, buried in a mountain of paperwork.

Darling's contributions to wildlife began, not surprisingly, in his home state of Iowa. He pressed government officials for the establishment of a state agency for wildlife and, in 1931, was named as one of the original board members of the newly formed Iowa Fish and Game Commission. He donated $9,000 of his own money to establish at Iowa State College (now Iowa State University) the nation's first college program of wildlife research. Years later, he convinced industrial leaders from Dupont, Remington Arms, and other corporations to fund the expansion of the Iowa wildlife research model nationwide.

Although Darling did not draw environmental cartoons frequently, he was noted for them, and this, combined with his role in Iowa conservation, made him an environmental figure of national stature. When the FDR administration established a three-person committee to recommend a program for the restoration and conservation of migratory fowl, they asked Darling, along with John C. Merriam, of the Smithsonian Institution, and Thomas Beck, editor of *Collier's* magazine and chair of the Connecticut State Board of Fisheries and Game, to serve. When Merriam declined, he was replaced by Aldo Leopold. The recommendations of this committee included the purchase of 12 to 50 million acres of submarginal land for wildlife habitat and appropriation of $50 million to accomplish that end.

The agency responsible for carrying out the committee's recommendations was the Bureau of Biological Survey, forerunner of the U.S. Fish and

Wildlife Service. While the service now is part of the Department of the Interior, the bureau of the 1930s was in the Department of Agriculture. The secretary of agriculture was Henry Wallace, future vice president and also fellow Iowan and friend of Darling. When the bureau failed to move on the committee's recommendations, it led to the resignation of Paul G. Redington, bureau chief, and the subsequent offer of the job to Darling.

On March 10, 1934, Darling was appointed chief of the Bureau of Biological Survey. Only a few days earlier, FDR had signed the Duck Stamp Act into law, and Darling would design the first federal duck stamp. During his tenure, Darling freed over $10 million from a tight federal budget and purchased 4.5 million acres of land for the nation's young wildlife refuge system.

Darling held the position of bureau chief for less than two years, but shortly before his resignation, he convinced Roosevelt to convene a North American wildlife conference in Washington. Darling recognized, as did many others, that popular support for wildlife conservation in the United States suffered from lack of coordination. He recommended at the conference that the hundreds of small conservation organizations unite under a single federation. Immediately a constitution for the General Wildlife Federation (now the National Wildlife Federation) was adopted, and Darling was elected the federation's first president.

Darling retired from the *Des Moines Register* in April 1949. In later years, poor health restricted Darling's conservation efforts to personal letter-writing campaigns, most notably opposing various dam projects of the Army Corps of Engineers. In 1960, he received the Audubon Medal for Distinguished Service. This accompanied the (Theodore) Roosevelt Medal for Conservation personally awarded him by Gifford Pinchot nearly twenty years earlier. He also was honored, posthumously, with the naming of the J. N. Ding Darling National Wildlife Refuge on Florida's Sanibel Island, an area he had supported protecting since the 1930s.

STEVEN V. SIMPSON

"Ding Darling: He Can Orate, He Can Snarl." *National Wildlife* 24 (April–May 1986): 8–9.

Lendt, David L. *Ding: The Life of Jay Norwood Darling.* Ames, IA, 1979.

McKerns, Joseph P., ed. *Biographical Dictionary of American Journalism.* New York, 1989.

Derleth, August William (February 7, 1909–July 4, 1971). Derleth once said that he was probably the most versatile and voluminous writer in quality writing. The claim has merit. His enormous bibliography includes poetry, essays, history, biography, juvenile books, mysteries, science fiction, and fantasy as well as edited works and extensive contributions to newspapers and magazines. He sometimes has been classified as a regionalist because many of his works have Wisconsin and the upper Midwest as a setting. This limiting label, though, points to another facet of his many interests. Derleth is also a nature writer in those works that deal with Sauk City, Wisconsin, where he was born, reared, and lived most of his life. These include *Village Year* (1941), *Village Daybook* (1947), *Walden West* (1961), and *Countryman's Journal* (1963). All of these mingle his thoughts on nature with other observations, descriptions, memories, and character sketches. The most significant of these is *Walden West*, which develops most completely his responses to nature.

Derleth's interest in nature grew primarily out of his boyhood experiences in the hills, woods, and fields surrounding Sauk City and from his extensive familiarity with the Wisconsin River, which becomes a central image in *Walden West*. The abiding memories of these early scenes provided him with a lifelong sense of continuity and led him to cherish his sense of belonging and attachment to the land he knew so well. In *Walden West*, he says that "a man belongs to where he has roots—where the landscape and the milieu have some relation to his thoughts and feelings, by virtue of having formed them."

As the title *Walden West* suggests, Derleth also owes a great deal to Henry David Thoreau's masterpiece about nature in New England. His introduction to Thoreau came early in grade school when a favorite teacher encouraged his interest in Transcendental literature: "The ideas that unfolded before my youthful eyes found fertile soil in which to grow." Derleth acknowledges that *Walden* became "a sort of Bible—not one to be preached from, but one to be kept in the recesses of my mind, not so much thought of as lived."

Derleth's primary contribution to nature writing is manifested in his exploration of environmental attitudes and themes in a literary form. Although he does not use the word *ecology*, he viscerally understands the important connections between human beings and the rest of the natural world, particularly in those passages when he almost mystically records the images of his native turf: the "primal music" of the spring peepers and cicadas, the emergence of Arcturus in the night skies, and the smells of his favorite slough on the Wisconsin River. Furthermore, he understands that if this close relationship is not acknowledged or acted on, human beings run the risk of cutting themselves off from their deepest roots.

His sharp awareness of the vastness of time also leads him to put the human species in perspective. Describing, for example, the mating dance of the woodcocks, he notes in *Walden West* that it is "something which went on before the arrival on this scene of man and his works, something which may last beyond man." The realization that human existence is transitory is humbling but not necessarily disheartening when viewed in the context of nature, because in the rhythms of the seasons, in the endless cycle of life and death, he recognizes a fate linked to natural cycles of which he is but a small part.

Although Derleth's contributions to American nature writing are relatively small in the context of the vast body of his other works, they are significant in his lyrical and memorable depiction of the landscape of a small corner of this native Wisconsin and in his ability to see in this microcosm environmental issues of importance to the larger world beyond.

Derleth married Sandra Evelyn Winters in 1953; they were divorced in 1959. They had two children, April Rose and Walden William.

WALTER HERRSCHER

"August Derleth" [obit.]. *New York Times*, July 6, 1971.

"August Derleth" [obit.]. *Publishers Weekly*, August 2, 1971, 44.

"Derleth, August William." *Who Was Who in America*. Vol. 5 1969–1973.

Devoto, Bernard Augustine (January 11, 1897–November 13, 1955). Born in Ogden, Utah, DeVoto established himself as a leading literary critic, a prominent historian of the American West, an able editor, and a dedicated conservationist.

DeVoto attended the University of Utah and Harvard University—the former for one year; from the latter he received his degree and won a Phi Beta Kappa key. His studies at Harvard were interrupted when he enlisted in the army in April 1917 and was commissioned a lieutenant in the infantry. He resumed his studies in September 1919 and graduated in 1920. For one year he taught history at a junior high school before becoming an

English instructor at Northwestern University, where he remained from 1922 until 1927. On June 30, 1923, he married Helen Avis MacVicar.

As a literary critic, DeVoto became Mark Twain's champion. Responding to adverse criticism, DeVoto contended that Twain should be honored for his unique contributions to American literature. For *Harper's Magazine*, he conducted "The Easy Chair" from November 1935 until January 1936. From 1936 until 1938 he served as editor of the *Saturday Review of Literature*. Though he considered himself a novelist, having written six in that genre, his interest in the West directed much of his life. He became convinced that mercenary eastern interests were destroying the West.

Interest in the West, perhaps because of his own origins, made DeVoto aware of the conservationist movement that occupied much of his time and energy. This interest found an outlet in his journalistic endeavors, especially the column "The Easy Chair," through which DeVoto attacked government policy that permitted the exploitation of the natural resources of the West, particularly grazing practices in the national forests, which he maintained damaged the forests, and particularly the watersheds to a critical point. Partly to blame was the Forest Service, assisted by legislators who had their own agendas. When the Grazing Service was created by the Taylor Act, it made a valiant attempt to correct the out-of-control practices by trying to reduce the overstocking of ranges through grazing fees.

Reforms, nonetheless, were short-lived. Senator Patrick McCarran of Nevada succeeded in creating the Bureau of Land Management by merging the Grazing Service and the General Land Office. Cutting the appropriations to this newly created bureau successfully made it impotent, and practices DeVoto fought against resumed. He aptly commented in an August 1954 article: "An APHORISM of the Chinese philosopher Mencius declares that the problem of government presents no difficulties: it is only necessary to avoid offending the influential families." One would suspect that reformers are often "more sinned against than sinning." Such was true for DeVoto. His fight for conservation of western lands was least appreciated in the West. He was attacked for being anti-West, for writing nonsense in his *Harper's* columns (his criticism of exploitation was called "Cow Dung" by a Nevada paper), and worst of all, for being a "paid propagandist." Perhaps the animosity directed at him by western newspapers stemmed from the manner in which he viewed the problem: it was not a struggle between the East and the West but, rather, one between big business and the individual, between the cattlemen and the settlers.

In a *Harper's* article of February 1944, DeVoto reaffirmed large corporations, not the pioneers, plundered the West, that the oil industry destroyed forests to make barrels for their oil. Nevertheless, it was the cattle industry that did the most damage with its unbridled use of the grazing facilities. In another *Harper's* article of August 1934, he reiterated his oft-repeated phrase that the West was "the plundered province." Initially, it was eastern interests that capitalized on the rich fur resources, mainly beaver. The second incursion into western assets was a consequence of mining interests, and these, too, were based in the East. Contrary to some western mythology, those who came to settle and build farms did not plunder but added to the resources of the West. "The interests of these people, the permanent inhabitants," he wrote, "have always been in conflict with the interests of transients, of those who were liquidating the West's resources." DeVoto found mining particularly repugnant and labeled it "liquidation." What was left, after the mines were decimated, were pathetic ghost towns. "You clean out the deposit, exhaust the lode, and move on."

A third infringement on the West's natural resources by eastern interests was of concern to DeVoto. Oil and natural gas became new targets for corporate investors. Though now controlled by the national government, who leased lands instead of issuing patents, power was still maintained by eastern investors who exercised monopolistic control over a burgeoning industry. DeVoto conceded that some wildcatters in the West did, on occasion, operate without help from the East; but independents had to submit, eventually, to the system controlled by eastern capital. Ironically, DeVoto noted, "The West does not want to be liberated from the system of exploitation that it has always violently resented. It only wants to buy into it."

Additionally, the cattle industry came under DeVoto's attack, but he was not quite clear about the origin of the offenders. Certainly, they were men of the West, in that they lived there. He called them "cattlemen from Elsewhere." In an even cloudier accusation, he called them "peons of their Eastern bankers." These pillagers of the West's grazing lands were constantly in conflict with the "nesters" and tried to force them off their lands through various nefarious schemes that included attacks by hired gunmen, the seizing of water rights, and foreclosing on landholders whenever possible. The outrage felt by DeVoto and other conservationists concentrated on the ownership of grazing land. In fact, only a minute portion of that land was owned by the cattlemen; the overwhelming share actually belonged to the American people. The public fees paid by the cattlemen were nominal compared to the private tariffs they would have had to pay to individuals.

Finally, the lumber industry denuded the land with little attempt at conservation. Here, DeVoto blamed western interests as well as eastern concerns. Not only were trees destroyed, but the naked ground presented an ecological danger—the Dust Bowl that occurred in the 1930s in the Great Plains was a result of overgrazing and the destruction of plants and trees.

DeVoto was pleased with government's attempts to reclaim the West through tree planting and dams that halted disastrous flooding. He fought tirelessly against the exploiters of our national parks. DeVoto feared that other attempts to exploit national parks would resume with the help of the West.

In an essay titled "The Citizen," which Arthur M. Schlesinger Jr. contributed to a memorial text honoring DeVoto, he wrote about DeVoto's 1946 trip through the West where once again he noted that the exploitation of nature was still rampant. Unfortunately, it was not only the East that was guilty of this "rape," as DeVoto termed it—the West was a willing ally. "If the East had not destroyed the resources of the West," he held, "the West would do so by itself."

DeVoto was a fighter for conservation and for civil freedom, and he was an astute commentator on the democratic process. "His broader view of democracy," Schlesinger maintained, "remained tempered and sardonic," but the battle to maintain its ideals "vindicates democracy by producing men of compassion, of courage and of faith. . . . Bernard DeVoto was such a man."

ROSS BRUMMER

Bowen, Catherine Drinker. "The Historian." *Four Portraits and One Subject: Bernard DeVoto*, ed. Julius P. Barclay and Elaine Helmer Parne. Boston, MA, 1963.

Mirrielees, Edith R. "The Writer." *Four Portraits and One Subject: Bernard DeVoto*, ed. Julius P. Barclay and Elaine Helmer Parne. Boston, MA, 1963.

Sawey, Orlan. *Bernard DeVoto*. Boston, MA, 1969.

Schlesinger, Arthur M., Jr. "The Citizen." *Four Portraits and One Subject: Bernard DeVoto*, ed. Julius P. Barclay and Elaine Helmer Parne. Boston, MA, 1963.

Stegner, Wallace. "The Personality." *Four Portraits and One Subject: Bernard DeVoto*, ed. Julius P. Barclay and Elaine Helmer Parne. Boston, MA, 1963.

Wilson, Edmund. "Complaints: II, Bernard DeVoto." *New Republic*, February 3, 1937, 21–23.

Dilg, Will H. (1867–March 27, 1927). A sportsman, writer, editor, and environmental activist, Dilg was the primary founder and the first president of the Izaak Walton League of America. The league was created in Chicago at a January 1922 gathering of fifty-four fishermen and hunters troubled by the destructive effects of industry on the nation's water and wildlife. By March 1927, at Dilg's death, there were over 300,000 members with more than 2,000 chapters. One of the league's more forceful advocates of environmental legislation, Dilg served as its president from its founding to 1926.

Dilg had long been a fishing enthusiast. Prior to the founding of the league, he had been editor of *Outers' Recreation*. There is a story that his environmental activism stemmed from the death of his young son. He seems to have been driven by a fear that the world of natural beauty that he and his contemporary fishermen had so enjoyed would be lost to future generations: "the American boy, who represents the future of the United States, is in danger of losing his heritage of sport afield and astream."

Although at its founding the Izaak Walton League was mainly concerned with water pollution, under Dilg's energetic guidance, it became the first conservation organization in the United States to address the full range of environmental problems.

As editor of the league's monthly magazine *Outdoor America* (known as the *Izaak Walton League Monthly* from August 1922 to September 1923), Dilg championed a variety of environmental causes. To add prestige to the magazine and its goals, he actively wooed and won such popular, mass-market writers as Zane Grey, Mary Roberts Rinehart, Harold Bell Wright, and Gene Stratton Porter as unpaid contributors. One of his supporters and an occasional contributor was Secretary of Commerce Herbert Hoover.

With a limited amount of success, Dilg attempted to attract women to the league and its work. Women writers frequently contributed pieces to *Outdoor America*.

Through his magazine and through the persistent courting of powerful politicians, Dilg was able to achieve noteworthy successes. In 1924, he and the league were responsible for saving a portion of the Superior National Forest from exploitation, and they were significant in the creation of the Upper Mississippi Wildlife and Fish Refuge. He also received nationwide attention by establishing a fund to save the starving elk of Jackson Hole,

Wyoming. In 1926 President Calvin Coolidge asked the league to conduct the first nationwide testing of water quality.

Dilg reveled in the achievements of himself and of the Izaak Walton League, declaring at its 1925 convention, "In the world of conservation, our League stands forth like a towering mountain set in the center of a vast prairie." A fellow league official described Dilg as "caught up in an ecstasy no less real, no less ardent than that of any religious evangelist."

According to Dilg's friends, he even shortened his life for his struggle. Suffering from throat cancer, he chose to forego an operation that would have prolonged his life but taken away his voice. He felt that he needed a voice so that he could more effectively campaign for his cause of the moment—the establishment of a national conservation department.

Dilg's combativeness had long annoyed certain other conservationists. Eventually, his noncompromising manner and his aggressiveness even antagonized many members of the Izaak Walton League. He was ousted as the league's president the year before his death.

In spite of Dilg's seeming ease at making enemies, he has been remembered as an effective environmental activist. After his death, an Izaak Walton League speaker noted, "It was because he was an extremist, an evangelist-crusader type, that he was able to do so much in a few years." In 1952 the league dedicated the Will H. Dilg Memorial, located in Prairie Island City Park, Minnesota.

KATHRYN HILT

Dilg, Will H. *Tragic Fishing Moments.* Chicago, IL, 1922.

Fox, Stephen R. *John Muir and His Legacy: The American Conservation Movement.* Boston, MA, 1981.

Paehlke, Robert, ed. *Conservation and Environmentalism: An Encyclopedia.* New York, 1995.

"Will H. Dilg Dies After Long Illness." *New York Times,* March 29, 1927.

Dobie, J. Frank (September 26, 1888–September 18, 1964). A famed Texas folklorist, Dobie was also a naturalist whose writings proclaim the abiding good of the natural world. He was born on a ranch in the brush country of southern Texas. He once said that the ranch (his "plot of earth"), his parents, and English literature had been the principal influences on his life. Happily for Dobie, his mother taught her children to read early and even found time to read to them. The young Frank became acquainted with such classics as *Ivanhoe* and *Idylls of the King.* Dobie's father, a cowman who sang hymns to the cows, also insisted on regular readings from the King James Bible.

Much of the frontier still survived in the life Dobie knew as a boy, but even though he himself was always a hunter, he rejected the American frontiersman's inclination to destroy land and animals. After describing the killing of his own first deer—a doe—he remarks that he takes "more pleasure in seeing wild things alert and alive than in seeing them fall." His sympathy extends also to plants. Live Oak County, where he lived, was often parched, and the live oaks sometimes died. Lon Tinkle quotes Dobie as saying that the thirst of the trees for water made him "want to weep for them."

Dobie entered Southwestern University at Georgetown, Texas, in 1906. There a course in English poetry "transmuted the world" for him. He discovered the Romantics, who articulated his own feelings about nature in sensuous, passionate language. Dobie's spiritual kinship with William Wordsworth, in particular, is striking.

Dobie's enthusiasm for poetry and his sincere desire to communicate that enthusiasm led him into high school teaching. He also worked as a reporter during summer vacations.

In 1913, Dobie entered graduate school at Columbia University. In New York, he educated himself in art by visiting the Metropolitan Museum on weekends. His appreciation of art shows up in his specific descriptions of nature. As a New Yorker, Dobie also attended the theater two or three times a week but still managed to write a master's thesis on John Heywood's *The Golden Age.*

In the fall of 1914, at twenty-six years of age, Dobie launched on his sometimes tempestuous teaching career at the University of Texas. A young colleague soon asked him to join the newly founded Texas Folklore Society. Difficult though it may be to believe, Dobie says that he had never before heard the word *folklore.* It came to have great significance for him.

In 1916, Dobie married Bertha McKee, a talented woman who functioned as his critic, frequently met his classes, and edited his work after his death. Meanwhile, unhappy about teaching only freshmen and chronically annoyed with "pedants," Dobie considered turning from teaching to journalism. But in 1917 he enlisted in the army and served as a lieutenant in field artillery. In 1919, he returned to the university and published his first article—a how-to-teach piece in the *English Journal.*

Dobie resigned from the university in 1920 to accept a job managing his uncle's ranch. His year as a professional cowman turned out to be one of the richest experiences of his life, and one special evening at the ranch altered his life. After listening to a Mexican vaquero tell his tales of the Southwest, Dobie experienced a moment of enlightenment, an "epiphany," as Tinkle says. "I resolved," Dobie writes, "to collect the traditional tales of Texas." He felt that in listening to those tales since childhood he had been "listening to a great epic," an epic that he would now record.

His uncle's financial failure hastened Dobie's return to the University of Texas, where, except for an interlude as head of the English Department at Oklahoma A&M, he remained until 1947. He returned to teaching a "fanatic," as he said, over "the cultural inheritance of a people." The collection of folklore became his vocation in almost a religious sense. He viewed the university and the Texas Folklore Society as tools to aid his mission. After reorganizing the society, he became editor of its publications.

During the two years (1923–1925) Dobie spent in Oklahoma, he began writing for magazines, notably the *Country Gentleman.* As he gained something of a literary reputation, a flamboyant figure who called himself "the Vaquero" asked his help in writing an autobiography. The result was *A Vaquero of the Brush Country* (1929). The story is told in the first person, as John Young, the vaquero, and Dobie become one persona.

Although he was almost forty when *Vaquero* was published, Dobie became a remarkably prolific author of books, articles, prefaces, entries in books by various hands, and hundreds of feature stories for Sunday newspapers. He was, moreover, an active editor.

Dobie's method was to collect tales, write short articles for the Texas Folklore Society or such magazines as the *Southwest Review,* then rewrite the same stories for book publication. The same story may also appear in more than one book. Dobie made maximum use of his files.

In 1924, Dobie had edited *Legends of Texas* for the Texas Folklore Society. Many legends recalled lost mines and treasures. Treasure hunters far and near were subsequently eager to share their tales, and the result was *Coronado's Children* (1930), which brought Dobie national fame and the Literary Guild Award in 1931. In 1939, Dobie published *Apache Gold and Yaqui Silver,* describing his own searches. The noted American author and critic Carl Van Doren placed high literary value on both treasure books as fiction. Another value, as in any book by Dobie, is his portrayal of nature.

Dobie, who was fond of travel, eagerly accepted a Guggenheim Grant in 1932 to collect the legends of Mexico. The result was *Tongues of the Monte* (1935), reissued as *The Mexico I Like* (1942). Always more than a scientific folklorist, Dobie is an imaginative storyteller in his own right, and *Tongues of the Monte* is his closest approach to a novel.

Dobie seems to be at his best writing about treasures or animals. *The Longhorns* (1941), one of his major books, combines folklore with scientific research.

While Dobie was publishing, he was also teaching. His most famous course led to the publication of his annotated bibliography, *Guide to Life and Literature of the Southwest* (1943; 1952), which is still a standard reference work. The chapter on wild life, in particular, is one of special interest to environmentalists.

In 1943, during World War II, Dobie became visiting professor of American history at Cambridge University. A photograph shows the Texan in his oversized Stetson hat, holding forth to a group of gowned Cantabrigians. Cambridge awarded Dobie an honorary master's degree complete with citation in Latin proclaiming that what he did not know about longhorns was not worth knowing. Dobie admired the civility of the English and associated that quality with their success in conservation. *A Texan in England* (1945) is an extraordinary travel book of interest to environmentalists. The chapter titled "The Lark at Heaven's Gate" reveals Dobie the romantic. Like Wordsworth and Percy Bysshe Shelley, Dobie delights in the skylark, but he will also celebrate the roadrunner and see beauty in the buzzard's soaring flight.

Dobie came home in 1944 but returned to lecture to American troops in Europe at the end of the war. Finally returned to Texas, Dobie found the state dominated by political conservatives whom he despised. In the era of McCarthyism, Dobie was perceived as "the leading liberal in Texas." In his usual abrasive fashion, he also mixed in politics. The Board of Regents dropped him from the faculty in 1947.

After leaving the university, Dobie published more books about animals. *The Voice of the Coyote* (1949) forcefully condemns the wanton slaughter of coyotes. *The Mustangs* (1952) is Dobie's most scholarly work. In 1948, he received a grant from the Huntington Library that enabled him to do research on, for example, the history of mustangs. Dobie speaks of his "overpowering" interest first in *The Longhorns,* then in *The Mustangs.* The wild horses more than any other animals symbolize spiritual freedom for Dobie.

Shortly after his separation from the university, Dobie also began work on *The Ben Lilly Legend* (1950). Lilly was a famous and impressive hunter whom Dobie had first met in the 1920s. He was attracted to Lilly's knowledge of the wilderness and to his love of freedom. Originally he thought that the old bear hunter lived in harmony with nature. But one who spent his life killing for the sake of killing fails to represent Dobie's ideal revealed in the lore of the Indian who loves the deer he slays.

Dobie's publications include anthologies made up of short pieces from his other books and essays. *On the Range* (1931) and *Up the Trail from Texas* (1955) are designed primarily for young readers. *Tales of Old-Time Texas* (1955) and *I'll Tell You a Tale* (1960) are revised versions of his best stories.

The morning mail on the day of Dobie's death in 1964 brought the first copies of *Cow People,* a collection of stories about southwestern ranchers and stockmen, including his father.

Following Dobie's death, his wife Bertha published several posthumous volumes, including *Rattlesnakes* (1965), *Some Part of Myself* (1967), *Out of the Old Rock* (1972), and *Prefaces* (1975). For decades Dobie had collected stories about rattlesnakes. Although he rarely anthropomorphizes wild things, Dobie does often compare them to human beings, and he praises the rattler for its total lack of hypocrisy. The snake never lies, and for that reason, Dobie says, he prefers its company to that of the governor of Texas. Although Dobie disparages autobiography, much of his wisdom as well as some of his best writing appears in the autobiographical essays collected in *Some Part of Myself.*

Throughout his work, Dobie suggested the primary value of the earth, its animals, and plants for human beings. With all his obvious concern for the environment, however, he avoided preaching. Henry Alsmeyer, nonetheless, notes many conservationist views in Dobie's newspaper columns and other articles. Dobie, for example, enthusiastically supported parks, sanctuaries, and botanical gardens. He disapproved of zoos.

Over the years, Dobie alternately criticized and praised the government agencies concerned with conservation. He recognized the need for laws governing the environment. As a humanist, however, he also believed that a civilized sympathy for nature is more important to the conservationist cause than any governmental regulations. In one column, Dobie spoke of the "cultivated gentleness" toward nature in Thoreau, Saint Francis, and Thomas Jefferson. Such an attitude, foreign to the frontier, is, Dobie implied, necessary to the future. He himself seemed to feel an almost mystical connection to the land, sky, plants, animals, birds, and reptiles of his southwestern environment.

LAWRENCE H. MADDOCK

Abernathy, Francis Edward. *J. Frank Dobie.* Austin, TX, 1967.

Alsmeyer, Henry L., Jr. "J. Frank Dobi." *A Literary History of the American West*, ed. J. Golden Taylor. Fort Worth, TX, 1987.

———. "J. Frank Dobie's Attitude towards Physical Nature." Ph.D. diss., Texas A&M University, 1973.

McVicker, Mary Louise. *The Writings of J. Frank Dobie: A Bibliography.* Lawton, OK, 1968.

Tinkle, Lon. *An American Original: The Life of J. Frank Dobie.* Boston, MA, 1978.

———. *J. Frank Dobie: The Makings of an Ample Mind.* Austin, TX, 1968.

Douglas, Marjory Stoneman (April 7, 1890–May 14, 1998). It was in her capacity as an environmentalist that Douglas, an author and journalist, first raised public consciousness about the dire condition of the Everglades, through her book *The Everglades: River of Grass* (1947). She was the founder of the influential organization Friends of the Everglades and was known for her unflagging efforts to save and restore the Glades.

Marjory was born in Minneapolis, Minnesota, the daughter of Florence Lillian Trefethen Stoneman, who went by the name of Lillian, and Frank Bryant Stoneman, a businessman, lawyer, judge, and newspaper editor. When Marjory was three, her father's business failed, and the family moved to Providence, Rhode Island. Further business reverses took a toll on Lillian's mental heath and resulted in a nervous breakdown. Not long thereafter, Lillian separated from her husband and, with her six-year-old daughter, traveled to Tauton, Massachusetts, to live with her parents and unmarried sister.

Marjory was educated in the Tauton schools and graduated from high school in 1908. With funds scraped together by her aunt and grandmother, she was able to attend Wellesley College, where she majored in English. She enjoyed her college experience and formed lasting female friendships. She also showed an interest in women's rights; in her junior year, with some of her classmates, she joined a club to support a woman's right to vote.

Marjory graduated in June 1912. The happy occasion turned grim, however, when she learned after graduation ceremonies that her mother was dying of cancer. She returned home to be with her mother, who died a few weeks later. Lonely and unsure of herself, she took a job in a department store in St. Louis to be near her closest friend, and after a year, she moved to Newark, New Jersey. There she met Kenneth Douglas, an editor on the *Newark Evening News,* and their three-month courtship led to marriage. But the union proved to be an unhappy one. At the outset, Marjory knew little about Douglas, who was some thirty years her senior. After learning that he was forging bank drafts in her name and drinking heavily, she followed the advice of her uncle and left her husband. (Although she later divorced Douglas, she retained his surname.) In September 1915 she departed for Miami, Florida; she reunited with her father, whom she had not seen since she was six years old.

Frank Stoneman had moved to Florida in 1896 and in 1908 founded the *Miami News Record*, named the *Herald* in 1910. Douglas began working for her father's paper as the society editor and occasionally took assignments as a reporter. After the United States entered World War I, she joined the naval reserve, becoming the first woman from Florida to do so. She resigned in 1918 to volunteer for the American Red Cross, serving in France, Italy, and the Balkans. On her return to the *Herald* after the war, Douglas edited a literary column and wrote editorials. She was one of Florida's foremost advocates of women's suffrage and civil rights. She joined her father in campaigns against slum lords. She signed up with an interracial committee lobbying for an ordinance requiring toilets and running water in all houses. Again, Douglas established the Herald Baby Milk Fund, the first charity in Miami unsupported by a church, to distribute milk to the less fortunate. Moreover, a project she supported in print was the proposal of Edward F. Coe, a landscape architect and Miami resident, to create an Everglades National Park. For several years, she also served on a committee formed to advance the park proposal.

Having grown tired of the grind and mental stain (she suffered a mental breakdown) of newspaper work, Douglas left the *Herald* in 1924 to become a freelance writer. With the income from her short stories, which appeared in popular publications such as the *Saturday Evening Post*, she was able to support herself, and in 1926 she moved into a small house she had built for herself in Coconut Grove.

In 1942 Hervey Allen, an old friend and author of the bestselling *Anthony Adverse,* (1933) asked Douglas if she wanted to contribute to a series of books he was editing on the rivers of America. When she asked if she could write about the Everglades, Allen assented. Despite her familiarity with the area, she knew little about its history or dynamics. Douglas, who at one time favored the development of the Everglades, changed her mind as she researched the book. In 1947 she published *The Everglades: River of Grass,* which was soon recognized as a classic and altered the public perception of wetlands as useless swamps. Over a period of five years, she quizzed the region's outstanding scientists, including those in the U.S. Department of Agriculture.

The book opens with a justly renowned and brilliantly evocative paragraph;

> There are no other Everglades in the world. They are, they have always been, one of the unique regions of the earth, remote, never wholly known. Nothing anywhere else is like them; their vast glittering openness, wider than the enormous visible round of the horizon, the racing free saltiness and sweetness of their massive winds, under the dazzling blue heights of space. . . . The miracle of the light pours over the green and brown expanse of the saw grass and of water, shining and slow-moving below, the grass and the water that is the meaning and the central fact of the Everglades of Florida. It is a river of grass.

The *Everglades* was, first of all, a natural history—with descriptions of its hydrology, geology, and biology—and a history of the Glades. This unique southern Florida saw grass wetlands was home to hundreds of species of plants and birds (including ibises, spoonbills, herons, egrets, and the rave wood stork), manatees (the famous plant-eating sea cow), and saltwater crocodiles and freshwater alligators (the only place in the United States where both species may be found).

But if much of the book is a colorfully rendered description of the flora and fauna of the Glades, the tone and purpose of the *Everglades* change sharply with the last chapter, titled "The Eleventh Hour." The chapter details the shocking story of the waste of the area's natural resources—from the slaughter of its wildlife to the building of dams and canals to drain the wetlands for flood control, ranching, farming, and sugarcane growing. "The Indians, [who like the Seminoles, inhabited the region] before anyone else, knew that the Everglades were being destroyed," Douglas observed. "Where there had been the flow of the river of grass, there was only drying pools and mosquitoes." She further observed,

> The saw grass dried, rustling like paper. Garfish, thick in the pools where there had been watercourses, ate all the other fish, and died and stank in their thousands. The birds flew over and far south, searching for fresh water. The lower pools shrank and were brackish. Deer and raccoons traveled far, losing their fear of houses and people in their increasing thirst. The fires began.
>
> The whole Everglades was burning. What had been a river of grass and sweet water that had given meaning and life and uniqueness to this whole enormous geography through centuries in which man had no place here was made, in one chaotic gesture of greed and ignorance and folly, a river of fire.

But the elegiac tenor passes toward the end of the chapter. Clearly, denying water to the Everglades would destroy it. Still, despite the evidence of greed and irresponsibility forcefully presented in what a *New York Herald-Tribune Weekly* reviewer called a "beautiful, and bitter, sweet and savage book" *The Everglades* closes with the hope that the public would at last act to salvage this "vast, magnificent, subtle and unique region, even in this last hour."

The year *The Everglades* was published, President Harry S. Truman formally opened Everglades National Park. In *The Swamp* (2006) Michael Grunwald quotes Truman expressively conveying the dimensions and heterogeneity of the Almighty's creation;

> Here are no lofty peaks seeking the sky, no mighty glaciers or rushing streams. . . . Here is land, tranquil in its quiet beauty, serving not as the source of water but as the last receiver of it. . . . For the conservation of the human spirit, we need places such as Everglades National Park, where we may be more keenly aware of our Creator's infinite variety, infinitely bountiful handiwork. Here we may draw strength and peace of mind from our surroundings.

The new park, which Edward Coe had proposed more than twenty years earlier, embraced only 10 to 15 percent of the original Everglades. Nonetheless, Douglas was pleased, calling it a "great accomplishment." The U.S. Army Corps of Engineers, however, subsequently built levees for flood control, causing the water level in the park to shrink. In the late 1960s, plans to build a jetport near the park's northern edge further threatened the ecological health of the park.

The projected Big Cypress jetport elicited strong opposition from environmentalists and galvanized Douglas to act in 1969. At the urging of Joseph Browden of the National Audubon Society, Douglas, then seventy-nine years of age and with failing eyesight, helped found Friends of the Everglades. As the group's first president, she traveled throughout central and southern Florida to deliver speeches against the jetport. Carrying a cane, and decked out in one of her floppy hats, Douglas was a diminutive and delicate person. Once in action, however, this intrepid independent-minded woman displayed a rapier wit and an acid tongue. She did not hesitate to chastise her allies if she thought they had gone adrift. But she reserved her combative stance for her opponents, particularly the Army Corps of Engineers (with whom she had an ongoing feud) and the Sugar Barons. Responding to pressure from environmental groups, including Friends of the Everglades, the state decided in 1970 to situate the jetport elsewhere.

In the 1980s, efforts were made by the state government to restore the water flow to the Everglades, and the U.S. Congress authorized the acquisition of over 100,000 acres to expand the park. But another serious threat to the park's ecosystem emerged: the quality of the water redirected to the Everglades was being compromised by pesticides and other agricultural pollutants. To address this issue, the state passed the Marjory Douglas Everglades Protection Act in 1991, which financed the construction of facilities to treat the wastewater from farms. The official signing ceremony was held in Douglas's front yard. She subsequently expressed displeasure with the law's ineffectiveness and asked that her name be removed from the law.

Douglas was repeatedly honored for her work. In 1975 she was named Conservationist of the Year by the Florida Audubon Society; a year later she received the same honor from the Florida Wildlife Federation. In 1989 she was named honorary vice president of the Sierra Club, and in 1993 President Bill Clinton presented Douglas with the Presidential Medal of Freedom for her work in preserving the Everglades. "Long before there was an Earth Day," the New York Times of May 15, 1998 quoted Clinton, "Mrs. Douglas was a passionate steward of our nation's natural resources and particularly her Florida Everglades."

In addition to her environmental activities, Douglas continued to pursue her literary career, writing novels, a drama, and short stories until she was unable to do so because of failing health. Toward the end of her days, she was bereft of both her sight and hearing. "I think death is the end," she averred. "I'm happy not to feel I'm going on. I don't want to." She decided, finally, that "this life has been plenty." She died in the Coconut Grove English-style cottage she had lived in for seventy-two years. She was cremated, and her ashes were scattered over the Glades.

As a longtime leader of the Florida chapter of the Sierra Club observed: "The Everglades wouldn't be there for us to try to continue to save if not for her work through the years." Douglas herself knew how to get to the heart of the issue: "The Everglades is a test," she was wont to say. "If we pass it, we get to keep the planet."

RICHARD P. HARMOND

Doherty, Kieran. *Marjory Stoneman Douglas: Guardian of the Glades.* Brookfield, CT, 1987.

Douglas, Marjory Stoneman, and John Rothchild. *Voice of the River.* Englewood, FL, 1987. Her autobiography.

"Marjory Stoneman Douglas" [obit.]. *New York Times*, May 15, 1998.

"Marjory Stoneman Douglas" [obit.] *Tampa Tribune*, 15 May 1998.

"Marjory Stoneman Douglas" [obit.]. Washington Post, 16 May 1998.

Douglas, William Orville (October 16, 1898–June 19, 1980). A Supreme Court justice, author, outdoorsman, and advocate of wilderness preservation, Douglas was born in Maine, Minnesota, where he experienced a difficult childhood. At age three, he was stricken by a nearly fatal case of polio; his father, the Reverend William Douglas, died when Douglas was six. His mother, Julia Bickford Fisk, then moved her impoverished family to Yakima, Washington, in the shadows of the Cascade Range. In Yakima, the young Douglas took to hiking the mountains as a way to regain strength in his polio-ravaged legs, and these early outdoor experiences built his sense of self-reliance and gave him an appreciation of the wonders and dangers present in the natural world. The writings of John Muir and of Aldo Leopold, as well as his own experiences in the Cascades, led him to believe that wilderness and wilderness preservation were essential to the preservation of the better elements in the human spirit.

Douglas graduated Phi Beta Kappa from Whitman College (Walla Walla, Washington) in 1920, and he received a law degree from Columbia University in 1925. He went on to practice law, teach at Columbia and Yale, and serve as chairman of the Securities and Exchange Commission before being appointed to the U.S. Supreme Court by President Franklin D. Roosevelt in 1939. Douglas remained on the Court until his retirement in 1975, and during his long tenure, he published more than twenty books on law, politics, travel, conservation, and wilderness preservation. Many scholars believe that his writings on wilderness values and wilderness preservation have been as important as his writings on civil rights.

Of Men and Mountains (1950), Douglas's first book dealing with the outdoors, is an autobiographical volume that recounts his childhood in the Pacific Northwest, details his efforts to conquer the effects of polio by hiking in the mountains near his home, and explains how his wilderness experiences shaped his personality.

My Wilderness: The Pacific West (1960) describes Douglas's experiences in Alaska and in the Olympic Mountains. *My Wilderness: East to Katahdin* (1961) details his adventures in the wilderness areas east of the Mississippi River. These companion works alerted the public to the fact that America's wilderness areas were under siege; both books served to make Americans aware of the challenges they faced in the realm of conservation. Douglas also hoped that *Muir of the Mountains* (1961), a biography for children, would help educate young people about nature and the conservation movement.

Douglas's most detailed argument for wilderness preservation appears in *A Wilderness Bill of Rights* (1965), which discusses the causes of the destruction of America's wilderness and defends the wilderness ethic against detractors. Douglas notes that commercial interests, population growth, water pollution, and road building are major threats to wilderness. To preserve remaining wilderness areas, Douglas proposes that America develop legal techniques to prevent land developers and others from destroying wilderness areas, and he advocates severe restrictions on the fencing of public land and the use of motorized vehicles in natural areas. The main focus of the book is to outline statutory protection for wilderness areas; Douglas knew that those seeking to preserve America's remaining wilderness would gain a powerful and permanent weapon if wilderness areas were granted formal legal status.

Douglas was the general editor of the McGraw-Hill Wilderness Series, which published his *Farewell to Texas: A Vanishing Wilderness* (1967). *Farewell to Texas* describes his experiences in the wilderness regions of that state and catalogs the factors that led to the destruction of most of the wilderness in Texas. He explains how public utilities, the Army Corps of Engineers, lumber barons, oil companies, and ranchers damaged the environment, and he urges Texans to campaign for the creation of more parks such as Big Bend.

The Three Hundred Years War: A Chronicle of Ecological Disaster (1972) is Douglas's last book on conservation and the environment. In contrast to his other works, which were heavily anecdotal, *The Three Hundred Years War* makes frequent use of scientific authorities. Douglas details the problems that result when a society has total faith in technological solutions to its problems at the expense of an ecological ethic. In *The Three Hundred Years War*, he observes that "technology and the profit motive have carried us far down the road to disaster. It is indeed a desperate race to institute preventive controls that will save the ecosystem." Douglas concludes the book with a chapter outlining the types of political action that should be taken to avoid ecological destruction. He calls for reform of government bureaucracy, for public hearings on environmental matters, and for citizen self-help environmental action on the local level.

In addition to his books on the environment and wilderness preservation, Douglas wrote articles on conservation and outdoor recreation for popular magazines such as *Field and Stream* and *Mademoiselle*. Like his books, his articles promote the wilderness experience and advocate preservation of wilderness areas.

During his long tenure on the Supreme Court, Douglas became one of the Court's most liberal justices. He was an activist who believed that judges had a duty to shape the law, and he sought to protect civil liberties and to defend the rights of privacy and dissent. In his legal writings, Justice Douglas consistently placed environmental concerns before those of development and commercialism. In addition to advocating legal rights for wilderness areas, he suggested that legal rights should be extended to all of nature. In a dissenting opinion in *Sierra Club v. Morton* (1972), Douglas wrote that there should "be assurances that all of the forms of life which it [the Court] represents will stand before the court—the pileated woodpecker as well as the coyote and the bear, the lemmings as well as the trout in the streams."

Douglas did not limit his commitment to conservation to his books and legal opinions. He had frequently warned of the destruction caused by bulldozing and road building, and in 1954, he led a 185-mile hike to protest the proposed building of a highway alongside the historic Chesapeake and

Ohio Canal. The protest was successful, and plans to build the highway were abandoned. In 1971, Congress recognized his service as a conservationist and his efforts to preserve and protect the canal and towpath from development by dedicating the Chesapeake and Ohio Canal National Historical Park to Douglas.

Outlining and promoting formal legal rights for America's wilderness areas through his books and other writings were Douglas's most important legacies to the conservation movement. His books, which were widely read, educated the public about the value of wilderness and the wilderness experience and alerted Americans to the crisis facing their environment.

Douglas married four times, first in 1923 to Mildred Riddle, with whom he had a son and daughter; Douglas and Riddle divorced in 1953. In 1954, Douglas wed Mercedes Hester Davidson; they divorced in 1963, and he married Joan Carol Martin that same year. In 1965, Douglas divorced again and married Cathleen Heffernan. He suffered a stroke in 1974 and retired from the Supreme Court in 1975. He was eighty-one when he died in 1980 at Walter Reed Army Medical Center, Bethesda, Maryland.

DAVE KUHNE

Countryman, Vern. *The Judicial Record of Justice William O. Douglas.* Cambridge, MA, 1974.

Duram, James C. *Justice William O. Douglas.* Boston, MA, 1981.

Frank, John P. "William O. Douglas." *The Justices of the United States Supreme Court, 1789–1969: Their Lives and Major Opinions*, Leon Friedman and Fred L. Israel. New York, 1969.

Mosk, Stanley. "William O. Douglas." *Ecology Law Quarterly* 5 (1976): 229–232.

Murphy, Bruce Allen. *The Legend and Life of William O. Douglas.* New York, 2003.

Dubos, René Jules (February 20, 1901–February 20, 1982). A distinguished microbiologist, environmentalist, and writer, Dubos was profoundly interested in how the world could be made better for humankind. He wanted science to integrate knowledge of the external world and of human nature to meet the fundamental needs of the individual and society. Dubos was not concerned with the environment per se, nor with environmental conservation and preservation as usually construed, but rather with nature as a garden to be tended and dressed in accord with a people's genius. He conceded that some of nature must be kept in its primeval state so that we may keep in touch with our biological origins. However, he held that most of nature should be recreated through human toil to serve our needs in a way that is ecologically harmonious and sustainable; Dubos referred to this idea as the "wooing of the Earth." The author of numerous articles and more than twenty books, Dubos is best known, perhaps, for the aphorism, often paraphrased by others: "The general formula for the future then might be: Think globally and act locally!"

Dubos was born in the village of Saint-Brice-sous-Forêt, France, just north of Paris and near Sarcelles, to Georges Alexandre and Adeline Madeleine de Bloedt Dubos. His parents ran a butcher shop in Saint-Brice and later in Henonville, another farming village of the Île-de-France region. At eight years of age, he succumbed to rheumatic fever, which damaged his heart valves. As he recovered, his excursions around his home led him to appreciate the surrounding countryside. During the illness and recovery, Dubos developed a lifelong fascination with history and literature, which he later often cited in his writings.

When Dubos was thirteen, his family moved so that his father could open a butcher shop in Paris. Several months later, Georges was called to military service; he died in 1919 from injuries sustained in the war. In the meantime, Dubos and his mother tried to run the butcher shop while he attended high school at College Chaptal. He had wanted to study history in the university, but his father's untimely death necessitated more practical plans. Dubos then decided to attend the École de Physique et Chimie, but another bout of rheumatic fever forced him to miss the entrance exam. After recovering, he took an entrance exam in economics and did so well (fourth out of 400) that he was admitted to study agricultural science at the Institut Agronomique in Paris, where he earned his Bachelor of Science degree in 1921. In his third year, he won a scholarship to the École d'Agriculture Coloniale in Paris that would have required him later to serve in the French Army in Indochina, but he was discharged from officer training because of heart problems.

In 1922, Dubos was offered and accepted the position of assistant editor of the journal *International Agriculture Intelligence*, published by the International Institute of Agriculture, part of the League of Nations, in Rome. In this job, he read an article by Sergei Winogradsky, a Russian microbiologist then at the Pasteur Institute in Paris. Winogradsky argued that microbes should be studied in association with other microbes in their own natural habitats. Dubos identified this article as the turning point in his scholarly career, and he once told an interviewer, "I have been restating [Winogradsky's] idea in all forms ever since." In Rome, Dubos met the American delegate to the International Institute, Asher Hobson, an economics professor at the University of Wisconsin, who encouraged him to pursue graduate studies in the United States.

While sailing to New York in 1924, Dubos ran into a couple that he had met previously as a part-time tour guide in Rome. The husband was Selman Waksman, a professor of microbiology at Rutgers University. As Dubos had had no fixed plans of what he would do after arriving in New York, Waksman took the young man under his wing, introducing him to colleagues and helping him to obtain a research assistantship in soil microbiology at Rutgers. His doctoral dissertation, completed in 1927, is a study of how soil bacteria decompose cellulose.

Dubos accepted a teaching position in microbiology at the University of North Dakota in Fargo, but after arriving there, he received a telegram offering a fellowship in the department of pathology and bacteriology at the Rockefeller Institute for Medical Research in New York City. He immediately packed his bags and left for Rockefeller Institute (renamed Rockefeller University) to work with bacteriologist Oswald T. Avery. Avery had been searching for a cure for lobar pneumonia by trying to find a way to break down the polysaccharide capsule enveloping the pneumococci bacteria. Using knowledge acquired during his doctoral studies, Dubos succeeded in finding a soil bacillus that produced antibodies lethal to pneumococcus. Noting that this bacterium produced the necessary enzyme only in a medium containing the capsular polysaccharide, Dubos concluded that cells have multiple potentialities that become manifest only where they are compelled to utilize them.

Later in the 1930s, Dubos tried to isolate a microbe that would be toxic to other bacteria. Using soil samples, he found that *Bacillus brevis* contained two chemicals, gramicidin and tyrocidine, that were toxic. The former could be applied only externally to kill bacteria in human and bovine-udder infections. This breakthrough led other scientists to search for antibiotics in the natural environment, and they soon succeeded in isolating penicillin (discovered earlier in bread mold) and a number of other modern antibiotics.

Hoping that a change of environment would help his first wife, Marie Louise Bonnet, whom he married in 1934, recover from tuberculosis, Dubos accepted an endowed professorship at Harvard Medical School in 1942.

Marie died before the move to Harvard, and Dubos decided later to return to Rockefeller to investigate tuberculosis and its causes. Tuberculosis had proved elusive in laboratory study due to its ability to mutate, but by 1947 Dubos found that by adding a common household detergent (Tween®) to the culture medium he could raise the bacilli so quickly they had little chance to mutate before being studied. This clever discovery allowed researchers to develop a vaccine for tuberculosis. Never satisfied with just laboratory analysis, Dubos published *The White Plague: Tuberculosis, Man, and Society* in 1952. The volume, which he wrote with his second wife, Letha Jean Porter, whom he married in 1946, shows how tuberculosis spread from the industrial, urban poor to all of society during the nineteenth century. The Duboses concluded that the white plague was the first penalty that nineteenth-century capitalist society paid for the ruthless exploitation of labor.

Dubos was no less critical of modern technological society. In *Man Adapting* (1965; and an enlarged edition in 1980), he discussed the biological and social problems of human adaptation to environmental challenges. In *So Human an Animal* (1968), he delivered a broadside against our willingness to adapt to a polluted, overly technical, and psychologically stressful environment. He warned that mismanaged prosperity could destroy human life unless we developed a scientific humanism that supported a sophisticated ecology, one that studies the responses of the entire organism to the total environment. In this volume, he once again demonstrated the rare ability to wed science and humanism in the manner of the great natural philosophers of earlier times; as a result, Dubos received a Pulitzer Prize in 1969 for the category of general nonfiction.

Dubos retired as an emeritus professor of pathology in 1971, culminating a forty-four-year association with Rockefeller University interrupted only by a two-year stint as an endowed professor at Harvard Medical School and School of Public Health from 1942 to 1944. Retirement freed him to devote the last decade of his life to lecturing on, writing on, and otherwise addressing environmental issues. In his writing, his focus gradually expanded from disease and medicine to the quality of all of human life. In public lectures, he was a passionate and eloquent speaker. In the early 1970s, with the help of Ruth and William Eblen, he was able to establish the René Dubos Forum (later called the René Dubos Center for Human Environments). The forum was to bring together interested parties to debate a current ecological controversy and to search for a concensual solution to any environmental problems identified. In recognition of his sterling achievements, he received forty-one honorary doctorates, of which thirty were from American institutions and eleven from foreign institutions. Many of these honorary degrees were in fields other than science. In addition to his Pulitzer Prize, he also won many other awards, including the Modern Medicine Award (1961), the Phi Beta Kappa Award in Science (1963 and 1965—for *The Unseen World* and *Man Adapting*, respectively), and the Tyler Ecology Award (1976). He also was a member of the National Academy of Sciences.

Despite the many honors, Dubos remained a man of simple pleasures. When he was not working in New York City, he spent most of his weekends at his estate in Garrison, New York. He and his wife Jean loved to plant trees and to garden there, and they liked to walk along the lush Hudson River valley. Although Dubos became an American citizen in 1938, he cherished his French roots and remained grateful for his rigorous French education. Dubos died on his eighty-first birthday in New York Hospital, the cause listed as heart failure.

In some respects Dubos was a typical environmentalist, but in important ways he was unique. Like many other environmental writers, he urged that people adopt a spirit of sacredness toward the earth. But Dubos rejected the Leopoldian philosophy that there is some invisible hand bestowing ecological harmony on nature as long as humans do not interfere. Rather, Dubos contended that each place has one or more vocations, which must be identified and developed for their potential and for the sustainable future of the human species. He supported human intervention in nature as long as we do not base the future on past and present events (a "logical" future) but instead on conscious use of science and technology for agreed-upon goals (a "willed" future). He further realized that ecosystems, if left alone, often evolve in response to random events impossible to predict. So if change is inevitable, Dubos reasoned that humans should marshal their knowledge of science and technology in support of environmental creation that respects the spirit and vocation of place. Dubos liked to refer to the English landscape, where, for example, the hedgerows of East Anglia support many plants and animals not found there originally but now considered part of the region's "natural" environment. Dubos was optimistic that our willed future could be ecologically complex, stable, and even desirable, provided that we adopted a theology of the earth, an awareness of the divine mystery that earth is our mother and that as we modify nature, we also determine the path of our own evolution.

It can be stated that Dubos was as concerned with people's relationships with each other as he was with their symbiotic relation with the environment. As Dubos broadened the scale of his analysis from soil ecology to that of global ecology, he also wrote about the city. More concerned with human interaction than with the built environment per se, Dubos cherished the intimate and public spaces of large cities such as New York, Paris, and Rome, and he dismissed suburbs as "second-class urban settlements" not offering a true urban way of life. Dubos wanted shelter to be more than for protection or display; he held that it must stimulate the development of human potential. According to Dubos, before we design shelter, we must ask, What is the genius of the place where the building is to be located? How do we wish to live? And what do we want to become?

Like many observers of modern life, Dubos also kept a wary eye on the rapid growth of the world's population, which he referred to as the "population avalanche." He rejected the usual causes for such growth that had been identified by others: medical advances, improved public health, religious beliefs, lack of sufficient contraception, and increased food supply. He reasoned that the population avalanche (as opposed to the "population explosion") reflected a continuous accelerating process that had reached a numerical level sufficient to cause public alarm. Yet he was sanguine (perhaps overly so) that industrialization and urbanization would check rapid population growth via economic and social progress and that the historical demographic trends of Western Europe would be repeated in the rest of the world.

The genius of Dubos lay in his ability to elucidate the relationships among humanness, society, culture, science, technology, and environment. Far more than almost all of his contemporaries in science, he was able to develop and share a holistic view of what it means to be human and of how all the external factors that make up our environment affect us. The frequent references in his works to history and geography provide a chronological depth and spatial breadth that make his arguments more timeless and universal. His frequent allusions to literature and poetry create an intertextual dialogue that frames his writing in a lofty architecture. Most important, Dubos's belief in our capacity to better understand nature and to manage it creatively, knowing and respecting the special character of each place, gives us optimism about the future and a sound philosophy by which to live.

STEVEN L. DRIEVER

Culhane, John. "En Garde, Pessimist! Enter René Dubos." *New York Times Magazine*, October 17, 1971.

Piel, Gerhard, and Osborn Segerberg Jr. *The World of René Dubos: A Collection from His Writings*. New York, 1990.

"René Dubos." *Current Biography*. 1973.

"René Dubos" [obit.]. *New York Times*, February 21, 1982.

Dutcher, William (January 20, 1846–July 1, 1920). A businessman, ornithologist, bird lover, and conservationist, Dutcher was born in Stelton, New Jersey, to Reverend Jacob Conklin Dutcher, a Dutch Reformed minister, and his wife Margaretta Ayres Dutcher. At the age of twenty-three, Dutcher embarked on a career in life insurance that lasted until his retirement in 1910. An active businessman, he enjoyed bird hunting along the Long Island and New Jersey shores, a habit that eventually led to his lifelong interest in ornithology and his dedication to the preservation of birdlife throughout the world. His published work in ornithology includes about one hundred titles devoted to notes on the birds of Long Island, notes and reports on bird protection, and popular leaflets. Among the most significant are his *History of the Audubon Movement from 1883 to 1904* (1905), *Some Reasons Why International Bird Protection Is Necessary* (1910), and a *Contribution to the Life History of the Herring Gull (Larus argentatus) in the United States*, which he published in connection with W. L. Baily in 1903.

Dutcher's interest in ornithology and bird protection was born of a hunting experience on the shores of Shinnecock Bay on Long Island in 1879. He shot a bird considered rare for that part of the country. That bird, a Wilson's plover, was commonly known to reside along the south Atlantic and Gulf coasts but not north of Delaware Bay. In October 1879, Dutcher wrote his first published note on the species in the *Bulletin of the Nuttall Ornithological Club*. This was the first of his notes on birds of Long Island, about which he continued to write throughout his lifetime.

Soon after his first publication, Dutcher became a member of the Linnaen Society of New York, marking the beginning of his active work in ornithology and bird protection. In September 1883, he was elected an associate member of the newly founded American Ornithologists' Union (AOU), and the following year he was appointed to the committee on Protection of North American Birds. It was with this committee that Dutcher helped found the original Audubon Society and draft a model law for the protection of nongame birds. The law, which did in fact serve as the model for subsequent legislation, came to be known as the AOU, or Audubon law.

In 1886, Dutcher was elected an active member of the AOU at the union's first meeting in Washington, DC. From this point on, he remained in the political spotlight as an advocate for bird protection. Among his triumphs in the political arena are the establishment of the first National Bird Reservation, Pelican Island in Florida, and the passing of the AOU's model law in seven U.S. states. Dutcher had traveled to the legislatures of several states, advocating better bird protection laws, many of which were passed owing to his efforts.

Dutcher was elected treasurer of the union in 1887 and held this position for sixteen years, during which time he met with remarkable success. Within two years of his election, he effectively collected dues from a record 90 percent of the union's membership.

Serving as chairman of the Committee on Protection of Birds of the AOU in 1886, and again in 1887, Dutcher was instrumental in the organization of many Audubon Societies in the closing years of the nineteenth century. The need for more societies came primarily from the increased destruction of birds resulting from the growing millinery trade. In 1901 a National Committee was organized to provide greater uniformity to the societies, and this committee was incorporated in 1905 as the National Association of Audubon Societies. Dutcher was elected president at the first meeting and held the office until his death.

After nearly thirty-three years of writing about and studying birds, he came to be regarded as an authority on many subjects within ornithology. The Camp-Fire Club of America awarded him its gold medal for his efforts toward better protection of wild life, and the Royal Society for the Protection of Birds in England made him an Honorary Fellow.

The final year of Dutcher's active career, 1910, brought with it the passing of landmark legislation in bird protection. During the spring of that year, the Shea-White bill prohibiting the sale of aigrettes was introduced in the New York state legislature. After a successful effort on behalf of the bill, it was approved and passed by Governor Charles Evans Hughes on May 7. The new law provided for the first time adequate restrictions on the traffic in aigrettes for millinery purposes. Dutcher then went to Berlin to represent the National Association of Audubon Societies at the Fifth International Congress of Ornithology. At this meeting he was appointed to the International Committee for the Protection of Birds.

In October 1910 Dutcher suffered a stroke that left him paralyzed on his right side and deprived him of speech. Although he could no longer work actively on behalf of bird protection legislation and ornithology, Dutcher retained his interest in birds and was kept informed of the workings of the National Association of Audubon Societies, as well as the continuance of his efforts by others.

Dutcher married Catherine Oliver Price in 1870; they resided in Bergen Point, New Jersey. One year later, their only son, Basil, was born on December 3. Dutcher died in 1920, survived by his wife and son.

ROBERT P. BIEHL

Palmer, T. S. "In Memoriam: William Dutcher." *Auk* 38 (1921): 501–513.

———. "The Life and Work of William Dutcher." *Bird Lore* 22 (1920): 317–321.

"William Dutcher" [obit.]. *New York Times*, July 4, 1920.

Edge, Rosalie (November 3, 1877–November 30, 1962). Once Described as "the only honest, unselfish, indomitable hellcat in the history of conservation," Mabel Rosalie Barrow was born into a decidedly upper-class family in New York City, where she resided most of her life. She married Charles Noel Edge, an English engineer, in 1909 and spent the first three years of her marriage in the Far East, where she became especially familiar with the China coast.

After returning to New York City when her husband decided to go into business, Edge worked assiduously in the vote-for-women movement until 1920 when the women suffrage amendment became law. Her work as secretary of the New York State Woman's Suffrage Party undoubtedly did much to provide the political savvy she demonstrated in her later conservation work.

Edge's entry into conservation came primarily through the avenue of ornithology for which she had acquired a taste as early as 1915 and perhaps as early as 1909 when she resided in China. It was during a bird walk in Central Park that she met Dr. William G. Van Name, a zoologist with the nearby American Museum of Natural History and a formidable force in conservation. The two became friends, and eventually Van Name felt comfortable enough to ask her to distribute a polemical pamphlet, "A Crisis in Conservation," he had written about the rapid disappearance of some American bird species. A not-very-veiled attack on the inactivity and complacency of many bird protection groups such as the Audubon Society, the pamphlet provoked a firestorm of protest from these groups; but it eventually resulted in two important consequences: a shake-up in the leadership of the Audubon Society and the establishment of the Emergency Conservation Committee (ECC) in 1929 with Edge as its crusading leader.

Over the next thirty years the ECC was responsible for the distribution of thousands of copies of about one hundred conservation-related pamphlets, nearly all of which were funded by Edge and by small individual contributions. Topics of the pamphlets ranged from the need for waterfowl hunting regulation (Edge was for it) to predator poisoning by the U.S. Biological Survey (she was against that practice) to the establishment of national parks in the American West (she favored creation of the Olympic and Kings Canyon National Parks). Equally as important as Edge's early application of mass-mailing techniques to conservation issues was her influence on a generation of environmentalists. David Brower, later founder of Friends of the Earth, and Roger Tory Peterson both spent time in Edge's New York apartment, addressing hundreds of envelopes containing ECC pamphlets.

Probably the most notable event associated with Edge was her work to acquire Hawk Mountain, near Kempton in southeastern Pennsylvania, in an effort to protect the birds of prey migrating past it. For decades, gunners had annually killed thousands of hawks at Hawk Mountain and at other sites along the Kittaninny Ridge, which are sometimes collectively referred to as the "bloody ridges." In 1934 Edge obtained a lease on about 1,200 acres along the top of Hawk Mountain, declared the leased land a sanctuary (much to the anger of local gunners), and appointed Maurice Broun, a young Bostonian then working as a research associate at the Austin Ornithological Research Station at Cape Cod, as the first caretaker of the sanctuary. Edge later bought Hawk Mountain and eventually vested ownership in the Hawk Mountain Sanctuary Association, of which she was a director and president for nearly thirty years. The Hawk Mountain Sanctuary Association has gone on to become the foremost raptor research organization in the world today, thanks largely to Edge's early vision of the mountain's biologic and symbolic importance.

There is no doubt that Edge was "grass-roots environmentalism's godmother." Her influence in a wide variety of conservation arenas remains to this day because she, like Rachel Carson and other environmentalists, saw the interconnectedness of the natural world and the need to maintain its integrity.

STEPHEN J. STEDMAN

Brett, James. *The Mountain and the Migration: A Guide to Hawk Mountain*. New York, 1986.

Broun, Maurice. *Hawks Aloft: The Story of Hawk Mountain*. New York, 1947.

Harwood, Michael. *The View from Hawk Mountain*. New York, 1973.

Stroud, Richard H., ed. *National Leaders of American Conservation*. Washington, DC, 1985.

Taylor, Robert. "Profiles: 'Oh Hawk of Mercy!'" *New Yorker*, April 17, 1948.

Zaslowsky, Diane. "In Profile: Rosalie Edge." *Hawk Mountain News*, no. 81 (Fall 1994): 20–23.

Eiseley, Loren Corey (September 3, 1907–July 8, 1977). An anthropologist, educator, essayist, and poet, Eiseley was born in Lincoln, Nebraska, and, except for brief residences in nearby Fremont and Aurora, spent his early days there. Eiseley experienced a financially poor and unhappy childhood. According to Eiseley, his mother, Daisy Corey Eiseley, was "stone deaf," "paranoid, neurotic, and unstable." His father, Clyde Edwin Eiseley, possessed a "great genius for love" and encouraged him to read widely but was an economic failure, and the parents argued frequently. On one occasion, when his parents were arguing late into the night, Eiseley, only a few years old, arose from bed and pleaded with them to stop. Such experiences may have contributed to Eiseley's lifelong bout with insomnia. Eiseley's struggle to come to terms with his lonely childhood would be the recurring theme of many of his later essays and poems. As he writes in his autobiography *All the Strange Hours* (1975), "Men should discover their past. I admit to this. It has been my profession. Only so can we learn our limitations and come in time to suffer life with compassion." Early on, Eiseley found solace and enjoyment in exploring the natural world of the ponds and salt flats around Lincoln and in inspecting the collection of mammoth bones at the University of Nebraska.

In the late 1920s and early 1930s, Eiseley vacillated between attending the University of Nebraska, drifting via freightcar around the West and Midwest, and doing menial work. Beginning in 1927 and continuing for the next decade, Eiseley published some thirty poems and short essays, mostly in the University of Nebraska literary journal *Prairie Schooner* (which he also helped to edit) and in *Poetry*, *Voices*, and the *Midland*. He would publish no more poetry until 1964. Diagnosed with tuberculosis in 1929, Eiseley retreated to the Mohave Desert, where he worked on a turkey ranch. A year later he was almost fully recuperated and returned to school. During his first few years at Nebraska, having demonstrated a considerable ability for writing, Eiseley pursued a degree in English, but as his segmented undergraduate years progressed, he became more interested in anthropology. During the summers of 1931–1933, Eiseley gained field experience by digging for bones in the Midwest for the Morrill Museum South Party. In addition to providing Eiseley with a practical expertise in archaeology, geology, and paleontology, his work with the South Party would become the basis for many of his later personal essays.

After some eight years of restlessness, Eiseley finally graduated from Nebraska in 1933 with a double major in English and sociology (with a concentration in anthropology). At the urging of his professors, Eiseley decided to continue to pursue his growing interest in anthropology, and he received a Harrison Scholarship for graduate study at the University of Pennsylvania. Although he initially found the urban world of Philadelphia "unendurable," Eiseley eventually adapted and received an M.A. in anthropology in 1935. After a year spent as a sociologist at the University of Nebraska, he returned to Pennsylvania under a Harrison Fellowship and became a student of ethnohistorian and anthropologist Frank Speck. As Andrew Angyal writes, "What Loren particularly gained from [Speck] was an appreciation of the poetic qualities of the Indian mind as revealed in its animistic beliefs and an understanding of the function of the Indian shaman as a visionary medium between man and animal." Eiseley and Speck were united by a mutual distrust of scientific reductionism and a need to survey the past. Both of these ideas would be central to Eiseley's later essays.

Completing his dissertation, which is an examination of the utilization of invertebrate and vertebrate faunal successions and pollen analytical methods as measurements of glacial and postglacial (Quaternary) time, Eiseley was awarded a Ph.D. in anthropology in 1937. On August 29, 1938, he married Mabel Langdon, whom he had known since his early undergraduate days at the University of Nebraska. The marriage continued for the rest of his life and would remain childless. Later that year he began his lifelong academic career by taking a position in the Sociology Department at the University of Kansas.

Eiseley moved to Oberlin College in 1944, where he became the department chairman and professor of anthropology and sociology. In 1947 Eiseley changed residences for the last time by becoming the chairman of the Anthropology Department at the University of Pennsylvania. Up to 1947, most of Eiseley's published writings had been more or less "purely" professional articles, published in scientific journals. Marked by the publication in October 1947 of his essay "Obituary of a Bone Hunter," and continuing with pieces written in 1948 (while experiencing an extended period of deafness caused by severely distended Eustachian tubes), most of his essays were characterized by the fusing of personal experience and reflection with scientific knowledge. Eiseley would come to call these writings "concealed essays"—essays "in which personal anecdote was allowed gently to bring under observation thoughts of a more purely scientific nature." Many of Eiseley's colleagues were agitated by his deviation from the convention of impersonal scientific writing, but Eiseley had found his niche. Although himself a scientist, an evolutionist, and a popularizer (as well as a critic) of Charles Darwin, Eiseley rejected science as a means of discovering the secrets of life. "I have come to believe that in the world there is nothing to explain the world," he writes near the conclusion of *All the Strange Hours.* In an attempt to circumvent the "confining walls of scientific method," Eiseley combines the metaphorical language and vision of a poet with his considerable scientific expertise in creating informative, reflective, and deeply humane writings. *The Immense Journey,* published in 1957, took Eiseley almost a decade to write and revise, but subsequent books came easier and needed little revision.

In 1948 Eiseley was appointed curator of early man at the University of Pennsylvania Museum, and in 1961, after a two-year term as provost of the university, he was named the first Benjamin Franklin Professor of Anthropology and History of Science, a position he held for the rest of his life. Eiseley was bestowed with numerous honors throughout his career, including election to the American Philosophical Society in 1960, a Guggenheim Fellowship in 1963–1964, election to the National Institute of Arts and Letters in 1971, the Bradford Washburn award for contributions to the public understanding of science in 1976, and by the time of his death, some thirty-six honorary degrees. But Eiseley suffered some ambivalence over his membership in academe.

Possessed of a deep, mellow speaking voice, Eiseley was much in demand as a lecturer at institutions across the country and even hosted an NBC-TV science program for children, *Animal Secrets,* in the mid-1960s, but he found the most enjoyment in later life by spending time with his wife in their large apartment near the Pennsylvania campus and in prowling secondhand bookstores for additions to his book collection. After a long illness, Eiseley died of cancer in a Philadelphia hospital at the age of sixty-nine.

A prolific writer, Eiseley is the author of numerous scientific articles, but he is best known for his books of essays, which include *The Immense Journey* (1957), *The Firmament of Time* (1960), *The Unexpected Universe* (1969), *The Invisible Pyramid* (1970), *The Night Country* (1971), *All the Strange*

Hours: An Excavation of a Life (1975), and *The Star Thrower* (1978), and his four volumes of poetry: *Notes of an Alchemist* (1972), *The Innocent Assassins* (1973), *Another Kind of Autumn* (1977), and *All the Night Wings* (1979).

The essays (and to a lesser extent, the poems) collected in these books constitute Eiseley's principal legacy to environmentalism. His essays are comparable to those of Aldo Leopold and Edward Abbey in that they are autobiographical and celebrate the natural world, but Eiseley's writings find, perhaps, a deeper kinship with the contemplative English and American naturalists (example, Gilbert White, Richard Jefferies, Ralph Waldo Emerson, and Henry David Thoreau) of the late eighteenth and nineteenth centuries. While Eiseley's writings do not explicitly invoke political action for the cause of environmentalism, almost all his writings carry profound ecological/environmental implications. René Dubos wrote that Eiseley "was one of the very first scientists to proclaim publicly that mankind must reinsert itself into nature. His words created a new kind of poetical literature based on objective Scientific knowledge and his warning helped start a new social movement."

Nature presents a paradox to Eiseley, for although after long observation the natural world can be revelatory, it remains unknowable, a mystery: "Nature remains an otherness which incorporates man, but which man instinctively feels contains secrets denied to him." In virtually all of his writings from *The Immense Journey* on, Eiseley attempts to demonstrate that the universe is not humankind centered. Modern man, which in *The Invisible Pyramid* he denotes ("neither in derision nor contempt" but in facing "a simple reality") as the "world eater," is only one species out of many, occupying a minute part of the geological time scale. Although man possesses the unique capability to restructure the natural world, he is indelibly located *within* nature. Man must, therefore, reorder his industrial and technological exploitation of the earth's finite resources and work to preserve the natural world: "The green world is his sacred center. In moments of sanity he must still seek refuge there."

BRYAN L. MOORE

Angyal, Andrew J. *Loren Eiseley*. Boston, MA, 1983.

Carlisle, E. Fred. *Loren Eiseley: The Development of a Writer*. Urbana, IL, 1983.

Dubos, René. Review of *The Invisible Pyramid*, by Loren Eiseley. *Smithsonian*, November 1970, 70–71.

Errington, Paul Lester (June 14, 1902–November 5, 1962). An animal ecologist, conservationist, wildlife manager, writer, and teacher, Errington was born in Bruce, South Dakota, the son of Joseph Errington and Lillian (Johnson) Errington. A scientist who never lost sight of the ethical implications his work held for environmental management, he has been characterized as "one of the pioneers of wildlife research" and "a naturalist in the best modern usage of this title."

As a child, he was stricken with both polio and rheumatic fever. He worked hard to overcome the physical impairments he suffered from those illnesses and as a young man was able to work as a professional trapper—employment that sprang naturally from his early interest in the outdoors and sharpened the skills he would later bring to bear on his scientific fieldwork. With a B.S. degree from South Dakota State College, he went on to do graduate work at the University of Wisconsin, where he studied under Aldo Leopold, receiving his doctorate in 1932. That same year, he joined the staff of Iowa State University, where he eventually became professor of zoology. He remained on the Iowa State faculty until his death.

Errington was recognized internationally for his work in the population dynamics of the higher vertebrates. In 1958 and 1959, with assistance from the Guggenheim and National Science Foundations and the Swedish government, he conducted research in northern Europe. (He spoke several Scandinavian languages and was instrumental in facilitating exchanges between researchers in those countries and those in the United States.) Errington stressed the close study of animals in their native habitat, with the least possible human interference. By thus observing and measuring the interdependent fluctuations in natural vertebrate populations, he gave a more precise meaning to the idea of the "balance of nature," with two major themes arising out of his work: the "oft-misunderstood and exaggerated role of predation in nature" and the "complicated mechanisms of compensation, interplay, interruptions, deflections, and successions that characterize free-living populations."

Errington's scientific work demonstrated the need to revise contemporary land and wildlife management practices. In 1947, he wrote that there had thus far "been two chief mistakes in management practice," first, "the 'cleaning up' and 'doctoring' of places that should be left alone" and, second, "the campaigning against those native vertebrates which are predators or competitive species and classified as vermin." He was particularly vigorous in defense of the latter, which, he argued, "comprise a resource demanding sane administration as much as any class of wildlife."

A member or fellow of numerous professional and conservation organizations—including the Ecological Society of America, the American Association for the Advancement of Science, the American Society of Naturalists, the Wildlife Society, the American Ornithologists' Union, and the Society of Zoologists—Errington was awarded the American Wildlife Conference's Aldo Leopold Medal in 1962.

As he made clear in *Of Men and Marshes: A Question of Values* (1987), and other writings intended for a lay audience, Errington considered ecology to be more than just a scientific discipline and a basis for sound management policies. It was also a philosophical stance, a "framework of universal order, of permanent physical realities, of evolutionary trends, and of the great phenomena of Life" from which the individual could "better see himself, his life, and his problems."

DAVID MAZEL

"In Memoriam: Paul Lester Errington." *Proceedings of the Iowa Academy of Sciences* 70 (1963): 40.

Scott, Thomas G. "Obituary: Paul L. Errington, 1902–1962." *Journal of Wildlife Management* 27 (April 1963): 321–324.

Fernow, Bernhard Eduard (January 7, 1851–February 6, 1923). A professional forester, teacher, and author, Fernow was born into an aristocratic family (who also spelled its name "Ferno") in Inowrazlaw, Germany. His grandfather, Graf Karl Frederick Leopold von Ferno, owned a large, forested estate in East Prussia, and young Bernhard (so family tradition holds) was promised he would inherit this land if he would systematically prepare himself to manage it. In consequence, at the age of nineteen, he went to the Hanover-Munden Forest Academy to begin his formal studies in forestry, which in Europe was already a well-developed academic discipline. During the Franco-Prussian War, he volunteered and served as an army lieutenant; afterward he spent a year studying law at the University of Konigsberg. He applied his combined knowledge of silviculture and law in his subsequent work as a professional forester at various locations in Prussia.

While working in Germany, Fernow became engaged to an American, Olivia Reynolds, and in 1876, much to his family's disappointment, he followed her to America. They married in 1879. In this country, in 1878, he was employed to manage a tract of land owned by the Cooper-Hewitt mining firm, and it was at this time that he began publishing professional articles outlining his systematic application of forestry principles that were then virtually unknown in this country. (That same year, he became a member of the American Association of Mining Engineers and also discovered an electrical process for stripping the tin from "tin cans," leaving behind just the iron. Fernow tried to turn this early recycling method into a profitable business, but the attempt failed in an unstable iron market.)

Beginning around 1879, he became sympathetic to the growing outcry for the conservation of natural resources and began publishing his views in such articles as "Forest Protection" (1881). His conservation interests were largely but not wholly utilitarian. He felt, for example, that just as important as the preservation of natural beauty were such practical issues as securing the future of the charcoal iron works industry. His call for public action was partly based on the absence of the sort of private conservation he had known in Europe. "[T]he time has not yet arrived," he wrote in 1881, "when the profit from forest-growing will induce private capital, at least in the East, to arrest the devastation of our timber lands. The protection, therefore, of this factor of national wealth must be, at present, mainly expected from a sound, protective legislation."

In 1882, Fernow helped found the American Forestry Congress (later renamed the American Forestry Association) and served as its secretary for the next twelve years. In 1886, Grover Cleveland appointed him chief of the Division of Forestry (later the U.S. Forest Service). As division chief, Fernow was instrumental in promoting the 1891 legislation that laid much of the groundwork for today's national forests.

Retiring from government work in 1898, he joined the faculty at Cornell University to organize America's first collegiate forestry school. This was the beginning of a long and productive academic career. In 1906 he helped begin a forestry program at Pennsylvania State College; in 1907 he left Cornell to begin a forestry department at the University of Toronto, where he remained until his retirement in 1919. During this time, he began *Forestry Quarterly* as the journal of the Society of American Foresters and wrote such books as *Economics of Forestry* (1902) and *A Brief History of Forestry in Europe, the United States and Other Countries* (1907).

Preceding Fernow in government work had been his friend and fellow German immigrant Carl Schurz, secretary of the interior under Rutherford B. Hayes (1877–1881). Schurz had advocated scientific forestry based on the European model, and though his specific forest conservation efforts made little headway in the Hayes administration, he can be credited with at least some of the preparation for Fernow's more successful forestry efforts.

Following Fernow at the Division of Forestry was Gifford Pinchot, whom he had known earlier. (At one point, seeing "no prospect for a young forester in the United States," Fernow had advised Pinchot "to go into landscape gardening or the nursery business, and perhaps work in some forestry as a side line.") Though similar to Pinchot in terms of his European training and utilitarian approach to forest management, Fernow had a more restricted vision of the division's policy role. "Convinced that neither the public nor the forest industry would yet support scientific management," he felt the division "should merely dispense information and technical advice to those who sought it, and not promote sustained-yield practices." But even though Fernow did not share the expanded vision, missionary zeal, and political acumen of his successor, he still must be credited not only with establishing the forest reserves themselves but also with establishing a level of professionalism that assisted Pinchot in building the Forest Service into a genuine force in American conservation.

DAVID MAZEL

Clepper, Henry. *Origins of American Conservation.* New York, 1966.

Highsmith, Richard M., Jr., J. Granville Jensen, and Robert D. Rudd. *Conservation in the United States.* 2nd ed. Chicago, IL, 1969.

Journal of Forestry 21 (April 1923): 306–348. Special section devoted to Fernow; includes complete bibliography.

Rodgers, Andrew Denny. *Bernhard Eduard Fernow: A Story of North American Forestry.* Princeton, NJ, 1951.

Forbes, Stephen Alfred (May 29, 1844–March 13, 1930). A museum curator, professor of zoology and entomology, state entomologist, and founder and first chief of the Illinois Natural History Survey, Forbes was born in a log cabin at Silver Creek, Stephenson County, in northern Illinois. He was the son of pioneer parents Isaac Forbes and Agnes (Van Hoesen) Forbes. One of six children whose father died when he was ten, Forbes's formal education was intermittent. He attended district schools until the age of fourteen and then studied at home, learning to read French, Italian, and Spanish. Forbes briefly attended Beloit Academy to prepare for college, but the family's limited means forced him to give up this plan.

In September 1861 the seventeen-year-old Forbes and his older brother Henry borrowed money to buy horses and joined the 7th Illinois Cavalry to fight in the Union Army during the Civil War. Entering the service as a private, Forbes rose to the rank of captain by the age of twenty. He was captured at eighteen while on a dispatch mission outside Corinth, Mississippi, and subsequently was held for four months as a prisoner. The hardships of war sharpened and focused Forbes's will to make something of his life. While a prisoner of war, he undertook the study of Greek and wrote later: "Anyone who had kept the solitary flame of his separate intellectual life steadily burning through all the blasts and storms of war might reasonably believe that nothing that should happen to him thereafter could possibly extinguish it." On his release from Confederate imprisonment, Forbes recuperated from malaria and scurvy in a Union hospital. He then rejoined his regiment and returned to the war, leaving the army in November 1865.

Following the war, Forbes briefly studied medicine in Chicago at Rush Medical College but did not complete his studies. From 1867 to 1872 he pursued a variety of occupations, raising strawberries in southern Illinois, studying and practicing medicine under a preceptor, and briefly studying at Illinois Normal University in Normal. Forbes began his publishing career in 1870 with botanical notes made in southern Illinois.

Forbes's lifelong interest in the natural sciences led him in his late twenties to a formal career in biology. His work in this field began in 1872 when he became curator of the Museum of the Illinois State Natural History Society in Normal. For three years beginning in 1875, Forbes served as an instructor of zoology at Illinois State Normal University. In 1877 he founded the Illinois State Laboratory of Natural History (and served as its director until 1917). In 1882 he was appointed state entomologist.

Forbes was awarded the Ph.D. "by thesis and examination" by the University of Indiana in 1884, although he had never received a bachelor's degree. That same year, he moved to the University of Illinois in Urbana to become a professor of zoology and entomology, serving until 1909. He became dean of the College of Sciences at the university in 1888, continuing in that position until 1905. When the Illinois State Laboratory of Natural History and the Office of the State Entomologist were combined in 1917, Forbes became the first chief of the new agency, the Illinois Natural History Survey, remaining in that position until his death in 1930.

His professional career as a field researcher, scientific writer, teacher, and administrator was marked by accomplishments in several fields: ornithology, entomology, ichthyology, aquatic biology, and ecology. Forbes's early research studies beginning in 1880 on the foods of birds, fishes, and insects represent pioneering investigations into the interrelationships between organisms and their environment. His awareness of this as well as the expectation that publicly supported scientific inquiry has practical and economically relevant uses for the people of the state is well illustrated by the following statement from his "The Food of Fishes":

> It is through the food relation that animals touch each other and the surrounding world at the greatest number of points, here they crowd upon each other the most closely, at this point the struggle for existence becomes sharpest and most deadly; and, finally, it is through the food relation almost entirely that animals are brought in contact with the material interests of man.

His work on the food of birds was extended later to a series of censuses of Illinois birds in various of the state's natural habitats throughout the year.

In 1887 he published his classic paper "The Lake as a Microcosm," in which he reported on the biological variation associated with changes in water levels in a number of Illinois lakes. Perhaps Forbes's greatest achievements are connected with the Illinois River. In 1894 the University of Illinois established a field station on the Illinois River at Havana with Forbes as its first director. It was at this station that the world's first studies of the aquatic biology of an inland river system were launched. Forbes believed that the detailed and long-term study that such a field station afforded would provide "a vivid idea of the individuality of a river as an organism, and of the complexity of its structure and sensitiveness of its physiological reactions."

Forbes was the author of over 400 publications, most on entomology, the remainder on zoology other than entomology, botany, ecology, general natural history, and science education. He was the recipient of the first-class medal of the Société Nationale d'Acclimatation de France in 1886 for his scientific publications. Forbes held a number of professional posts throughout his life. At various times he served as president of the Illinois Academy of Science, chairman of the International Congress of Zoologists, president of the American Association of Economic Entomologists, president of the National Society of Horticultural Inspectors, president of the Entomological Society of America, and president of the Ecological Society of America, among others. In 1918 he was elected a member of the National Academy of Sciences.

Forbes married Clara Shaw Gaston in 1873. They had five children: Bertha Van Hoesen, Ernest Browning, Winifred, Ethel, and Richard Edwin.

SUZANNE ROSS

Howard, L. O. "Biographical Memoir of Stephen Alfred Forbes." *Ecological Investigations of Stephen Alfred Forbes.* New York, 1977.

Mills, Harlow B., et al. "A Century of Research." *Illinois Natural History Survey Bulletin* 27 (1958): 95–103.

Scott, Thomas G. "Stephen Alfred Forbes: May 29, 1844–March 13, 1930." *Audubon Bulletin* 119 (September 1961): 1–5.

Fossey, Dian (January 16, 1932–December 24, 1985). A primatologist, writer, and conservationist, Fossey was born in San Francisco, California. Her parents' divorce and indifferent treatment by her stepfather contributed to the sense of social isolation that is evident in her later personal and professional writings. Following a natural love of animals, she attempted to train for a career in veterinary medicine, transferring to occupational therapy due to difficulty with chemistry and physics courses. Graduated from San Jose State College in 1954, Fossey took a position at the Korsair Children's Hospital in Louisville, Kentucky, where she remained for eleven years. It was at this time that her interest in African wildlife began to develop, first through contact with a journalist from the *Louisville Courier Journal* who eloquently described his experiences as a tourist and subsequently by accounts from a close friend who had made a safari in 1960.

By June 1963, Fossey had accumulated sufficient funds to cover the costs of six weeks on safari, mortgaging her wages for three years to obtain the loan. In answer to inquiries from incredulous friends and relations, she stated that she planned to recover the investment by selling her photographs, publishing an account of her journey, and possibly filming wildlife as well. As preparation for the trek, she read voraciously anything in print on Kenya, Tanganyika, Uganda, southern Rhodesia (Zimbabwe), and the Belgian Congo (Zaire). The description of his fieldwork with the mountain gorillas of the Congo conducted at Kabara from August 1959 to September 1960 by zoologist George Schaller in his books *The Year of the Gorilla* (1964) and *The Mountain Gorilla: Ecology and Behavior* (1963) impressed and intrigued her. As she later noted in her diary:

> Neither destiny nor fate took me to Africa. . . . I had a deep wish to see and live with wild animals in a world that hadn't yet been completely changed by humans. . . . Almost at the end of my trip I found the place I had been looking for. . . . Right in the heart of central Africa . . . are great old volcanoes towering up almost fifteen thousand feet, and nearly covered with rich green rain forest—the Virungas.

Her initial introduction to African wildlife began on the plains of Kenya in September 1963 and extended through the Serengeti, Tsavo, and the Ngorongoro Crater. Of lasting import to her future was a visit to Olduvai Gorge, where she became acquainted with renowned paleoanthropologist Louis Leakey. In the course of their conversation, Fossey noted her intention to visit the Virunga region in hopes of seeing the mountain gorillas. Leakey responded that, apart from Schaller's research, no long-term study of this species had yet been carried out. This was the beginning of the association that would later result in the foundation of the Karisoke Research Centre and Fossey's widely known studies of the mountain gorillas of Rwanda. Departing from Olduvai, Fossey and her guide traveled to Kisoro, near the borders of Uganda, the Congo, and Rwanda, within hiking distance of the western slopes of the Virunga chain. She was able to visit Mount Mikeno in the Congo in the company of wildlife photographers Alan and Joan Root and experienced the thrill of meeting a group of six adult gorillas. Taken together with the then-recent discussion with Leakey, this event crystallized a determination to return to study them more deeply.

On her return to Louisville and her hospital position, Fossey made several attempts to publish accounts of her trip and to interest the National Geographic Society in her still photographs of the Karamojong people. By the time Leakey arrived in the city on a speaking tour in March 1966, these efforts had not been notably successful. Following his presentation, Leakey inquired regarding the success of her trip to Kabara and her plans for the future. In answer to her statement that she would like to work with animals and was fascinated by the gorillas of the volcano rain forests, Leakey requested that she meet with him further. The end result of these interviews was an offer to Fossey (the twenty-third candidate so evaluated) to serve as principal investigator in the first extended field study of the remaining mountain gorilla populations. At this time, Jane Goodall had been in the field for some six years, working with chimpanzees at the Gombe Stream Reserve in Tanzania, and Birute Galdikas had not yet begun her work with the orangutans of Borneo's Tanjung Puting National Park. According to Fossey's account, Leakey voiced his opinion that women were more suited to field ethological studies than men due to their greater patience and capacity to immerse themselves in the task. With a mixture of elation and unreality, she accepted the challenge and, with support from the Wilkie Foundation and the National Geographic Society, departed for Africa on December 15, 1966.

Following a brief period of staging in Nairobi, Fossey (accompanied by Alan Root) drove to the Ugandan border, intending to resume observation of the gorilla groups at Kabara. By January 15, 1967, she had successfully established a base camp in the Parc des Virungas of Zaire and begun to perfect the skills in habituation, such as mimicking contentment vocalizations and grooming mannerisms, necessary for close approach. Her work was abruptly cut short on July 9, 1967, when the eastern Congo was declared under a state of siege due to the insurrection led by Moise Tshombe. Fossey returned from the forest to find a party of soldiers sent by the park's director to ensure her safe removal from Kabara and eventual exit from the Congo. By the end of the month, she had lost access to the Parc des Virungas area and determined to continue her interrupted studies on the Rwandan side of the range, having refused Leakey's suggestion that she shift venue and species and take up orangutan studies in Indonesia or observe the lowland gorillas of West Africa. On September 24, 1967, accompanied by Alyette de Munck, a longtime resident of the Virungas, Fossey and a train of porters climbed to a saddle between the volcanoes Visoke and Karisimbi to found the research base of Karisoke, from which all of her subsequent work (and that of the numerous students and assistants who were to join her over the years) would be conducted.

The unique contribution of Fossey to modern primatology lies in her extension of the habituation techniques utilized by Schaller and others with several mountain gorilla groups within the Parc des Volcans for the most extended study of the species carried out since its formal naming in 1902. Through gradually making the dominant silverback males, attendant females, and offspring accustomed to her presence by copying patterns of vocalizations and familiar patterns of behavior such as pretending to feed on wild celery and other food plants, Fossey was eventually accepted by group members to the degree that she could openly observe interactions between family members and track the histories of individual animals. Her emotional relationship to the Virunga gorillas is very evident in her popular monograph *Gorillas in the Mist*, published in 1983 to both summarize her work to that point and call attention to the effect the pressures of poaching, overpopulation, and the unceasing demand for agricultural land within Rwanda were having on the Parc des Volcans and its wildlife in general. While acknowledging the place of international agreements restricting trade in endangered species and similar legislations in conservation work, Fossey was convinced that what she termed "active conservation" would be more effective in protecting the gorillas. Following the death of one of her primary specimens, Digit, at the hands of poachers in 1977, she founded the Digit Fund, whose goal was to support proactive anti poaching tactics in which she herself often participated. This activist approach to wildlife preservation was looked at askance by many and contributed much to the numerous lurid and highly charged stories that came to be told about her.

Her lasting legacy lies in the fact that she thrust and held the issue of the survival of the mountain gorillas as a species into international awareness and tenaciously continued to publicize their plight for over a decade at a period of time that was crucial to their existence. Her widely publicized murder in December 1985 at Karisoke only served to call further attention to the cause for which she had labored, spoken, and written. The transformation of *Gorillas in the Mist* into a widely distributed film with actress Sigourney Weaver brought the story of her conservation efforts to an audience far wider than she had ever reached during her lifetime.

ROBERT B. MARKS RIDINGER

Fossey, Dian. *Gorillas in the Mist.* Boston, 1983.

Hayes, Harold. *The Dark Romance of Dian Fossey.* New York, 1990.

Montgomery, Sy. *Walking with the Great Apes: Jane Goodall, Dian Fossey, Birute Galdikas.* Boston, MA, 1991.

Mowat, Farley. *Woman in the Mists: The Story of Dian Fossey and the Mountain Gorillas of Africa.* New York, 1987.

Fuertes, Louis Agassiz (February 7, 1874–August 25, 1927). Born in Ithaca, New York, into an academic family, Fuertes was an artist and ornithologist. As a child, he and his sister collected all sorts of wildlife. His special interest in birds was noted, and arrangements were made for him to have access to one of the original elephant folio sets of John James Audubon's monumental work *Birds of America* (1917), which had been donated to the public library in Ithaca. Exposure to these masterworks of naturalistic art spurred him to further develop his own artistic gifts, beginning with a drawing from life of a male red crossbill done in 1888 when he was fourteen. From at least age fifteen, he kept lists of birds he had seen, cross-referencing them with Audubon's volumes.

In 1891, he became a member of the American Ornithological Union. Entering Cornell University in 1893 as a student in the College of Architecture, he continued to draw and was introduced by one of his fellow students to his uncle Elliott Coues, the most prominent American ornithologist of the day. With his assistance and encouragement, Fuertes decided to be a bird illustrator. An exhibition of fifty of his drawings at the American

Ornithological Union in late 1895 led to requests for his skills from Florence Merriam (then engaged with the manuscript of her *Handbook of Birds of the Western United States* (1902), for whom he made several drawings that appeared in her 1896 work *A Birding on a Bronco*. His next publishing project was a commission from Coues and Mabel Osgood for 111 pictures for the children's book *Citizen Bird* (1897). At the 1896 meeting of the union, Fuertes made the acquaintance of prominent artist and naturalist Abbott Thayer, with whom he was to have a lifelong friendship. Thayer's experiments on protective coloration were highlighted at that meeting, as were fifty of Fuertes's drawings.

After graduating from Cornell in June 1897, he undertook his first field expedition of bird collecting and illustrating, a six-week trip to the Indian River region of Florida with a party led by Thayer. The appearance of *Citizen Bird* had immediately given the young man a solid reputation as a leading ornithological artist, and his skills were quickly sought after. On returning from Florida, he was recruited for the 1899 Harriman Alaska Expedition and collecting trips to Texas, the Bahamas (to study flamingoes), and California with Vernon Bailey between 1901 and 1903. Following his marriage to Margaret Sumner on June 2, 1904, he traveled to Jamaica and began to give a series of lectures at Cornell, illustrated with his own work, which lasted until 1907. At this time, he was also engaged in preparing illustrations for Thayer's volume *Concealing Coloration in the Animal Kingdom* (1909). His fieldwork resumed in 1908, with a visit to the Cape Sable region of southwestern Florida, followed by trips over the next five years to the Magdalen Islands in the Gulf of St. Lawrence, the Yucatan and Mount Orizaba regions of Mexico, and the Cauca Valley of Colombia.

Between 1913 and 1920, he was in residence in Ithaca, working on eight series of paintings commissioned by the National Geographic Society. These paintings are notable because they were the most widely disseminated of all Fuertes's works outside the ornithological community and because they also included illustrations of large and small North American mammals and dogs in their subject matter. His popular legacy, begun with *Citizen Bird*, continued in 1919 and 1920 with *The Burgess Bird Book for Children* and *The Burgess Animal Book for Children*, respectively.

In belated recognition of his contributions to the university, he was appointed resident lecturer at Cornell in 1922. The lure of the field still drew him, first to Holland in 1924 and then to Wyoming in 1925, capped by a nine-month expedition to Abyssinia in 1927 with Wilfred Osgood, curator of zoology at Chicago's Field Museum, to study and record the unique birdlife there. A more pleasing tribute was offered that same year when the Boy Scouts selected him as one of seventeen great American outdoorsmen whom they named as Honorary Scouts. The planned record of the Abyssinia trip and completion of plates for the manuscript of the *Birds of Massachusetts* were never carried out. On August 25, 1927, when he was returning to Cornell, his car was struck by a train at a grade crossing at Unadilla, New York, and Fuertes was killed instantly.

The delicacy and power of Fuertes's depictions of any living thing mark him as the successor to Audubon as the premier wildlife artist of the United States during the late nineteenth and early twentieth centuries. The scope of his work varied enormously, ranging from book illustrations to permanent habitat groups for the American Museum of Natural History and the New York Zoological Society and murals for the Flamingo Hotel in Miami Beach. Both as a contributor to field research and as an independent lecturer, he offered the reading public and scientific community images of birdlife and animal life that brought home the message of their fragility and worth.

ROBERT B. MARKS RIDINGER

Boynton, Mary Fuertes. *Louis Agassiz Fuertes: His Life Briefly Told and His Correspondence.* New York, 1956.

Chapman, Frank M. "Memories of Louis Fuertes." *Bird Lore* 51 (January–February 1939): 3–10.

Peck, Robert McCracken. *A Celebration of Birds: The Life and Art of Louis Agassiz Fuertes.* New York, 1982.

Gabrielson, Ira Noel (September 27, 1889–September 7, 1977). Born in Sioux Rapids, Iowa, Gabrielson was a biologist. As a child, one of his favorite interests was collecting various types of bird's nests from the fields and woods of the area, the beginning of a fascination with living things of all species that was to become the dominant interest of his life.

Following his completion of high school and graduation from Morningside College in 1912, he taught biology in Sioux City, Iowa for three years, leaving teaching in 1915 to accept the post of assistant economic ornithologist with the Bureau of Biological Survey, at that time part of the Department of Agriculture. After initial field research on the food habits of birds in New England, the Mississippi Valley, and the Great Plains, he was transferred from Washington to Portland, Oregon, where he served as supervisor of rodent and predatory animal control work in the Pacific region and regional game management director for the northwestern states. Returning in 1935 as chief of the bureau with a well-established reputation as an expert on the birds, plants, and animal life of the West, he was appointed the first head of the Fish and Wildlife Service upon its creation in the Department of the Interior in 1940. In 1946, he resigned this appointment to accept the presidency of the Wildlife Management Institute (founded in 1911), which he served in this capacity until 1970, then taking the post of chairman of the board. During this time, he also became deeply engaged in the cause of international conservation, working as an organizer of the World Wildlife Fund (U.S.) and cofounding the International Union for the Conservation of Nature and Natural Resources. This global focus did not prevent him from taking an active role in state and regional conservation work in the Washington, DC, area, where he chaired the Regional Park Authority of Northern Virginia from 1959 to 1976 and served on the Virginia Outdoor Study Commission from 1966 to 1974.

While his organizational and administrative gifts played an important role in the establishment of planetary conservation as an aspect of local American environmental preservation efforts, his writings in both scientific and popular journals and his books popularized the limitations and benefits of wildlife conservation. Shortly after his arrival in Oregon, Gabrielson began to coauthor several studies for the Biological Survey on the economic value of the starling and the food habits of winter migratory birds. These two publications, completed in 1921 and 1924, respectively, inaugurated a writing career that would eventually encompass dozens of articles and several notable books on regional ornithology of the Northwest and Alaska, wildlife refuges, and wildlife management. An expansion of his initial roles as ornithologist and field biologist is clearly visible in the works published prior to the entry of the United States into World War II. Among them are *Birds of the Portland Area* (1929), *Western American Alpines* (1932), *Game Management on the Farm* (1936), and *Birds of Oregon* (1940). A series of articles in 1934 presented the richly diverse plant ecosystems of Washington, discussing landscapes as varied as the rain-shadow desert east of the Cascades and the Olympic and Siskiyou ranges. These pieces set the pattern for much of his later work, combining an emphasis on natural detail with the message that such beauties must be preserved. Massive flooding in the spring of 1936 prompted an article on the interrelationship of efforts at environmental management and natural cycles essential to proper functioning of riparian ecosystems. The proposed solution offered by the Corps of Engineers, a system of dams and levees, would be severely tested in the Great Flood of 1993.

From 1935 to 1939, wildlife conservationists in the American West shifted their focus to the restoration and rehabilitation of overgrazed range lands and other marginal habitats, while continuing to work for further funding and expansion of the system of wildlife refuges inaugurated by President Theodore Roosevelt in 1903. At the First North American Wildlife Conference in 1936, seven basic concerns were set forth: more land for wildlife, closer cooperation of federal and state agencies, closer coordination of federal activities, a wider recognition of wildlife values by those who manage lands, efforts to correct stream pollution, adequate research programs, and protection regulations based entirely on the needs of wildlife. The threat posed to endangered species of birds and animals from the commercial exploitation of limited habitats was a problem vehemently opposed by Gabrielson at this time as well. In the summer of 1940, he made an extended journey to the then–Territory of Alaska to inspect newly established wildlife refuges intended to safeguard major nesting colonies of seabirds. The trip resulted in an abiding interest in the environment and wildife of the area, later culminating in his 1959 volume *The Birds of Alaska*.

Upon the entry of the United States into World War II, all national resources (and the agencies that managed them) were assigned new priorities in the mobilization effort. Even the Fish and Wildlife Service found itself facing the decimation of its personnel through conscription and restricted budgets. Gabrielson's writing from this era highlighted the natural splendors and wild glories of such unique sites as the Okefenokee Swamp, emphasizing that protecting and preserving such places was an essential contribution to the maintenance of national spirit and morale. He also found himself co-opted to the post of deputy coordinator of fisheries and spent 1943 through 1945 attempting to balance the needs of the disrupted fishing industry with federal demands for ships and marine products.

His nongovernmental involvement in activism for wildlife began in 1947 with a review of the progress of the service from 1935 to 1946 and returned with undiminished vigor to the increasing pressure being put on wild lands by the rapidly expanding postwar economy, particularly the threat posed by unmonitored use of pesticides. The population explosion and massively increased access to Alaska occasioned by the system of bases and airfields created during wartime was also the subject of discussion in his writings. Much of Gabrielson's work during the last thirty years of his life was devoted to continuing to educate the public on the spiritual necessity and economic realities of managing the vast resource represented by the wild species of America. His widely read volume *Wildlife Conservation* (1944) explored the interdependence of erosion control, water conservation, forest and grasslands, and their role in regional ecologies and profiled the specific cases of fur-bearing animals, predators, and the place of refuges and breeding areas in the national scheme of conservation. The latter theme was expanded into the 1949 companion work *Wildlife Management*. Its purpose was to illustrate the need for a balanced approach to the creation and operation of wildlife reserves by offering a picture of their contemporary status.

In all his writings, he argued for the sane and sensible use and control of the wild things of the continent, much as the popular writer Emerson Hough had done earlier in the century. He communicated a fascination with the wild to a popular audience while making the otherwise abstract issues and values of conservation comprehensible and vital. Professional recognition peaked with the awarding of a bronze medal by the National Audubon Society in November 1949. The citation accompanying it stated that "his chief monument will ever be the vast chain of national refuges from arctic Alaska to the Florida Keys, where he has proved that our vanishing wild life and game resources can be enjoyed and maintained from generation to generation." He died in Washington, DC, at the age of eighty-seven.

ROBERT B. MARKS RIDINGER

Corey, Herbert. "Piscatorial Solomon." *Nation's Business* 31 (November 1943): 36, 44–47.

Griscom, Ludlow. "Dr. Gabrielson Honored." *Audubon Magazine* 52 (January–February 1950): 2, 68.

Halle, Louis J. "Gabrielson." *Audubon Magazine* 48 (May 1946): 140–146.

"Ira Gabrielson, Wildlife Expert and Leading Conservationist." *National Parks and Conservation Magazine* 51 (December 1977): 18–19.

Smith, Llyod W. "Ira N. Gabrielson, 1889–1977." *American Forests* 83 (November 1977): 4, 37.

Gore, Albert Arnold, Jr. (March 31, 1948–). Primarily known as an American politician, former Vice President Gore has been hailed as an important, even prophetic, voice on environmental issues. According to the *Concord Monitor* (February 27, 2007), "Gore was one of the first politicians to grasp the seriousness of climate change and call for a reduction in emissions of carbon dioxide and other greenhouse gases." He ordered congressional hearings on toxic waste in 1978–1979 and began investigations into the dangers of global warming in the 1980s. He has written three best-selling books; starred in an Academy Award–winning documentary; and traveled extensively, giving public lectures offering information and solutions to the climate crisis. Respected worldwide as a leading spokesman for the environment, Gore was awarded the prestigious Nobel Peace Prize in 2007.

Gore was born in Washington, DC. His father was serving his fifth term in the House of Representatives, representing Tennessee's Fourth Congressional District. Pauline, his mother, worked with her husband in his congressional office. Gore was the couple's second child; his sister, Nancy, was ten years old when he was born. Politics was the family business, and it was assumed that Gore Jr. would follow in his father's footsteps.

Gore Sr. had an illustrious political career, serving in the House of Representatives from 1939 to 1953 and in the Senate from 1953 to 1971. He enjoyed a good reputation in Congress, known as an independent thinker and a hard worker. Decidedly liberal, he was one of the first southerners in Congress to vote for civil rights legislation. He refused to sign the Southern Manifesto, a document declaring opposition to the Supreme Court's decision to desegregate public schools (*Brown v. Board of Education*, 1954). One of his finest achivements was the massive federal highway program that he supported. To keep his Tennessee constituents informed about the progress of the program, he prepared radio broadcasts similar to Franklin Roosevelt's "fireside chats." Gore Sr. was certainly an influential powerhouse of a man who passed his son a baton weighted with presidential aspirations.

When Congress was in session the Gores lived in the far-from-luxurious Fairfax Hotel in Washington, DC. During summer vacations the family returned to the farm in Carthage, where hay and tobacco were grown and cattle were raised. The farm was a fabulous place for children, with horses, cows, and the Caney Fork River for swimming, fishing, and canoeing. Gore attended elementary school in Carthage until fourth grade, when he was enrolled in the prestigious St. Albans School for Boys in Washington, DC. Although he was a good and serious student, his teacher sensed that the boy really wanted to be back on the farm he loved. Years later, Gore said in an interview, "I'm happy my parents gave me a good education . . . but if you're a boy and you have the choice between the eighth floor of a hotel and a big farm with horses, cows, canoes, and a river . . . it was an easy choice for me."

As much as Gore loved the farm, his life there was not all play; he had backbreaking chores to complete every day. It was on the Carthage farm that he first became aware of the need to preserve the environment. His father taught him to fill a gully—or it would become a ditch, and precious topsoil would be washed away. His mother made him aware of the lessons of Rachel Carson's *Silent Spring* (1962), which warned about the dangers of DDT and other pesticides. As a result Gore developed ecological concerns at an early age.

During his high school years Gore's time continued to be divided between Washington, DC, and the family farm. While school was in session, he worked hard, earned excellent grades, and managed to be active in sports and student government. When summer arrived, it was back to Carthage to work the farm by day and meet by night with a close-knit group of friends including his steady girlfriend, Donna Armistead. Curiously, he kept these two parts of his life completely separate and never invited Donna or any of his Carthage friends to visit him at St. Albans. His classmates heard nothing about his agrarian life, and his Carthage friends heard very little about his school life.

In 1965 Gore, a National Merit Scholarship finalist and three-letter varsity athlete, was accepted at Harvard, the only university to which he applied. Two of his roommates in Dunster House were actor Tommy Lee Jones and John Tyson, son of black well-to-do parents and a defensive back of sufficient skill to be drafted by the Dallas Cowboys. Tyson and Gore became close friends. Years later Tyson would credit Gore for making him less antagonistic toward other races. In 1969 Gore graduated from Harvard cum laude with a Bachelor of Arts in government.

Although as opposed to the Vietnam War as his father, Gore enlisted in the U.S. Army on August 7, 1969. Gore Sr. was up for reelection, and whether his son's decision was one of conscience or political savvy remains debatable. In 1970, before he had to report for duty, Gore married Mary Elizabeth Aitcheson (Tipper Gore). He and Tipper first met at his high school prom and dated throughout college. The newlyweds shipped off to Fort Rucker, Alabama, where Gore was assigned as a military journalist. Every weekend they could get away, the couple traveled to Tennessee to campaign for Gore Sr. Senator Gore's opponent blasted him for his position on the war and was successful in his attacks; he defeated Gore Sr. by a wide margin. His father's defeat left Gore Jr. bitterly disappointed and soured on politics. He shipped out to Vietnam on Christmas Day of 1970, where he served four months in Bien Hoa and another month in Long Binh. Since his unit was standing down, he received an honorable discharge two months early to attend divinity school at Vanderbilt University. For the next few years, Gore worked as a reporter for the *Tennessean* in Nashville and attended law school part-time.

In 1976 Gore entered the political arena when Congressman Joe Evin announced his retirement after thirty years as Tennessee's Fourth District representative. Gore won the Democratic primary and was elected to his first congressional post. He was reelected three times (1978, 1980, 1982) and successfully ran for a seat in the Senate in 1984, where he served until 1993 when he became vice president under Bill Clinton. During his first term in Congress in 1978, Gore received a letter from a Tennessee family complaining of illnesses they suspected were being caused by a nearby toxic waste dump. This letter prompted Gore to hold congressional hearings on the subject of toxic waste; he focused on the Tennnessee site and one at Niagara

Falls, New York. The second site, called the Love Canal, would become infamous as one of the most terrible examples of the hazards of toxic waste dumping. The canal, named after its creator William Love, was intended to carry water for electricity, but the scheme fell through, and the canal was left unfinished. Years later the Hooker Chemicals & Plastics Corporation used the site for its toxic waste. The canal was filled in when Niagara Falls needed the land; homes as well as an elementary school were built right on top of fearsome toxic waste. The water supply was contaminated, and the results were devastating. Children and adults suffered serious illnesses; the number of miscarriages, stillbirths, and infants born with defects was staggering. The federal and state governments were forced to spend $28 million on the evacuation and relocation of the residents of Love Canal. The legislative outcome of the hearings was the Superfund Law of 1980. Gore was the prime mover behind this measure, which provided a fund of $6.1 billion to clean up toxic waste sites.

He continued to be actively involved in environmental issues throughout his years in the House and the Senate. He sponsored or authored legislation that called for pesticide-free imports, encouraged recycling, and fostered cooperation between military and civil research scientists. His fact-finding trips to the South Pole, the Amazon rain forest, the North Pole, and the Caribbean were all efforts to continue to gain firsthand information on the ecological effects of climate change.

Gore's life changed drastically on April 3, 1989, when the youngest of his four children, Albert Gore III, was hit by a car, thrown thirty feet in the air, and scraped along the pavement for another twenty feet until he finally landed in a gutter. The six-year-old hovered between life and death for days; for the next month Gore and his wife stayed at the hospital by their son's side. The outpouring of sympathy from people in all walks of life made Gore deeply self-aware. He began to examine relationships: within himself, with his family, and between people and their environment. It seemed he finally knew what he wanted to do with all the information he had gathered traveling around the globe. At a nurse's station near his son's hospital room, Gore began writing what would become *The Earth in Balance: Ecology and the Human Spirit* (1992). The book called for a worldwide cooperative effort to discard the notion that human actions have no lasting effect on the earth's environment. A global Marshall Plan was proposed with five strategic goals to help humanity restore the balance missing from its relationship with the earth. In the conclusion of the bestselling book,

Gore wrote that "we do have a future. We can believe in that future and work to achieve and preserve it, or we can whirl blindly on, behaving as if one day there will be no children to inherit our legacy. The choice is ours."

As the 1992 presidential election approached, Gore had to decide whether to leap into the political fray. After losing the Democratic nomination in 1988, he knew he would have to win this time to keep any presidential hopes alive. There were, however, larger considerations. Albert III had survived his dreadful accident, but the trauma and long recuperation had taken their toll on the entire family. Gore put his responsibilities as a father and husband above his political ambition. After the decision was made, Gore said he knew it was the right one—and he and his family enjoyed the campaign-free year. When Bill Clinton secured the nomination, Gore was selected to be his running mate for several reasons: he and Clinton were in agreement on most important issues; he was physically strong enough to wage a vigorous campaign; he had served in Vietnam; and he had a strong environmental record. Gore felt the three-month vice-presidential campaign would not disrupt his family the way a presidential bid would have, so he accepted.

On January 20, 2001, Gore was inaugurated as the forty-fifth vice president of the United States. During his two terms as vice president, Gore continued to be attentive to environmental issues. He made sure that $1 billion was included in the budget for new technology initiatives, established a forest management plan in the Pacific Northwest, set up a Clean Car Initiative, and worked with the president on numerous executive orders. These orders made the administration lead by example by cutting toxic waste by 50 percent, purchasing energy-efficient vehicles and computers, phasing out ozone-depleting chemicals, and installing water-saving technologies in all federal buildings. He was also responsible for the creation of a whale sanctuary in the South Pacific Ocean. On Earth Day 1994, he launched the GLOBE (Global Learning and Observations to Benefit the Environment) Program, a school-based education and science activity that made use of the Internet to increase student ecological awareness and to generate research data for scientists. In addition, he was instrumental in the passage of the North American Free Trade Agreement (NAFTA), which eliminated tariffs for U.S. companies, thereby encouraging them to open markets for environmental technologies overseas.

After losing to George W. Bush in the controversial 2000 presidential election, Vice President Gore rechanneled his energy to focus on saving the earth from irrevocable change. In his bestselling book *An Inconvenient Truth* (2006) Gore elevated the problem of global warming from a political issue to a moral one. "Humanity," he said "is sitting on a ticking time bomb. If the vast majority of the world's scientists are right, we have just ten years to avert a major catastrophe that could send our entire planet into a tail-spin of epic destruction involving extreme weather, floods, droughts, epi-demics and killer heat waves beyond anything we have ever experienced." Amid the harrowing facts and chilling predictions is the personal story of Gore's own journey from an idealistic college student who recognized the looming environmental crisis to a young senator facing a life-changing family tragedy to the man who lost the presidency and returned to the most important cause of his life, the threatened environment.

Gore starred in the film version of the book, which was released on May 24, 2006, and received rave reviews from critics who saw him as he had never been seen before—funny, engaging, open, and on fire about getting the truth of the planetary emergency out to ordinary citizens before it is too late. Supporters bestowed "green knighthood" status on Gore for his fervent crusade to stop global warming in its tracks by exposing the myths and misconceptions that surround it. The documentary film was a huge success and resulted in two Academy Awards, Best Documentary Feature and Best Original Song. Like Gore himself, *An Inconvenient Truth* has its detractors who challenge the science of his premise and accuse him of evangelistic hyperbole. Some feel that Gore went beyond the evidence and was overselling the scientific certainty about knowing the future. However, some of the most virulent critics were well known for anti-Clinton/anti-Gore sentiments and attacked the global warming issue with a fervor remi-niscent of the town elders in Henrik Ibsen's *An Enemy of the People* (1882). More rational criticisms have come from the scientific community, who claim that the former vice president's work may hold imperfections and technical flaws, but most agree with the fundamentals—that warming is real, and humans are the cause. Gore's third book, *The Assault on Reason* (2007), was less about the environment and more about the incompetence of the current administration. An impassioned Gore presented a well-documented brief that accused the Bush White House of ignoring expert advice on the Iraq war troop levels, global warming, the deficit, and other issues. The book revealed a fiery Gore asserting that the president is out of touch with reality and given to using the language of fear to "drive the pub-lic agenda without regard to the evidence, the facts or the public interest."

Presently, Gore and a team of specialists are involved in training more than 1,000 volunteers to give informative climate change presentations throughout the United States. On February 9, 2007, Gore and Richard Branson announced the Virgin Earth Challenge, a competition offering a $25 million prize for the first person or organization to produce a process for the removal of atmospheric greenhouse gases. The Gore family resides in Nashville, Tennessee, and owns a small farm near Carthage. He and his wife have four children and two grandchildren.

Throughout his career as an elected official, Gore has worked to make the earth we inhabit a safer place. His writings have been directly responsible for drawing worldwide attention to the threatened environment. He was awarded the Nobel Peace Prize in 2007 for his extraordinary efforts to expose the facts and fictions surrounding the climate change issue. Gore's belief in our ability to make changes and ensure a future for ourselves and our planet is both inspirational and challenging. In his Nobel Prize acceptance speech, Gore issued the challenge once again:

> We are standing at the most fateful fork in that path. So I want to end as I began, with a vision of two futures . . . and with a prayer that we will see with vivid clarity the necessity of choosing between those two futures, and the urgency of making the right choice now. . . . We have everything we need to get started, save perhaps political will, but political will is a renewable resource. So let us renew it, and say together: "We have a purpose. We are many. For this purpose we will rise, and we will act."

ADRIENNE GLYNN

Broad, William J. "From a Rapt Audience, a Call to Cool the Hype." *New York Times*, May 13, 2007.
Burford, Betty. *Al Gore: United States Vice President.* Hillside, NJ, 1994.
Dunham, Richard S. "The Trashing of Al Gore." *Businessweek*, May 24, 1999.
Gore, Al. *Earth in the Balance: Ecology and the Human Spirit,* Boston, MA, 1992.
———. "Nobel Prize Acceptance Speech." *Huffington Post*, December 10, 2007.
Guggenheim, Davis. *An Inconvenient Truth.* Paramount Pictures, 2006.
Kakutani, Michiko. "Al Gore Speaks of a Nation in Danger." *New York Times*, May 22, 2007.
Turque, Bill. *Inventing Al Gore.* Boston, MA, 2000.
Zelnick, Bob. *Gore: A Political Life,* Washington, DC, 1999.

Grant, Madison (November 19, 1865–May 30, 1937). A sportsman, lawyer, naturalist, and eugenicist, Grant was born and raised in New York City. He was the oldest of four children born to Newark physician Dr. Gabriel Grant and his wife Caroline (Manice) Grant. Dr. Grant was the first physician in the Civil War to win the Congressional Medal of Honor for his bravery at the Battle of Fair Oaks in June 1862. During the war this physician from Newark, New Jersey, met the wealthy Caroline Manice, and they were married in 1864 while Grant was the medical director of an army hospital in Madison, Indiana.

Dr. Grant was a member of the Order of United Americans and had been a member of the Newark Know-Nothing Lodge before becoming a member of the Republican Party. After his retirement from the army in January 1865, he returned temporarily to Newark and shortly thereafter moved to New York, where he became a member of the exclusive Union Club. His oldest son, Madison, as well as De Forest, Norman, and a daughter Katherine were all born in New York.

Madison was raised in a privileged setting. He traveled extensively with his family to the Middle East, Africa, and Europe. He went to private schools in New York and spent his teenage years in Dresden, Germany, where he was tutored. At this time, Germany was a nation concerned with the preservation of the environment, especially the forests.

In his work as a naturalist and eugenicist, Grant had as his principal concern the need to warn his fellow countrymen of the dangers that society faced as a result of the destruction of its natural resources and the replacement of what he considered the normal, traditional American bloodstock with inferior non-Nordic immigrants. Obviously, as a man of wealth, he could have avoided the problems inherent in being a tocsin sounder, but he was consistent throughout his life in the belief that America was passing from greatness. He was not alone.

Others agreed with this pessimistic assessment. Those men were in a sense Malthusian alarmists. They believed the frontier was already closed. They were reformers. They wanted to use the government to protect the environment and maintain popular control.

Grant was graduated from Yale University in 1887 and received an LL.B. from Columbia University in 1890. He passed the bar and opened a law office in New York City. Rarely did he engage in legal practice. His office served as the center for his role as naturalist and eugenicist. He became politically active in 1894–1895 when his brother De Forest became a fervent supporter of fellow businessman William Strong, who ran for mayor of New York City in 1894. Madison helped that successful effort to try to undo the corruption they claimed had been created by Tammany Hall.

Grant, on the recommendation of George Bird Grinnell, the editor of *Forest and Stream* magazine, became a member of the Boone and Crockett Club in 1893, which Theodore Roosevelt had helped establish in 1887. For admission the club required the collection of three trophy heads. It encouraged big game hunting. But the group underwent a major transformation when it became obvious that hunting big game would lead to their extinction; Daniel Boone and Davy Crockett became interested in the preservation of these animals. They formed a committee on parks to build on the work begun with the congressional establishment of Yellowstone National Park in 1872; the organization of a National Forest Reserve Service in 1891; and the passing of the Park Protection Act in 1894.

With the success of Mayor William Strong, Grant now proposed to the leaders of the Boone and Crockett Club the establishment of a New York Zoo to provide local residents with park facilities for the city. Grant, Grinnell, Elihu Root, and two other members were able to get the New York State legislature to pass a bill in 1895, drafted by Grant, to set up a private group, the New York Zoological Society, for the maintenance of a zoo. Under Mayor Strong, the city then provided the society with 261 acres of public land, the Bronx Park.

The bulk of the credit for the passage of the legislation and the acquisition of the land was placed at Grant's door. Roosevelt was delighted with all that Grant had accomplished: initially, he was a first-class lobbyist in Albany; then, with his brother De Forest's help, he had been able to gain concessions from New York City.

Roosevelt subsequently asked Grant and Grinnell to have the state legislature in New York prevent the harassing of deer in the Adirondacks. Again Grant took on the assignment, went to Albany, and was able, in 1897, to help pass a bill that outlawed the hazardous procedure for five years. Other states followed New York's lead.

When the New York Zoological Society came into existence, Grant served the group as secretary (1895–1924), as chairman of the executive committee (1908–1936), and as president (1925–until his death in 1937). Most of those who served in nonpaid administrative positions had little time to devote to the zoo; Grant, however, was there and came to dominate the zoo and its personnel.

When it became obvious that if people were to reach the zoo, they would require a roadway, Grant became the initiator of a plan to build a road to the zoo, the Bronx River Parkway Commission, the first of its kind in the East. He served as the president of that commission from 1907 to 1925. Grant was not interested in providing public transportation to the facility. Instead, he opted for a middle- and upper-class clientele, those who could afford automobiles.

Grant's New York, as well as the nation, had truly expanded. Between 1840 and 1890 the country had doubled in geographic size, and the population had expanded almost fourfold, from 17 million in 1840 to 63 million in 1890. By 1890 New York had more than 1 million residents. Grant was not happy with this explosion of people.

But Grant did not restrict his energies to the zoo and the parkway. Dr. William Temple Hornaday, the director of the zoo, interested Grant in the need to save the American buffalo. Grant, with Ernest Harold Baynes, who aroused Hornaday's concern, was one of fourteen who in 1905 organized the American Bison Society. The society reacted to the establishment of a buffalo herd in Canada; they wanted to preserve the animals in the United States. The group undertook to stock the newly created buffalo range on the Flathead Reservation in Montana. Most of the funds came from Grant and his friends.

Similar concerns with the environment led him to join with Henry Fairfield Osborn, the socially prominent director of the American Museum of Natural History, and John C. Merriam, the naturalist, in the foundation of the Save-the-Redwoods League to preserve the remaining groves in California. Grant insisted on a policy that escalated demands for preservation of the redwoods. "*Ask for much*," he insisted. "We cannot save *too many Redwoods*. Every tree is worth more to the state and nation, than as timber, be the price of the timber ever so high."

In his earlier publications, he was mainly concerned with conservation. In 1894, for example, his article "The Vanishing Moose" appeared in *Century Magazine*. His "Distribution of the Moose," originally published in the *Seventh Annual Report of the Forest, Fish and Game Commission of the State of New York*, (1875) was given wider circulation in editor George Bird Grinnell's *American Big Game and Its Haunts: The Book of the Boone and*

Crockett Club (1904). In it, Grant discussed the history of the relationship between the elk and the moose from ancient times through the medieval period to colonial Vermont and New Hampshire and the present. The theme stressed the decline of the moose and its habitat.

In 1917, Grant's efforts on behalf of conservation were rewarded: the National Institute for Social Sciences presented him with a gold medal at a huge dinner at the Hotel Astor. He was elected a vice president of the group as well. Grant received a gold medal from the Society of Arts and Sciences in 1929 for his work on the Taconic Park Commission, which he had joined in 1924, as well as for his work in developing the Bronx River Parkway. He received a large number of letters of congratulations including one from Dr. Charles B. Davenport, who exulted, "What a wonderful satisfaction it must be to you to have played so large and successful a part in conservation."

Grant was a successful fund-raiser as well. He received over $1 million from the John D. Rockefeller estate, plus $25,000 from J. Coleman DuPont as well as several other large grants. Grant was willing to use both private and public funds to preserve the environment. In 1928, he wanted to purchase private property in the national parks. "What is left of our national heritage," he wrote in the *New York Times* (April 12, 1928), "must be carefully guarded. In the past, Americans have cheerfully squandered their national resources of timber and wild life and are now entering upon a period where millions must be spent to restore what has been needlessly destroyed."

But by 1917 Grant's priorities had begun to change. He continued his work at the Bronx Zoo; however, more and more of his time was taken up by his concern over the influx of what he termed "the inferior people of the world." That portion of his career has been given a great deal of attention, to the neglect of his role as a naturalist. In that arena, he was certainly an elitist, but he was also a most public-minded citizen.

His two major publications center on his role as an immigration restrictionist: *The Passing of the Great Race* (1916) and *The Conquest of a Continent* (1934). Both works were praised by none other than Adolf Hitler—not a high note of praise. There is a legend among the Scottish Grants that when a Grant who has lived an honorable life dies, the wailing of the "Pipes of Glory" is heard. When Grant died, there were no reports of wailing pipes.

THOMAS J. CURRAN
RICHARD P. HARMOND

Alexander, Charles C. "Prophet of American Racism." *Phylon*, 23 (1962): 73–90.
Cutright, Paul Russell. *Theodore Roosevelt: The Making of a Conservationist.* Urbana, IL, 1985.
Crandall, Lee. *A Zoomen's Notebook.* Chicago, IL, 1966. The curator of birds, Crandall had many contacts with Grant at the Bronx Zoo.
Grinnell, George Bird, and Charles Sheldon, eds. *Hunting and Conservation: The Book of the Boone and Crockett Club.* New Haven, CT, 1925.
Kurt, Stefan. *The Nazi Connection: Eugenics, American Racism and German National Socialism.* New York, 1994.
Nash, Roderick. *Wilderness and the American Mind.* New Haven, CT, 1967.

Grinnell, George Bird (September 20, 1849–April 11, 1938). A conservationist and ethnographer, Grinnell was born in Brooklyn, New York, the oldest of five children of George Blake Grinnell, businessman, and Helen Alvord Lansing. Growing up in an upper-class home, he lived in several locations during his early childhood: Brooklyn, lower Manhattan, and Weehawken, New Jersey. Grinnell was seven when, in January 1857, his family moved to Audubon Park, the thirty-acre former estate of artist-naturalist John James Audubon, on still-rural upper Manhattan. He would live in the "Park" for over a half century, until the expanding city overwhelmed the area.

There, Grinnell's first teacher was Audubon's widow, Lucy Bakewell. "Grandma Audubon," as he lovingly called her, played a central role in helping to shape his early interest in nature. Audubon's two sons, Victor and John Woodhouse, were also important in building in Grinnell an affinity for the natural world. Their nearby houses were full of the fascinating possessions of their renowned father, including western hunting trophies, the "ornithological biographies," and, of course, many paintings. Both men continued the work of their father and frequently received fresh specimens of strange birds and animals that Grinnell and the other neighborhood children stared at in amazement as they were taken from their shipping containers.

The many hours spent examining the study skins and mounts in the Audubon homes were complemented by visits to his grandfather's house in Greenfield, Massachusetts, where his uncle, Thomas P. Grinnell, had gathered together a large collection of mounted birds and mammals that he had bagged on his many hunting trips. With the reinforcement given by the teachings of the Audubons, it is easy to see how the young Grinnell would quickly acquire his uncle's sporting and natural history interests.

Those intertwined avocations would be strengthened still more on the eve of the Civil War, when, at about the age of twelve, he began to go shooting. The gun proved to be a liberating force. No longer was he limited to examining the study skins and mounts of his uncle or the Audubons or to sitting passively by while they told about the exciting hunt that ended in the specimen's capture. Now he could shoot his own specimens and begin his own collection, which eventually totaled several hundred, including the now-extinct passenger pigeon and heath hen.

In 1866, the same year Grinnell entered Yale, the college awarded Othniel C. Marsh the first chair of paleontology in the United States. Several years later, as Grinnell was about to graduate, he learned that Marsh was going to lead an expedition that summer to the unmapped West in search of fossils. Marsh's acceptance of Grinnell as one of a dozen volunteer assistants from the college would prove to be a turning point in his life.

The plan of the 1870 March expedition was to follow the recently completed transcontinental railroad, taking extended side trips north and south of the line, eventually reaching the Pacific coast. The "Great West," a land characterized by huge bison herds and free-ranging Native Americans, was about to pass away, and Grinnell was a witness. In eastern Nebraska, for example, the train was stopped by a bison herd that took three hours to cross the tracks!

Hostile Indians, trying to hold on to their land, were also an ever-present threat. The day the Marsh expedition arrived at its first jumping-off spot, Fort McPherson, Nebraska, hunters from the fort in quest of pronghorn-antelope meat were attacked by a small party of Cheyenne, one of whom was killed.

The fort's chief scout was William F. "Buffalo Bill" Cody. Grinnell became well acquainted with him and also with Major Frank North, leader of the famous battalion of Pawnee Indian scouts who fought with the whites against the Sioux, Cheyenne, and Arapaho. For the romantic Grinnell, who had thrilled in his youth to the western adventure stories of novelist Mayne Reid, the months he spent in the "Wild West" were the realization of a fantasy. A love affair with the land beyond the Missouri had begun, and he would come back to it almost every year of his life.

Grinnell returned to Yale in 1874 to assist Marsh at the Peabody Museum and to pursue graduate studies. Also in that year he acted as Marsh's scientific representative on George Armstrong Custer's reconnaissance to the Black Hills of the Dakotas. In that same capacity, he joined William Ludlow, an army engineer, on his 1875 survey of the recently created Yellowstone National Park. There, he was alarmed to find commercial hide hunters killing great numbers of big game, and he registered his protest against these activities in the report published by the government as part of the papers of the reconnaissance.

Continuing his graduate studies, he submitted a dissertation titled "The Osteology of *Geococcyx californianus*" (roadrunner) at Yale and was awarded the Ph.D. in 1880. Since 1876, he had been natural history editor of the hunting and fishing weekly *Forest and Stream.* On completion of his doctoral work, he decided not to pursue a scientific career. Instead, with the financial help of his father, he became owner, editor, and publisher of *Forest and Stream.* He would remain at that post until 1911, and the editorial battles he waged had a profound impact on the history of American conservation.

Grinnell used his weekly to call attention to the rising dissatisfaction among sport hunters and fishermen over the destruction of wildlife and channeled it into a crusade to husband both wildlife and habitat. Among the most important of many successful editorial campaigns was his effort to abolish the legal sale of game in America and, with it, the commercial hunter. Virtually all his muckraker-like editorials were unsigned because he felt that the viewpoint of a newspaper carried more weight than the opinion of a single individual. Other reasons for wishing to remain anonymous were his patrician's contempt for the ostentatious and his knowledge that allowing others to take credit for what he had initiated would keep them as allies in future battles.

Among Grinnell's editorial crusades was an effort, beginning in 1882, to expose the incompetence of the federal government in its handling of Yellowstone National Park, created, *on paper,* in 1872. His objective was to establish in law the concept that America's first national park should be an inviolate wildlife and wilderness sanctuary. This campaign succeeded in 1894 when Congress passed "An Act to Protect the Birds and Animals in Yellowstone National Park." Historians agree that Grinnell and *Forest and Stream* deserve much of the credit for the passage of this legislation, which established a crucial precedent for how future national parks would be conceived and managed.

Starting in 1885, Grinnell made regular trips to the mountainous St. Mary Lakes region of northwestern Montana, where he discovered the ice mass later designated the Grinnell Glacier in his honor. Between 1885 and 1892, he compiled a "sketch map" of this magnificent area, on which he named a number of the mountains, names they still retain. In an article that he submitted in 1892—but that was not published until September 1901—he initiated a campaign that finally resulted in the establishment of Glacier National Park in 1910.

Another of his sustained editorial endeavors, beginning in April 1882, was to have the European science of forestry adapted to American woodlands. In May 1884, for example, *Forest and Stream* demanded that the federal government immediately appoint "A Competent Forestry Officer," a "trained professional" to lead in "the inauguration of a system of forest conservancy." Editor Grinnell also encouraged the establishment of the New York State Forest Preserve in the Adirondacks in 1885, as well as legislation granting the president of the United States the right to set aside forest reserves.

Grinnell was always most active in wildlife conservation. In 1886, he established the original Audubon Society, the first national organization in American history for *nongame* conservation, and in 1887–1888, he joined with Theodore Roosevelt and others in founding the Boone and Crockett Club, named after two of America's most famous hunters. With influential, upper-class sportsmen as members and with Grinnell's *Forest and Stream* as its "natural mouthpiece," the club played a key role in bringing several nationally important conservation projects to fruition. Among these were the enactment of the Forest Reserve Act in 1891, the creation of Mount McKinley (now Denali) National Park in Alaska in 1917, and the passage of the Migratory Bird Treaty Act in 1918. Besides being one of its founders, Grinnell was a member of the club's first executive committee; vice president, chairman, or member of many of its action committees; and president from 1918 to 1927. So important was he that in 1926 a leader of the club, noted explorer and naturalist of Alaska Charles Sheldon, went so far as to declare: "The Boone and Crockett Club . . . has been *George Bird Grinnell* [emphasis in original] from its founding." That may be something of an overstatement, but there is no doubt that Grinnell played a central role in making it the first conservation organization in the United States to deal effectively with issues of national scope.

The Boone and Crockett Club is notable for another reason. It was through their close relationship in the club that Roosevelt was given a complete indoctrination in Grinnell's conservation philosophy. Although President Roosevelt's conservation program has been thoroughly examined by historians—who now consider it his most enduring accomplishment—they have generally overlooked the fact that the fundamentals of that program were foreshadowed in a series of books he coedited years before. These were the three Boone and Crockett Club volumes on hunting and conservation that he and Grinnell compiled during the 1890s: *American Big-Game Hunting* (1893), *Hunting in Many Lands* (1895), and *Trail and Camp-Fire* (1897). In one or more of these works can be found the demands that the federal government expand the national parks and forest reserves, establish a means for systematically administering them, and create a series of game refuges. All of these proposals had been called for earlier by him in *Forest and Stream*.

There is little doubt that Grinnell was the source of a number of the key concepts Roosevelt later established as national policy. For example, in the area of forest conservation, Grinnell *prepared* Roosevelt for Gifford Pinchot, who implemented ideas already familiar to Roosevelt. Pinchot was the "trained professional" Grinnell had first called for in 1884, who would lead the nation in the "inauguration of a system of forest conservancy." From the very first consultation, in 1899, between Roosevelt and Pinchot, the forester was impressed with the future president's understanding of forestry and his enthusiasm for suggestions on how the timberlands could be perpetuated and exploited simultaneously. Behind that receptivity were almost fifteen years of tutelage by Grinnell.

Roosevelt's contemporaries understood Grinnell's influence. After Roosevelt died in 1919, a group of his friends formed the Roosevelt Memorial Association and in 1923 created the Roosevelt Medal for Distinguished Service. In the spring of 1925, when the presentations were made in the outdoor-conservation field, it was no accident that Grinnell and Pinchot were given medals at the same time and coupled as *pioneers*—the word President Calvin Coolidge used in the White House ceremony. In the minds of the association's members, Grinnell, Pinchot, and the late president were inextricably linked.

In 1911, two years after Roosevelt left the White House, Grinnell retired from *Forest and Stream* after more than thirty years of directing the most-respected outdoor journal in the United States. Thousands of sportsmen and naturalists had read the publication each week, and the cumulative effect of his teachings about everything from the immorality of spring waterfowl hunting to the necessity for systematic forest conservation can only be surmised, but it must have been prodigious. With his departure from the weekly, *Forest and Stream* went into a rapid decline that Roosevelt personally tried to halt by handpicking the members of a "governing board" who would advise the journal's new owners. But the governing board was never allowed to govern, and with much sadness, Grinnell and Roosevelt watched *Forest and Stream*'s standing continue to deteriorate. It went from a weekly to a monthly in 1915 and finally ceased publication in August 1930 after first selling its subscription lists to the present *Field and Stream*.

A major reason for Grinnell's leaving *Forest and Stream* in 1911 was to be able to devote more time to research and writing on the Native Americans of the Plains, cultures that he had become increasingly fascinated with over the years. In addition to articles in the *American Anthropologist* and the *Journal of American Folklore*, his works include *Pawnee Hero Stories and Folk Tales* (1889), *Blackfoot Lodge Tales* (1892), *The Punishment of the Stingy and Other Indian Stories* (1901), *The Fighting Cheyennes* (1915), and *By Cheyenne Campfires* (1926). His major monograph is the two-volume *Cheyenne Indians* (1923).

His ethnographic studies helped him "see" nature more as Native Americans did and contributed to the development of his conservation philosophy. For example, his article "Tenure of Land among the Indians," which he published in 1907 in *American Anthropologist*, contains an early statement of Indians' communal notion of land "ownership" and their reverence for the land and its wild creatures, plus the idea that these temporary "possessions" must be passed on in good order to future generations.

In addition to his ethnographic contributions, Grinnell also wrote for young readers. Between 1899 and 1913, he published seven books in the so-called Jack series (for example, *Jack, the Young Ranchman* (1899) and *Jack among the Indians* (1900), which were based on his own (or friends') experiences. Along the same lines were his historical studies *Trails of the Pathfinders* (1911) and *Beyond the Old Frontier* (1913).

Despite his growing interest in Native American history and culture, Grinnell never lost his love for wildlife—or hunting. His *American Duck Shooting* (1901) and *American Game-Bird Shooting* (1910), which became classics, contain conservation messages as well as techniques on how to bag waterfowl and upland game birds. And though he never took up angling, he seemed to have had an interest in fish, as demonstrated by the data he compiled on *both* the salmon industry and the Natives while a member of the 1899 E. H. Harriman expeditions to Alaska.

Even after his retirement from *Forest and Stream*, he remained active in conservation, in the broadest sense of that term. In 1911, for example, he helped found the American Game Association. He also served as a director of the National Association of Audubon Societies; chaired the Council on National Parks, Forests, and Wildlife; and in 1925 succeeded Herbert Hoover as president of the National Parks Association. By 1929, when he suffered the first of a series of heart attacks, Grinnell had been a central figure in most of the environmental campaigns of his time.

At almost ninety, he died at his New York City home, pneumonia being the immediate cause of death. He was survived by his wife, Elizabeth Kirby Curtis, whom he married in 1902. There were no children. In cataloging his amazing array of accomplishments in a column-and-a-half obituary, the *New York Times* referred to him as the "father of American conservation." Surely no one deserved the title more.

JOHN F. REIGER

Diettert, Gerald A. *Grinnell's Glacier: George Bird Grinnell and Glacier National Park.* Missoula, MT, 1992.

Fox, Stephen. *John Muir and His Legacy: The American Conservation Movement.* Boston, MA, 1981. An analytical, wide-ranging history up to 1975 that has much on Grinnell.

Parsons, Cynthia. *George Bird Grinnell: A Biographical Sketch.* Lanham, MD, 1992.

Reiger, John F. *American Sportsmen and the Origins of Conservation.* 1975. Norman, OK, 1986. Covers Grinnell's conservation efforts before 1901.

———, ed. *The Passing of the Great West: Selected Papers of George Bird Grinnell.* 1972. Norman, OK, 1985. Based on Grinnell's unpublished "Memoirs," which trace his life from 1849 to 1883, this work blends his words with the editor's commentary to paint a picture of the "Great West's" last years.

———, ed. "With Grinnell and Custer in the Black Hills." *Discovery: The Magazine of the Yale Peabody Museum of Natural History* 20 (1987).

Trefethen, James B. *Crusade for Wildlife: Highlights in Conservation Progress.* Harrisburg, PA, 1961. Important for Grinnell's conservation activities after 1901.

Grinnell, Joseph (February 27, 1877–May 29, 1939). Born at Fort Sill, Oklahoma, Grinnell was the eldest of three children of Quaker parents.

During Grinnell's early years, his father served as physician at the local Indian agency. The elder Grinnell was restless, however, and soon moved his family to Tennessee, where he began a private practice. After a brief period, he abruptly moved the family to the Dakota Territory, where he again worked at an Indian agency. Then came an interlude in Pasadena, California, followed by a stint at the Indian School at Carlisle, Pennsylvania. At last, Dr. Grinnell took his family back to Pasadena and settled down.

While attending Pasadena High School, young Grinnell developed a serious interest in birds. After graduating, he matriculated at Throop Polytechnic Institute (later to become California Polytechnic Institute) in Pasadena. During his college years, Grinnell expanded his bird collection with specimens from California and a summer trip to Alaska. He won a degree in 1897, then embarked on an eighteen-month search for gold in Alaska. Though he returned with empty pockets, Grinnell brought back to California something quite valuable—hundreds of additional bird specimens and a firm commitment to ornithology. Soon, his first journal articles on the subject were published.

Graduate studies at Stanford led to completion of his M.A. in 1901, after which he taught for a year as an assistant instructor at Throop in 1897–1898. While continuing his graduate work, he became an assistant in embryology at Stanford University's Hopkins Laboratory in 1900 and an instructor in ornithology at Stanford in 1901–1902. He taught botany and zoology at Palo Alto High School from 1901 to 1903. Progress toward his doctorate was interrupted by a case of typhoid fever in 1903, but Grinnell nevertheless achieved a full professorship at Throop by 1905.

In 1906, Grinnell married Hilda Wood, who had been a student at Throop, and she soon became a research assistant to her husband. They would later have several children. In 1907, one of Grinnell's students introduced him to philanthropist Annie Montague Alexander (1867–1950), heiress to a Hawaiian sugarcane producer. Alexander was an enthusiastic big game hunter, pioneering conservationist, and occasional student at the University of California. Just the year before, she had conceived plans for a natural history museum at the Berkeley campus. With Grinnell in mind as director, she now proposed the establishment of a zoological museum to the university's trustees, pledging her ongoing financial support.

When the trustees accepted Alexander's terms, she persuaded them to name Grinnell as head of the new institution. During the building of the Museum of Vertebrate Zoology (MVZ), she sent Grinnell on a three-month trip in late 1907 to study the operating procedures of zoological museums in the East. In 1908, Grinnell took over as head of the MVZ and in 1909 gave 2,000 of his personal mammal specimens to the museum. Ten years later, he donated 8,000 of his bird skins. These represented about half the bird and mammal specimens he collected during his lifetime.

In 1912, Grinnell resumed work on his doctoral dissertation at Stanford, was awarded his degree in 1913, and became an assistant professor at Berkeley that same year. In 1914, Alexander informed Grinnell that she had revised her will and would eventually bequeath one-third of her estate to the MVZ, a vote of confidence in his leadership. The next year, Grinnell received an attractive invitation to join the U.S. Biological Survey staff in Washington, but loyalty to the museum and its benefactor prevailed. Promoted to associate professor at Berkeley in 1917, he was named full professor in 1920.

For over thirty years, Grinnell directed the activities of the Museum of Vertebrate Zoology and yet was deferential toward Alexander, who had the final say regarding museum expenditures and staff hires. She also retained the right to discuss with Grinnell the nonmuseum activities in which he engaged: conservation-related activities and his editorship of the ornithological journal the *Condor*. She preferred that he focus on research. For her part, Alexander managed her portfolio with canny ability, strengthening the museum's endowment. She continued to underwrite substantial parts of its program and, with an eye to the future, often provided Grinnell and other MVZ staff members with funds for their personal research. An active partner in management, she had charge of all financial and business dealings with the university trustees and administration.

Despite her personal commitment to the MVZ, Alexander was also investing money, time, and energy in the University of California's Museum of Paleontology, which she had helped establish. Thus, she encouraged Grinnell to solicit financial support for the MVZ from outside sources. Grinnell, however, pursued a taxing schedule. The pressure of his research, preparation of his publications, management of the museum and its myriad activities, and teaching preempted his attention. Besides, his temperament unsuited him for such a role. Reserved and modest, Grinnell was not well suited to the glad-handing of a capital campaign.

With the exception of fund-raising, the partnership between Grinnell and Alexander was productive and cordial. The museum's enviable record of research and publication, plus its expanding graduate and undergraduate programs, depended completely on this remarkable duo. Course offerings in vertebrate zoology and ecology continued to expand. Grinnell had originally considered Stanford a better setting for the museum but accepted Alexander's preference for Berkeley. From the beginning, it was decided that the MVZ would focus its attention on land mammals, birds, and reptiles, with ocean specimens almost completely lacking. The fact was that Stanford's first president, David Starr Jordan, was an authority on fish and other ocean vertebrates. Alexander had no desire to compete with him, and Grinnell concurred.

In another basic decision, outreach to the general public was minimized. Alexander initially underwrote creation of three MVZ specimen dioramas for public exhibition but was ultimately dissatisfied with the result. No other dioramas were commissioned, although the originals remained in place until a new museum building was constructed in 1930. Thereafter, no provisions were made for public exhibits.

One of the innovations introduced by Grinnell was the concept that each animal collected was a "voucher specimen," valued not only as an exemplar of its species but also as a representative of its habitat. Grinnell also developed the concept of the ecological niche, which essentially stated that no two animals in nature can occupy the same niche for long before one excludes the other. Another contribution was in methodology. Influenced by the fieldwork procedures of C. Hart Merriam and others, Grinnell followed a four-part procedure for working up observations and specimens acquired in the field.

The Grinnell method can be described as follows: first, the collector recorded in field notebooks his observations concerning where and when each specimen had been collected, the ground covered during each day of fieldwork, the weather, and other details. Next, he expanded on this information in field journals. Following that, the collector prepared species accounts. Finally, he constructed a catalog of the specimens. A further rule was that any field scientist should label each specimen carefully, with information on the animal's species or subspecies (insofar as this could be determined in the field), its measurements, sex, precisely location where collected, and the date. This protocol, employed by many of Grinnell's students, has since been adopted, in its essentials, by later generations of vertebrate zoologists.

To expand museum collections and as an educational tool, Grinnell recognized field trips as vitally important. By World War I, the director and his associates found that automobiles were a godsend for fieldwork. Grinnell used vehicles dubbed "Dipodomys" and "Perodipus," after jumping rats. Given their hard use, it was not surprising when the auto named Perodipus collapsed after just one season. During these trips, Grinnell's students later recalled that he was so focused on California fauna that he would require field parties to turn back if they happened to cross the border into a neighboring state.

By 1923, fifteen years after its founding, the MVZ's collection of mammal specimens stood third in size among all American collections. Only the Bureau of Biological Survey in Washington and the American Museum of Natural History in New York boasted larger collections. The MVZ was particularly known for its western species. By the end of the 1930s, Grinnell and his museum colleagues had amassed some 20,000 vertebrate specimens, 2,000 photographs, and 13,000 journal pages with detailed observations.

Grinnell's personal research was primarily focused on birds, but some of his 550 publications dealt with California fauna, especially mammals. Since Alexander felt unqualified to write her own species accounts, Grinnell also wrote papers based on specimens that she, often with assistance from her friends and museum staff, had collected while on expeditions to Alaska, Oregon, Baja California, Hawaii, and elsewhere.

Although Grinnell respected Alexander's prowess as a hunter and collector, he was reluctant to open his graduate programs to women. The first woman graduate student was not admitted until 1931; the second did not matriculate until 1933. During Grinnell's tenure at Berkeley, women were never permitted to participate in extended field trips or longer expeditions. The director believed them incapable of dealing with the stresses of fieldwork. Although Mrs. Grinnell had long accompanied her husband on expeditions, the wives of other faculty members were excluded from longer field trips. Strangely, Alexander did not interfere with Grinnell's policy.

Even after Grinnell's death in May 1939, the stricture continued, at least briefly, under E. Raymond Hall, one of Grinnell's former graduate students. Hall served briefly as acting director of the MVZ until ornithologist Alden Holmes Miller became Grinnell's successor in 1940. At last, in the spring of 1941, Alexander suggested that Mrs. Grinnell—who held a graduate degree in zoology—serve as chaperone and teacher for female graduate students on field trips. However, at least for another year, women graduate students were driven to their fieldwork sites by male students or faculty. Men also helped the women break up camp and drove them back to Berkeley when their projects were completed.

A productive ornithologist, Grinnell is also credited with introducing mammalogy as an academic subject in American universities. The high quality of research in mammalogy, ornithology, and herpetology by his museum colleagues was reflected in their excellent papers and books. A number enjoyed joint appointments as both MVZ staff and faculty in Cal Berkeley's Zoology Department. Many of Grinnell's former students launched outstanding graduate programs at other academic institutions. In doing so, they extended his philosophy of fieldwork, collecting, and research through later generations of American mammalogists and ornithologists.

Taxonomic fashions among zoologists between "splitters" and "lumpers" ebbed and flowed during Grinnell's career. He sometimes experienced frustration in trying to persuade colleagues to accept his reasoning. Grinnell contended that California's complex geography and environment demonstrated the existence of subspecies and the need for establishing them definitively. Other authorities disagreed regarding the degree of difference between individual organisms. These so-called lumpers considered subspecies as confusing impediments to understanding regional faunas. In fact, the subspecies issue has never been resolved; zoologists continue this debate today.

Grinnell's many publications included "Mammals and Birds of the Lower Colorado Valley with Especial Reference to the Distributional Problems Presented" (1914); "A Distributional List of the Birds of California" in *Pacific Coast Avifauna* (1915); *Game Birds of California* (with H. C. Bryant and Tracy Storer; 1918); *Vertebrate Animals of Point Lobos Reserve* 1934–1935 (1936); and *Fur-Bearing Mammals of California* (with J. S. Dixon and J. M. Linsdale; 1937). His "Bibliography of California Ornithology" appeared in three installments in *Pacific Coast Avifauna*, the first in 1909, the second in 1924, and the third in 1939. In addition, he edited fourteen volumes of the University of California *Publications in Zoology.*

Grinnell gave considerable attention to the biogeographical factors influencing the distribution of vertebrates in California, expanding on the earlier "Life Zone" research developed in the western United States by C. Hart Merriam (1855–1942), first chief of the U.S. Biological Survey, between 1885 and 1910. Although Merriam's theory was criticized elsewhere, Grinnell believed that it reflected the biogeographic realities in California. His "Life Zone Indicators in California" (coauthored by H. M. Hall; 1919), *Animal Life in the Yosemite* (coauthored by Tracy Storer; 1924), and A Revised Life Zone Map of California" (1935) all testified to his convictions about this subject. When Merriam, who had strongly encouraged the MVZ in its infancy, retired from active research in 1939, he donated his sixty-five years of research materials, photographs, and publications to the MVZ, a major addition to the museum's resources.

Grinnell served as president of the American Ornithologists' Union from 1929 to 1932 and as president of the American Society of Mammalogists in 1937–1938. He was a foreign member of the British Ornithologists' Union and a corresponding member of the Zoological Society of London. Various newly discovered forms of living and fossil mammals and birds have been named in Grinnell's honor. Grinnell Mountain, a 10,284-foot peak located in San Bernadino County, roughly one hundred miles north of Los Angeles, also commemorates the man and his many accomplishments.

KEIR STERLING

Eakin, Richard M. "History of Zoology at the University of California, Berkeley." *Bios* 27. 2 (May 1956).

Grinnell, Hilda Wood. "Joseph Grinnell, 1877–1939." *Condor* 42 (1940).

Hall, E. Raymond. "Joseph Grinnell (1877 to 1939)." *Journal of Mammalogy* 20 (1939).

Linsdale, Jean M. "In Memoriam: Joseph Grinnell." *Auk* 59 (1942).

Miller, Alden H. "Joseph Grinnell." *Systematic Zoology* 13 (1964).

Halle, Louis J. (November 17, 1910–August 13, 1998). Government official, educator, conservationist, and author of several books that explore social and political themes, Halle was celebrated for his writings on nature. A foremost expert in international affairs, he is a prime example of an individual not trained in the sciences whose avocation became nature and the environment. An important ornithologist, it may be said of him that as coal miners took birds down into the mines to monitor the quality of the air they had to breathe, so, too, did Halle demonstrate that a concentration on birdlife is one way to evaluate the quality of the environment.

Halle was born in New York, the son of Louis Joseph and Rita (Sulzbacher) Halle. He had a rather usual boyhood, but he was a bright teenager and gained admission to Harvard. On his graduation in 1932 he went to work for International Railways of Central America, El Salvador, and Guatemala. Two years later he accepted an editorial position with Longmans, Green & Company in New York. In 1937, he returned to Harvard for graduate study, after which he joined the Department of State as a staff member for Inter-American Affairs.

During these early and formative professional years, Halle wrote *Transcaribbean* (1936), *Birds against Men* (1938), and *Rivers of Ruin* (1944). Of these three books, *Birds against Men* gave special evidence of his keen reaction to nature. Although self-trained in ornithology, this work was so well received that it was chosen for the John Burroughs Award in 1941. Having been so honored, it would seem that Halle would concentrate on nature writing, but instead, in 1946, he accepted a position as assistant chief of Special Inter-American Affairs at the State Department. In the same year, he married Barbara Mark, with whom he had four children.

In 1950 he became an adviser in the Bureau of Inter-American Affairs, then shortly after, a member of its policy-planning staff. Despite all the time his positions demanded, he never abandoned his interest in birds and external nature. Accordingly, to follow up on the success of *Birds against Men*, Halle wrote another ornithological study, *Spring in Washington* (1947).

Spring in Washington originated from a journal he began writing in 1945. When compiled and edited, his observations made a book detailing the coming of spring in the nation's capital, on its birdlife in particular and on various manifestations of nature in the city, along its shores, creeks, and marshes. With considerable delight, Halle expatiates on avian life, trees, flowers, the wind, and weather, recording the natural emergence from the chill of winter through the beauty of spring into the splendor of summer.

During his years in Washington, Halle became an enthusiastic bird-watcher and an eager bicyclist. On many a morning, he pedaled ten or twelve miles to search out migratory birds before going to his office in the State Department. As a devotee of nature, he was captivated by budding flowers, cloud formations, the beauty of sunrises. "To snatch the passing moment and examine it for signs of eternity is the noblest of occupations," he wrote. "Therefore I undertook to be the monitor of the Washington seasons when the government was not looking."

Rhapsodically he noted the flowering of Japanese cherry trees. As wonderful as such an experience was, he added, it could not quite equal "the wave of warblers that passes through a countryside in mid-May, remaining sometimes only a day." The appearance of a familiar bird immediately awakened a train of forgotten associations, which made each spring transcend its predecessors. "Knowledge of spring gives me the freedom of the world," he contemplated. The discovery of spring each year after the winter's hibernation, he concluded, "is like a rediscovery of the universe."

Day after day, his eyes took in all sorts of birds, from swans to crows, from gulls to ducks, from eagles to vultures, from chickadees to great blue herons. A heavy favorite was the thrush. "I never heard a wood thrush until I was a grown man," he lamented, "though I must have been surrounded by them every spring. Each year I discover new sights and sounds to teach me how blind and deaf I must still be."

Halle's writing of *Spring in Washington* coincided to a considerable extent with World War II. The war does not figure in his narrative, though he did serve in the army in 1941. He was assigned to a pigeon company at Fort Monmouth, New Jersey. The objective of his unit was to teach falcons to bring down enemy carrier pigeons, but the experiment failed badly. Halle, ever humble, complained that he spent most of his time cleaning up pigeon droppings. Later, in 1943, he had more success when he joined the U.S. Coast Guard Reserve and served as a lieutenant junior grade.

Ever since its publication in 1947, *Spring in Washington* has been a nature lover's classic. (In 1988, the Audubon Naturalist Society sponsored its republication through the Johns Hopkins University Press.) *Spring in Washington* and his other books, plus his expertise in international affairs, earned Halle a research professorsip in the Woodrow Wilson Department of Foreign Affairs at the University of Virginia in 1954.

While at the University of Virginia, interest in world politics and the possibility of atomic destruction inclined Halle to write one of his most enthusiastic and provocative books, *Choice for Survival* (1958). On one level, the work deals with the dangers of nuclear annihilation; on another level, the book examines the effects advanced weaponry would have on the environment. Rebutting the prophets of doom, Halle argues in a calm fashion that humanity has a future.

His reputation was such that he was invited to join the faculty of the Graduate Institute of International Studies, Geneva; he taught there between 1956 and 1977 and was then appointed professor emeritus. During the period of his tenure at the institute, Halle wrote one of his more important books, *Dream and Reality: Aspects of American Foreign Policy* (1959), a diplomatic history meant to illuminate the general nature of international relations and American foreign policy problems.

Halle is important in the environmental movement not for a single major contribution but more for awakening thousands to ecological problems. With his twenty-two books, he molded public opinion and made his readers receptive to nature and conservationism. Among his books, the most significant are those that focus on the microcosm of birdlife, as he did in *Birds against Men* and *Spring in Washington.* In their own way, along with the books of other activist environmentalists, his efforts led to the passage of the Wilderness Act of 1964 and the passage of the National Wildlife Refuge System Act of 1966.

Much to his credit, Halle has been compared favorably to Henry David Thoreau. In their books, both demonstrate that a human being in contact with nature is a gifted, humane creature. Writing at a leisurely pace, they took the time required to express their philosophical positions in poetic prose. They differ, however, in their worldviews. The author of *Walden* (1854) was more of an individualist, a personalist encapsulated in his own localistic niche. Beyond Thoreau, Halle was expert in matters of international concern, and he even dwelt on the possibility of nuclear annihilation and survival. Always optimistic about humanity and the future, he refused to accept projections about cultural and moral decline. External nature remains about us for our appreciation, he held, and even for our redemption.

As for humankind's place in the cosmos, Halle liked to point out that population and cities grow exponentially, often with regrettable results. He dreaded a withdrawal of humans into "an environment produced by the mind and people living inside urban agglomerations." Increasingly, individuals might be "in the position of the embryo in the egg, the infant in an incubator, the mariner in a submarine, the space-traveler in a spacecraft." Nonetheless, Halle remained enthusiastic and amelioristic about human existence. "I could view the crumbling of our civilization with equanimity if violets were to spring up in the cracks," he once observed. "And they would, too."

GEORGE A. CEVASCO

Barnes, Bart. "Louis Halle, Jr., Naturalist and Author, Dies at 87" [obit.] *Washington Post*, August 22, 1998.

Harmond, Richard, and G. A. Cevasco. "Literary Environmentalism in the Generation before *Silent Spring*, 1945–1960." *Literature of Nature: An International Sourcebook*, ed. Patrick D. Murphy. Chicago, IL, 1998.

"Louis J. Halle." *Who's Who in America*. 1997.

Hornaday, William Temple (December 1, 1854–March 6, 1937). A noted naturalist, zoologist, taxidermist, and conservationist, Hornaday was born on his father's farm near Plainfield, in Hendricks County, Indiana, the youngest of four sons of William Hornaday and Martha (Varner) Hornaday. His paternal grandfather was one of four brothers who came to the United States from England soon after the American Revolution, while his maternal grandparents, the Varners, were of Pennsylvania Dutch and Scottish ancestry. In 1857, William Hornaday moved his family to a farm just south of Eddyville, in Wapello County, Iowa. Hornaday writes of this place, "I was mostly 'raised' on a fine, big farm in the beautiful rolling prairie country of south-central Iowa, three miles south of the Des Moines River." This area and its natural surroundings afforded Hornaday the opportunity to see wildlife in its most natural setting.

Orphaned by his mother's death in 1867 and his father's two years later, Hornaday attended Oskaloosa College in 1871 and the Iowa State Agricultural College in Ames in 1872, from which he learned the objectives of taxidermy and museology. In November 1873, Hornaday received an offer to work for Henry Augustus Ward's Natural Science Establishment in

Rochester, New York. Ward was, according to Hornaday, "a museum builder," and he utilized the growing talents of the young Hornaday by sending him on nature expeditions to Florida, Cuba, and the Bahamas in 1874 and to South America in 1875. From 1877 to 1879, Hornaday traveled to India, Ceylon (Sri Lanka), the Malay Peninsula, and Borneo and collected such natural riches that his name was made as a taxidermist and naturalist. In 1882, Hornaday was appointed chief taxidermist at the U.S. National Museum in Washington, DC, a post he held until 1890. In 1885, he published *Two Years in the Jungle*, followed two years later by *Free Rum on the Congo.* At the National Museum, Hornaday is credited with establishing a living animals department, an idea that extended to the founding of a National Zoological Park in Washington to be administered by Hornaday. However, a dispute arose as to the administration and size of land for the park, and he resigned his position.

For the next six years, Hornaday worked for a real estate firm in Buffalo, New York. On April 1, 1896, he was named the first director of the New York Zoological Park, now known as the Bronx Zoo. As director, he launched a series of articles and books for his "war for wildlife," in which he called attention to hunting and exploitation of animals. His 1898 report "The Extermination of Our Birds and Mammals" shocked many people into seeing the true side of the unending slaughter of wildlife during much of the nineteenth century. He raised some $100,000 to start the Permanent Wild Life Fund and was a founder of the American Bison Society, which sought to repopulate the western United States with buffalo. His 1913 study *Our Vanishing Wildlife* was one of the most well received works in the field of conservation. A lifetime of struggle to regulate hunting and the exploitation of wildlife is documented in his last work, *Thirty Years War for Wild Life* (1931). Hornaday died at his home in Stamford, Connecticut.

Hornaday married Josephine Chamberlain on September 11, 1879; they had one daughter, Helen Ross Hornaday.

MARK GROSSMAN

Dolph, James. "Bringing Wildlife to Millions: William Temple Hornaday: The Early Years: 1854–1896." Ph.D. diss., University of Massachusetts, Amherst, 1975.

Hornaday, William Temple. Papers. Library of Congress, Washington, DC. See especially his autobiographical memoirs "Eighty Fascinating Years: An Autobiography" (box 112) and a shorter essay titled "My Fifty-four Years with Animal Life," dated May 1929.

Hough, Emerson (June 28, 1857–April 20, 1923). A journalist, novelist, and conservation advocate, Hough was born in Newton, Iowa, at a time when that territory still retained much of its frontier character. The fourth of six children, much of his early education was taken in hand by his father, who introduced him to the skills of hunting while instilling in him a love of wild things and an appreciation of unspoiled country that would never fade. He also was exposed at a tender age to Henry Howe's 1851 volume *Historical Collections of the Great West*, whose colorfully romantic accounts of the already fading time of the cowboy and the sodbusters influenced his own later novels. Following a natural academic bent, he completed high school (a rare achievement for that day) and entered the State University of Iowa in 1876, graduating with a concentration in modern languages.

An unsuccessful attempt at teaching was followed by a transfer to the field of law, which admitted him to the bar in 1882. It was the lure of practicing law in the gold fields of New Mexico that sent him on his first trip to the West, armed likewise with a commission from *American Field*, a leading outdoor magazine of the day, for a series of pieces on the Southwest. Thus, the two interests that were to characterize the course of his life—a love of writing and a desire to be an effective voice for the sane management and assessment of the nation's wild creatures—were part of his career from its very beginning. His eighteen months' residence in the frontier town of White Oaks, New Mexico, brought him into contact with a world of direct and rough honor, whose life was to be reflected in many of his later works for the popular audience, such as his novel *Heart's Desire* (1905). Returning to the Midwest due to family illness, he tried his hand at further newspaper work, finally landing the post of editor for "Chicago and the West" at the widely circulated magazine *Forest and Stream* in 1889. It was in this capacity that many of his contributions to the conservation movement in the United States would be made.

Over the next thirty-five years, Hough inhabited the world of commercial journalism, traveling widely in Nebraska, Colorado, Wyoming, Minnesota, Texas, and New Mexico to gather material for his columns and feature articles. Steady employment freed him to pursue his own creative ideas, drawing heavily on his firsthand experiences of life in the frontier world. With the exception of *Madre d'Oro* (1889), a drama set in Aztec Mexico, and a book for children, *The Singing Mouse Stories* (1895), he did not embark on his formal career as a writer until he was approached to begin work on the volume that appeared in 1897 as *The Story of the Cowboy.* This work, part of the "History of the West" series, came to the attention of Theodore Roosevelt, resulting in the beginning of a lasting friendship. Over the first decades of the twentieth century, Hough created a body of fiction that served as a vehicle for the transmission to popular consciousness of the facts of frontier life as it had truly been lived and an ethic of conservation. Beginning with *The Girl at the Halfway House* (1900) and reaching through the successful *Heart's Desire* to later works such as *54–40 or Fight* (1909) and *The Covered Wagon* (1922), Hough placed his vision into the public mind at a time when the United States was becoming more nostalgic for, and interested in, the roots of its national character. Part of this interest centered on the value of wilderness as a formative influence that should be recognized and preserved.

His columns and stories (the latter often utilizing familiar characters drawn from his novels) spoke to an America awakening to environmental consciousness through the efforts of such public figures as George Bird Grinnell, founder of the Audubon Society. His most spectacular personal contribution to conservation came in March, 1894. As part of a project for *Forest and Stream,* he snowshoed into the decade-old preserve of Yellowstone National Park to verify the condition of the last large bison herd. His discovery of barely 100 emaciated animals where 500 were reported and his outraged descriptions of virtually unrestrained poaching were instrumental in sparking the 1894 passage of legislation protecting the park's game. Returning to writing for a younger audience, he published *The Young Alaskans* in 1908, beginning a series of four books that stressed good hunting practices and game conservation.

It was through the medium of his articles and columns in mass-circulation magazines and periodicals that Hough made his most significant contribution to educating the American public on various problems of conservation. His earliest effort in this vein was a two-part series done in 1905 for the *Saturday Evening Post.* Under the title "The Wasteful West" he graphically described the extermination of the plains bison and the damage inflicted on the region's forests by the contemporary timber industry. This feature was followed in 1907 by an article on bear hunting and in 1908 by "The Slaughter of the Trees," which called for the adoption of harvesting timber through thinning existing growth rather than wholesale clearing. Conservation clubs across the United States quickly had it reprinted in pamphlet form. Another of his interests was the protection of game birds from extinction through establishing protected breeding areas and prohibiting the shooting of migrating flocks in the spring, a step eventually taken by the United States and Canada with passage of the Weeks-McLean Law in 1916. His 1913 article "Wealth on Wings: The Wild Fowl of America" provided part of the publicity campaign leading to successful passage of protection legislation creating wildlife refuges on the flyways. In recognition of his support for their struggle, Hough was invited to a conference of superintendents and conservationists held in March 1915 among the great trees of Sequoia National Park. His opposition to unthinking exploitation of the environment continued almost to the day of his death in 1923 in articles such as "The President's Forest." His later writings also returned to the mystic image of the West of his first novels, with a series devoted to retracing the famous trails.

The chief significance of Hough for the field of conservation stems from the impact of his writings. While others spoke in favor of the preservation of wilderness as necessary to the health of the American soul, his personal vision of the broad majesty of the land infused every line of his works. In the pages of both his novels and his articles in popular magazines, he persistently placed issues of conservation's importance before the public eye in a style that removed their esoteric scientific distance and made them comprehensible needs requiring immediate attention.

ROBERT B. MARKS RIDINGER

Johnson, Carole M. "Emerson Hough and the American West." Ph.D. diss., University of Texas, Austin, 1976.
Stone, Lee. *Emerson Hough: His Place in American Letters.* Chicago, IL, 1925.
Wylder, Delbert E. *Emerson Hough.* Boston, MA, 1981.

Houle, Marcy Cottrell (August 1, 1953–). Documentaries, films, and glossy magazines have made wild nature part of everyday life for millions of Americans. Houle, biologist and author, has seen exotic things in her day, too: peregrine falcons defying extinction to raise chicks high in the Rockies, hawks and eagles taking advantage of abundant prey on Oregon's Zumwalt Prairie to nest together at greater densities than in any other place in the world. But Houle goes beyond simply communicating the beauty and wonder of the natural world. Houle's real subjects in her three books and numerous articles are the human communities on whose goodwill and understanding the survival of wild nature depends. Houle shows that no matter how well we understand the behavior of wild animals and the dynamics of ecosystems, we will never know how to protect nature until we also understand ourselves.

Houle's career in exploring the human dramas behind nature conservation programs was nurtured early on by the progressive political environment and abundant opportunities for exploring nature in her hometown of Portland, Oregon. Houle's family had lived in Oregon for four generations when she was born in 1953 on thirteen acres of old-growth forest in southwest Portland, and Houle's interest in biology began with exploring her own backyard. Houle and her two sisters also went frequently to Vancouver Island and spent weekends backpacking around Oregon's forests and parks. Later, Houle attended Colorado College, whose unusual "block plan" curriculum afforded students the opportunity to supplement their time on campus with frequent travel. For Houle, that meant she was able to see for herself many of the ecosystems she read about in class. For a native Oregonian who was practically "born with a land-use ethic," Houle's experience studying biology in Colorado offered further reason to orient her career around the natural world.

But Houle also had a choice to make. Along with her commitment to wildlife biology, she had another passion: writing. The question was which one to choose. Her professors in college were certainly less than enthusiastic about Houle's literary ambitions. As Houle put it in a 2007 interview for this entry: "Writing was not something that was encouraged, and certainly not looked upon as any kind of career path. Biologists I talked to worried that I was going to become an 'advocate' because lots of times advocates may not have the best science, even though they may have wonderful motives." Ultimately, Houle decided that she did not have to pick just one thing to be—that science and writing did not have to be mutually exclusive. She wanted to feed both the "objective" side of her soul and the "artistic, emotional, intuitive" side. She wanted both to do research into ecology and animal behavior and to write about it in a way that nonscientists could understand and enjoy. When Houle left college for her first real job, studying a pair of rare peregrine falcons that had just taken up residence on a spectacular outcropping in the Colorado Rockies, she strongly suspected that a report to the Colorado Division of Wildlife would not be the only piece of writing that would emerge from her time in the mountains.

Though Houle left college knowing she wanted to be a writer, the books did not flow from her fingers right away. Instead, in the ten years between college and her marriage to civil engineer John Houle, Houle threw herself into scientific research, taking copious notes that eventually formed the foundation for her first two books. Houle did biological fieldwork mainly in two places. For four summers, she monitored peregrine falcon activity up and down the Rocky Mountain spine for the Colorado Division of Wildlife. Much of her work kept her close to Chimney Rock, a dramatic spire, dotted with American Indian relics, thrusting up from the San Juan basin in remote southwest Colorado. A pair of peregrine falcons made this lonely eerie their home, and Houle spent the better part of two summers camped out several hundred meters away from the raptors at the nearby sister spire of Companion Rock, filling notebooks with observations on the falcons and on the people who lived in the valley. After four years of peregrines, Houle enrolled at Oregon State University for an M.A. in biology and swiftly found herself spending the next three summers on the Zumwalt Prairie in northeastern Oregon, one of the largest remaining grasslands in the United States. On foot, on horseback, and bumping around in rusty Forest Service trucks, Houle surveyed over a hundred nesting pairs of raptors in the Zumwalt. Houle's research ultimately dramatically contradicted conventional wisdom on the relationship of ranching to wildlife ecology. But once again, her three summers on the Zumwalt inspired Houle to write as much about the regular people—ranchers, foresters, and others—of the Zumwalt as about the birds she had ostensibly gone east to study.

Houle began to write in earnest after she married John in 1982 and settled into a farmhouse on Sauvie's Island in northwest Portland. Her first book, *Wings for My Flight: The Peregrine Falcons of Chimney Rock* (1991), vividly describes the high-altitude life of two peregrine falcons, Arthur and Jenny, as they struggle to raise three chicks in the face of DDT, dwindling habitat, and often-hostile human communities. Most important, Houle's first book highlights profound tensions surrounding efforts to bring endangered species back from the brink—tensions that can erupt into violence. Before the arrival of the falcons, surrounding communities had been counting on a tourist park based around Chimney Rock's American Indian ruins to inject life into the moribund economy. Elaborate plans complete with a gondola to the spire itself were on everyone's lips. The falcons changed all that. Even if the plans were pie-in-the-sky to begin with (Chimney Rock's ruins, while impressive, do not compare to nearby Mesa Verde

National Park), they were at least something to hope for, and *Wings for My Flight* deftly records Houle's public and internal struggle to justify efforts on behalf of wild nature when human communities are also suffering. Outraged citizens confront Houle on several occasions, including a woman who asks whether the falcons "are worth shutting down the million-dollar development that so many people have spent years in planning? Are they worth the thousands of dollars the local businessmen will lose? What you're doing is all right in theory, but it just doesn't work in reality." The secretary's views are shared by many of the people that Houle meets over the course of the book. Ultimately, Houle is bullied by an unsavory construction foreman, the female peregrine is shot and killed, and Houle's trailer is raided and all of her equipment stolen.

After Houle's trailer is raided, however, the tide in the town begins to turn, with more people expressing interest in her work. Relieved at this newfound support, Houle finds closure on some of her own private debates. Endangered species can be protected for many reasons, she concludes, from their ecological importance to the repository of genetic capital that they represent. But the most important reason to care for creatures like the peregrines is entirely subjective: it is just how falcons and other wild things make people feel. "Watching Arthur [the male peregrine], I knew for certain the secretary was wrong, though I'd never be able to explain it to her," Houle writes. "It was not a question of a simple value judgment that animals are to be held in higher esteem than people. . . . It was that we as human beings would somehow be made less each time a species was taken away." In the years since Houle left Chimney Rock, the prospects for the peregrine falcon have improved considerably. *Wings for My Flight* suggests that the reasons for this improvement can be found as much in the hearts and minds of regular people as in the science to which she and her partner contributed during their sojourn high above the San Juan valley.

Houle's second book, *The Prairie Keepers: Secrets of the Grasslands* (1995), balances tales of adventures in prairie life with a complex meditation on the ties that bind human and natural communities together. While a graduate student, Houle travels to the Zumwalt Prairie in eastern Oregon on assignment for the U.S. Fish and Wildlife Service to find out how many raptors are nesting there—and why they appear to nest there in such great quantities. As soon as she arrives, Houle discovers that her job carries a hidden catch: the ranchers who own most of the Zumwalt think of her employers as nosy, lazy, paper-pushing tree huggers, while her colleagues at

Fish and Game view private landowners as the bane of wildlife—and do not think much of female scientists, either. Over the course of her research, Houle's work ethic and open mind begin to win the trust of the Zumwalt's ranchers, and she warms to them in turn. Her eventual research findings do not endear her to Fish and Game, however. In fact, Houle finds just the opposite of what she and her employers were expecting: cattle on the Zumwalt, far from damaging the land on which raptors' prey subsists, in fact play a vital role in sustaining vast populations of rodents, which in turn feed over a hundred nesting pairs of red-tail, ferruginous, and Swainson's hawks. Houle's findings meet with incredulity in local government agencies, which suspect her of being paid off by ranchers—or at least of being biased toward them. But Houle's conclusions, grounded in rigorous scientific methodology, eventually win over the skeptics: without the cattle, who play the role on the Zumwalt that elk and bison do further east, primary growth would overwhelm the prairie, rodents would starve, and hawks would find a prime feeding (and breeding) ground much diminished.

The Prairie Keepers suggests a mysterious, heartening parallel between the ecology of the prairie and the human ecology of the people who live on and around it. Concerning the animals of the Zumwalt, Houle writes that "what seemingly clashed is really the reason for harmony"—grazing cattle support rodents, which in turn support hawks. But the same is also true for the people of the Zumwalt. Private landowners and public authorities trusted with protecting nature seem to be after completely different things—an age-old living off the land for one, a return to wilderness for the other. But the truth is that private land can be managed in a way that is good for ranches and wild nature both. And public authorities can forward their conservation mission by encouraging sustainable, responsible practices on the part of ranchers and farmers. Seemingly competing species actually constitute a healthy ecosystem on the Zumwalt, and seemingly very different humans can work together to accomplish mutually beneficial goals.

Houle's writing has won her significant acclaim. *Wings for My Flight*, published in 1991, won a 1992 Christopher Award and a 1991 Oregon Book Award and was named a 1992 "Best Books for the Teenage" by the New York Public Library. The impact of *The Prairie Keepers*, however, has been greater. According to Phil Shephard, Northeast Oregon Stewardship director for the Nature Conservancy, Houle's research laid the groundwork for the Conservancy's recent work in the region, culminating in a nature preserve that now spans fifty-one square miles. Houle's rapport with Zumwalt ranchers has also led her to play the part of unofficial intermediary between landowners, government, and private conservation groups since *The Prairie Keepers* was published in 1995. According to Professor Patricia Kennedy, who recently updated Houle's studies of the Zumwalt for the Nature Conservancy, "Houle's legacy of constructive working relationships with landowners on the Zumwalt" has been "essential" for conservation work in the region.

Houle has given talks all over the country on her two books to audiences ranging from aspiring writers to landowners' associations. She lives today in Portland, in the house where she raised her family, and continues to write full-time.

ETHAN SCHOOLMAN

Cade, Tom J., James H. Enderson, Carl G. Thelander, and Clayton M. White, eds. *Peregrine Falcon Populations: Their Management and Recovery.* Boise, ID, 1988.

Kennedy, Patricia. Interview with Ethan Schoolman. 2007.

Shephard, Phil. Interview with Ethan Schoolman. 2007.

St. John, Alan D. *Oregon's Dry Side: Exploring East of the Cascade Crest.* Portland, OR, 2007.

Ickes, Harold LeClaire (March 15, 1874–February 3, 1952). A journalist, attorney, author, secretary of the interior, and conservationist, Ickes was born in Altoona, Pennsylvania, on the eastern slopes of the Allegheny Mountains. His father, Jesse Williams Ickes, was a storekeeper and comptroller of Altoona; his mother, Martha Ann McCune, was the daughter of a county judge. Harold—he was called "Clair" by family and friends—frequently visited Judge McCune (the family had earlier spelled the name "McEwen"), who owned a farm on the Juniata River. Ickes's love of nature and the outdoors was sparked by these visits; Altoona was a grimy railroad town, and he found escape from the depressing conditions in the town and from a home life that was not always harmonious by hiking and horseback riding in the wooded hills near his grandfather's property. His appreciation of the national parks began in the early 1920s when he trekked across Yellowstone National Park on horseback.

When his mother died in 1890, Ickes moved to Chicago to live with his aunt, Ada Wheeler, and her family. He received a B.A. from the University of Chicago in 1897, and after graduation, he took a job as a reporter. He worked for more than a decade in numerous Chicago campaigns for political reform and social justice, and he was a supporter of Theodore Roosevelt's Progressive Party. He had first met Roosevelt while covering the 1900 presidential campaign, and he was impressed with, and influenced by, Roosevelt's commitment to conservation. In 1907, Ickes obtained a law degree from the University of Chicago, and he practiced law until 1933.

The turning point in his career came in 1932 when he formed an independent Republican committee supporting Franklin D. Roosevelt's presidential bid; after the election, FDR rewarded Ickes by appointing him secretary of the interior and public works administrator. Ickes served as secretary of the interior for more than twelve years, and although he was unable to realize his ambition of transferring the Forest Service from the Department of Agriculture to the Department of the Interior—a transfer that would have consolidated most federal land under the control of one cabinet office—he was able to make many important contributions to the conservation movement during his long tenure.

Aware that Interior's reputation had been severely damaged by the Teapot Dome affair and other scandals, Ickes took office intending to restore public confidence to the department and to transform it into a "Department of Conservation." He was a demanding administrator who took pleasure in calling himself a "Curmudgeon" and "America's number one Sour Puss."

As head of Interior, Ickes took personal interest in the National Park System. Great Smoky Mountains National Park, which had been authorized by Congress in 1926, officially joined the park system while Ickes was in office. The Everglades National Park, first proposed in the 1920s, was authorized in 1934, and Big Bend National Park was authorized in 1935; both parks would officially join the National Park System in the 1940s. Ickes fought to ensure that both Everglades and Big Bend would—at least for the most part—be preserved as wilderness parks. In 1936, he championed the creation of the 600,000-acre Joshua Tree National Monument in southern California as well as the first National Recreation Area, the 1.5 million-acre Lake Mead National Recreation Area.

In 1933, when FDR transferred control of all National Monuments from the Forest Service to the Department of the Interior, Ickes lobbied for the creation of a national park that would expand the boundaries of Olympus National Monument, a 620,000-acre preserve that had been established by President Theodore Roosevelt. For years, debate raged about the need for the proposed park. Finally, Ickes and his assistants prevailed, and in 1938 President Franklin Roosevelt signed into law a bill establishing Olympic National Park, a park that would eventually include nearly 900,000 acres of some of the most beautiful land in the United States.

Ickes also worked hard to create Kings Canyon National Park, 453,000 acres of wild country in the Sierra Nevada adjacent to Sequoia National Park. To preserve this spectacular wilderness, Ickes had to battle those who favored damming the Kings River to provide irrigation for the San Joaquin Valley. Typically, the Forest Service backed the irrigation projects and suggested the building of dams in the Kings Canyon mountain area. Eventually, opposition was overcome, and in 1940 FDR signed into law legislation creating the Kings Canyon National Park. Ickes had wanted the park to be treated as a primitive wilderness with little access except by foot or horseback, and his desires were largely achieved when Newton B. Dury, Park Service director, established a wilderness management program for both Olympic and Kings Canyon.

In addition to securing Olympic and Kings Canyon National Parks, Ickes helped establish Isle Royal National Park, a 133,000-acre wilderness island in Lake Superior; the Organ Pipe Cactus National Monument, a 331,000-acre desert preserve in Arizona; and the Cape Hatteras National Seashore—the first. During his first seven years as secretary of the interior, he saw the National Park System grow from 8.2 million acres to more than 20 million acres.

Ickes, however, was not content with merely expanding the National Park System; he was also concerned with how the parks should be developed and used. Above all else, he wanted to ensure that the parks would be preserved as wilderness areas. During his tenure at Interior, very few roads would be built in the national parks. In a speech to park authorities, Ickes said, "I am not in favor of building any more roads in the national parks. . . . I do not have much patience with people whose idea of enjoying nature is dashing along a hard road at fifty or sixty miles an hour." He also told his park superintendents that "I do not want any Coney Island. I want as much wilderness, as much nature preserved and maintained as possible. . . . I think the parks ought to be for people who love to camp and love to hike and who . . . [want] a renewed communion with Nature." As part of this effort to preserve the wilderness character of the parks, Ickes opposed the development of luxury hotels and resorts in park lands. To educate the public, Ickes initiated a series of weekly radio broadcasts over NBC extolling the virtues of the National Park System.

Management of the National Park System was only one of Ickes's responsibilities as secretary of the interior. Also under his direction were the Bureau of Indian Affairs, the General Land Office, the U.S. Geological Survey, and administration of lands in U.S. Territories (Alaska, Hawaii, Puerto Rico, and the American Virgin Islands). While Ickes was head of Interior, the department also gained control of the Biological Survey, which had authority over hundreds of wildlife refuges, and the Bureau of Fisheries, which had authority over commercial fishing operations. Aware of the environmental damage caused by the Dust Bowl, Ickes established a Soil Erosion Service within the Department of the Interior.

As part of his administration of the General Land Office, Ickes pushed for legislation that would empower Interior to regulate grazing on public lands and national forests. Passed in 1934 and amended in 1936 to include additional acreage, the Taylor Grazing Act put 143 million acres of public domain under government protection.

Ickes and his assistant Horace Albright were largely responsible for managing the Civilian Conservation Corps (CCC), a conservation project proposed by FDR during his first hundred days as president. The CCC, which operated from 1933 until 1942, fought forest fires, built trails, combated soil erosion, and constructed conservation projects.

As head of the Bureau of Indian Affairs, Ickes established more than 4.5 million acres of wilderness preservation on Indian Reservations. A civil libertarian, he also worked to preserve Indian societies and cultures. Many improvements on Indian Reservations were made by the Indian CCC, another of Ickes's creations.

During World War II, Ickes served as petroleum coordinator and championed energy conservation. After FDR's death in 1945, Ickes found himself at odds with President Harry Truman over the appointment of oilman Edwin Pauley as undersecretary of the navy. During Pauley's confirmation hearings, Ickes testified that Pauley offered him campaign funds in return for the federal government's dropping of claims to offshore oil resources in California. Truman maintained that Ickes was mistaken in this charge, and Ickes resigned in protest. Pauley's name was eventually withdrawn from nomination.

After leaving Interior, Ickes continued to aid the conservation movement by writing columns for the *New York Post* and the *New Republic*. In his columns, Ickes argued against efforts to abolish Jackson Hole National Monument (later made part of Grand Teton National Park), and he fought against the building of a dam that would have flooded much of Dinosaur National Monument. Notable among Ickes's other writing are his *Autobiography of a Curmudgeon* (1943), the three-volume *The Secret Diary of Harold L. Ickes* (1953–1954), and his eight-part series "My Twelve Years with F.D.R," written for the *Saturday Evening Post* and published in June and July of 1948. In addition to his autobiography and writings on politics, Ickes authored books on energy policy, freedom of the press, and the Public Works Administration. The Ickes Papers are housed in the Library of Congress.

Ickes's chief legacies to the conservation cause were the result of his accomplishments as secretary of the interior. He reformed Interior, restoring public trust and making the preservation of America's natural treasures the principal goal of the department. While he was at Interior, Ickes worked to expand the National Park System by millions of acres—bringing such magnificent parks as Olympic and Kings Canyon into the system—and he developed a philosophy of use for the parks that has helped to preserve their wilderness character.

Ickes married twice, first in 1911 to Anna Wilmarth Thompson, a political reformer who served three terms in the Illinois legislature; together they had a son, and Ickes adopted her son from a previous marriage. Anna died in an automobile accident in 1935. In 1938, he wed Jane Dahlman, with whom he had two children.

DAVE KUHNE

Harmon, M. Judd. "Some Contributions of Harold L. Ickes." *Western Political Quarterly* 7 (1954): 238–251.

Swain, Donald C. "Harold Ickes, Horace Albright, and the Hundred Days: A Study in Conservation Administration." *Pacific Historical Review* 34 (1965): 455–465.

Trani, Eugene. "Conflict or Compromise: Harold L. Ickes and Franklin D. Roosevelt." *North Dakota Quarterly* (Winter 1968): 20–29.

Watkins, T. H. *Righteous Pilgrim: The Life and Times of Harold L. Ickes.* New York, 1990.

White, Graham, and John Maze. *Harold Ickes of the New Deal.* Cambridge, MA, 1985.

Jackson, William Henry (April 4, 1843–June 30, 1942). A man whose life almost exactly spans the first century of photography, Jackson is regarded as the central photographer of the nineteenth-century American West. In the 1870s, for an eastern public skeptical of the stories of the fabled wonders of America's western frontier, Jackson's photographs, especially those of the Yellowstone area in the Wyoming Territory, presented direct, objective, and convincing visual proof of the grandeur of America's monumental natural scenery and helped to infuse the idea that it was America's national destiny to conquer, to inhabit, to control, and ultimately to protect its western lands.

Traveling with his darkroom wagon or his mule Hypo ("almost as indispensable to me as his namesake, hyposulphite of soda, was to dark-room chemistry"), Jackson also photographed and recorded the life of the American Indian, even then seeming a relic of the past, and in so doing endowed the Indian with an aura of myth. He photographed the ancient Indian cliff dwellings of the Mesa Verde, providing evidence that the Indian lived in harmony with his natural surroundings and subtly implying that the White Man could do the same. He chronicled the progress of what Walt Whitmam hailed as "type of the modern—emblem of motion and powers—pulse of the continent"—the railroad—paying tribute less to nature than to humans' technological prowess in subduing her.

Jackson's significance as a conservationist rests primarily on his service as official photographer to the U.S. Geological and Geographical Survey of the Territories led by surveyor and geologist Ferdinand Vandiveer Hayden every summer between 1870 and 1878. During these years, Jackson accompanied Hayden and his teams to the Yellowstone, the Grand Tetons, the Rockies, and other parts of the Wyoming, Utah, Colorado, and Southwest territories. His photographs were instrumental in providing crucial documentary proof that the wonders of the Yellowstone, sometimes called "Colter's Hell" after John Colter, a member of the Lewis and Clark Expedition who may have been the first white man to witness its wonders—geysers, hot springs, mud pots, fumaroles, the mighty Grand Canyon of the Yellowstone

River and its waterfalls—were not fabricated and might in fact be even greater than had been imagined.

The Yellowstone Park Act, signed into law by President Ulysses S. Grant on March 1, 1872, made Yellowstone the first national park, a "public park or pleasuring ground for the benefit and enjoyment of the people." Passage of the bill can be credited in large part to the albums of Jackson's photographs presented to members of Congress and to the exhibition in the Capitol Rotunda of Jackson's photographs and Thomas Moran's paintings. "[I]f any work that I have done should have value beyond my own lifetime," said Jackson in his autobiography *Time Exposure* (1940), it was the "happy labors of the decade 1869–1878," though modestly he claimed to have been "seldom more than a sideshow in a great circus."

Had Jackson photographed nothing else, the 400-odd negatives made annually for the Hayden Surveys would have assured his place in the history of the American West and American landscape photography. The energy and zeal with which Jackson focused what Hayden called his "uncompromising lens" has made him the quintessential frontier photographer. Others such as Eadweard Muybridge and Carleton E. Watkins photographed the American West, and three photographers besides Jackson participated in the Hayden expeditions (Jackson loaned a camera to one, whose own camera had fallen into Yellowstone Canyon; another later lost his negatives in a fire); but Jackson most strongly captured and catered to the public imagination. His energy and stamina exemplify the "rugged individualism" of the American frontier. Intentionally or not, Jackson's photographs confront the important ideas and themes of nineteenth-century America: the individual, the land, nature, and humankind's relation to and place in an awesome, though not hostile, virgin landscape; furthermore, his canny sense of business and publicity enabled him to package and present his photographs effectively for an appreciative public by means of stereographic views, splendid picture albums, and award-winning exhibitions.

Jackson was born in Keeseville, New York, to Quaker parents of British ancestry. He attended school in Troy, New York, but only through the eighth grade. An autodidact, he read a great deal and taught himself to paint and learn daguerreotypy, an early photographic process using silver-coated metallic plates sensitive to light and developed by mercury vapors. In such pursuits, he was helped and encouraged by his father, a carriage maker and blacksith, and his mother, a talented watercolorist. When he was nineteen, he served one year in the 12th Vermont Volunteer Infantry.

After his discharge, he operated a photographic studio in Rutland. In 1868 he left Vermont and settled in Omaha, Nebraska. Zealous in his photogaphy, he obtained a large camera to take shots of vast landscapes, small mining towns, and geological features found only in the Rocky Mountains. Over the next few years, he photographed Native Americans, in particular, members of the Pawnee, Winnebago, and Ponca tribes. Such photographs brought his work to the attention of Ferdinand Vandiveer Hayden.

Lack of funds ended the Hayden Surveys in 1878, but Jackson retained full rights to his negatives and began a long career as a commercial photographer, first independently in Denver in 1879 and later merging into the mask of twentieth-century corporate mass production in 1897 as a director of the Photochrom Company, a subsidiary of the Detroit Publishing Company, which bought his entire stock of 40,000 negatives. In 1894 and 1895, Jackson had briefly resumed the rigorous adventures of his youth by accepting Major Joseph Pangborn's offer to document the World Transportation Commission's "world tour" across Europe, Egypt, India, Australia, New Zealand, the East Indies, China, Japan, and Russia.

Despite financial setbacks (the Photochrom Company went bankrupt in 1924, and he lost his savings in the 1929 stock market crash), Jackson remained active, serving as research secretary of the Oregon Trail Memorial Association and painting thirty murals depicting the Survey years commissioned in 1936 by the Department of the Interior for its new building. In March and April of 1942 the Metropolitan Museum of Art saluted him, among others, in a photographic exhibition titled "Photographers of the Civil War and the American Frontier."

Jackson died from injuries suffered in a fall at age ninety-nine. His first wife Mollie died in childbirth with their infant daughter in 1872; his second wife, Emilie Painter, whom he married in 1873, died in 1918. They had a son Clarence and two daughters. Several landmarks in the Grand Tetons bear Jackson's name.

Photographer, painter, adventurer, explorer, topographer, businessman, and corporate executive, Jackson was also an author, enthnographer, and archeologist. He published numerous albums of photographs, such as *The White City (as it was . . .): The Story of the World's Centennial Exposition* (1894); *Wonder-Places: The Most Perfect Pictures of Magnificent Scenes in the Rocky Mountains. The Master-Works of the World's Greatest Photographic Artist, W. H. Jackson* (1894), containing twenty-one plates without text, published for $1,000 each; and albums of Mexico, Canada, Venezuela, as well as the more familiar American frontier. He wrote over a dozen articles for magazines such as the *Colorado Magazine* and *Harper's Weekly.* Jackson's photographs formed an integral part of the descriptive catalogs compiled annually by Hayden for the U.S. Geological and Geographical Survey of the Territories and of articles published in the *U.S. Geological and Geographical Survey Bulletin.* In 1875, Jackson contributed to the *Bulletin* a scholarly article titled "A Notice of the Ancient Ruins in Arizona and Utah Lying About the Rio San Juan." Between 1891 and 1906, when it became a reality, Jackson urged the creation of Mesa Verde National Park in southwest Colorado. Jackson's legacy is still apparent: fifty-six of his 1870s photographs illustrate *Rangeland through Time: A Photographic Study of Vegetation Change in Wyoming, 1870–1986*, by Kendall L. Johnson (1987).

The physical difficulties attendant on field photography using the wet-plate collodion process must be emphasized. The process required that a glass plate be sensitized immediately before exposure and developed immediately afterward. Referring to the Survey period, Jackson said,

> This was a period of great experimentation for me. The art of timing exposures was still so uncertain that you prayed every time the lens was uncapped, and no picture was a safe bet until the plate had been developed. Working in a fully equipped studio was hazardous enough. Going at it in the open meant labor, patience, and the moral stamina—or, perhaps, sheer phlegmatism—to keep on day after day, in spite of the over-exposed and underdeveloped negative, and without regard to the accidents to cameras and chemicals.

Jackson's own words are less effective in conveying the scope of the physical difficulties facing a frontier photographer than a photograph: "515. Photographing in High Places" (1872) shows Jackson squatting with his equipment atop a narrow cliff beside his assistant. Part of his success was undoubtedly due to his willingness to experiment, though experimentation carried its risks: Jackson's first attempt to use Warnerke's "sensitive negative tissue, supplied in bands," the predecessor of modern roll film, to photograph the Presbyterian Church Southwest missions in 1877 failed completely.

Jackson used larger and larger cameras to create panoramic landscapes, sometimes using negatives as large as twenty by twenty-four inches that when joined could record a radius of 150 miles. In 1873, to photograph the fabled Mountain of the Holy Cross, so called for the impression of a cross produced by snow in fissures near the mountain's summit, he and two assistants climbed Notch Mountain, carrying (as pack animals were useless at this height) one hundred pounds of equipment the last 1,500 feet and waiting through a chill night to take eight wet-plate exposures the next morning, which he developed by noon using melted snow as wash water. The most famous of these photographs inspired Henry Wadsworth Longfellow's sonnet "The Cross of Snow" and appeared the very emblem of God's blessing on America's western destiny.

Jackson's photography combined the objectivity of the scientist with the sensitivity of the artist. He gave much credit for the artistic quality of his photographs to his friend Thomas Moran, the official Survey artist, who taught him valuable lessons in composition and selection of viewpoint. Through his creation and mass dissemination of photographs that engage the viewer, drawing the individual into a monumental but unthreatening landscape, Jackson helped both to promote settlement of the West and to shape the response of the American public to the wonders of its wild spaces.

PAGE WEST LIFE

Fabian, Rainer, and Han-Christian Adam. *Masters of Early Travel Photography.* London, 1983.

Hales, Peter B. *William Henry Jackson.* London, 1984.

———. *William Henry Jackson and the Transformation of the American Landscape.* Philadelphia, PA, 1988.

Newhall, Beaumont, and Diana E. Edkinds. *William H. Jackson.* With a critical essay by William Broecker. New York, 1974.

Jewett, Sarah Orne (September 3, 1849–June 24, 1909). A leading figure in the "local color" school of writers, Jewett was born and died in the family mansion in South Berwick, Maine. Her paternal grandfather Captain Theodore Furber Jewett, a sailor and adventurer, made a substantial fortune in shipping and trade. Her father Theodore Herman Jewett, a graduate of Bowdoin College and the Jefferson Medical College in Philadelphia, was a doctor in South Berwick. Her maternal grandfather William Perry was a respected physician, surgeon, inventor, and community leader in

Exeter, New Hampshire. Her maternal grandmother Abigail Gilman descended from an eminent New England family—sired by "Edward the Emigrant" (Gilman), who arrived in America in 1638. Sarah, her two sisters, and her mother Caroline Frances Perry Jewett frequently visited Exeter, New Hampshire.

Her two grandfathers and her father particularly encouraged her innate interest in nature. Dr. Perry, who raced horses into his late eighties, taught her horseback riding. She learned a love of the sea and seafaring stories from Captain Jewett. Her father, a botanist and zoologist as well as an accomplished physician, regaled her with facts about plant and animal life while she accompanied him on his rounds through the New England roads and byways to visit his patients.

Jewett was educated at Miss Rayne's School and Berwick Academy. A desultory student, her real education took place in her frequent expeditions into nature near her home and on her trips with her father. Like many of her characters, she spent long hours exploring the Maine woods, rowing on the rivers, or horseback riding through the countryside.

Her hometown, South Berwick, Maine, exercised a formative and permanent influence on her character, interests, and career choice. Berwick, the ninth town in Maine, lies on the southern border, about ten miles up the Piscataqua River. It afforded Sarah exposure to a wide range of natural environments and human occupations. These included forests, rivers, ocean, and mountains, on the one hand, and sailors, fishermen, farmers, lumberjacks, hunters, merchants, and village professionals, on the other. Especially after she started writing, she became a frequent visitor to Boston. She traveled extensively and finally settled into a life equally portioned between Boston and South Berwick. Nevertheless, Maine remained central not only to her writing but to her sense of her self. She believed in an intrinsic relationship between character and place.

The importance she attached to place attracted her to the "local color" school, a literary movement that grew out of Realism in the late nineteenth century and emphasized specific portraits of particular locales and the lives and characters of ordinary people determined by these locales. Most of Jewett's more than 170 works depict life and nature in rural Maine. She is counted among the most prominent local color authors. In its emphasis on the uniqueness of specific regions and the formative influence that nature exerts on human character and human spirit, Jewett's interpretation of local color tenets anticipates the concepts of environmentalists' bioregionalism.

Jewett published her first story in 1868 and began publishing regularly in the *Atlantic Monthly* in 1869. William Dean Howells, the editor of the *Atlantic Monthly*, early perceived her gift for describing nature and capturing the spirit and tenor of rural and natural environs. Under his urging, in 1877 she assembled her first collection of stories, *Deephaven*, about life in a Maine seacoast town. *Country By-Ways* (1881)—a series of reflective essays including "A Bit of Shore Life," "River Driftwood," "A Winter Drive," and "An October Drive"—describes various regions in Maine and articulates some of her natural philosophy. The novel *A Country Doctor*, which draws on her childhood, was published in 1884. The collection of short stories *A White Heron and Other Stories* was published in 1886. "A White Heron" is probably her single most famous story and one of the most influential on environmentalists. The work that established her reputation was *The Country of the Pointed Firs* (1896), a series of interrelated stories about life in a village on the Maine coast. According to one critic, it "brought the local color novel to its highest degree of artistic perfection in nineteenth century America."

Jewett witnessed the unraveling of the fabric of seacoast life in South Berwick and other towns like it along the Maine coast. Shipping, sailing, and fishing were losing viability, and small farms were also in decline. More and more people were leaving the country for the greater opportunities that cities afforded. Industrialism and technology—with their factories, mills, and railroads—were radically altering people's styles of life. Jewett's regret for these changes accounts for the elegiac tone of most of her writing. Her deepest lament was for industrialism's destruction of humans' rapport with nature.

Jewett felt strongly about the natural and spiritual toll of industrialism and did not resist diatribe in some of her stories. "I think the need of preaching against this bad economy is very great," she stated. "Man has done his best to ruin the world he lives in." Many of her works include critical depictions of the decimation of the Maine forests, the pollution of the rivers, and the destruction of wildlife. Her international reputation brought widespread attention to New England's environmental problems, but her most enduring legacy for environmentalists resides in her theories about nature and humans' relationship to nature.

Jewett espouses a transcendentalism similar to Henry David Thoreau's or Ralph Waldo Emerson's and frequently refers to natural objects as if they have souls or spirits. Humans are at their best when they live in harmony with nature, "so close to nature that one simply is a piece of nature, following a primeval instinct with perfect self-forgetfulness" ("A Dunnett Shepherdess"). The egalitarian nature of her thought has important consequences for environmentalists. Its antihierarchical stance insists that environmental decisions should not be strictly anthropocentric or utilitarian. She believes that industrialism and mass production have resulted in an objectification of humans and nature that threatens the substance of our lives. When humans look at trees, streams, and wildlife for their monetary value instead of their intrinsic worth, they separate themselves from the world around them in a way that corrupts them morally and spiritually. Although she celebrates nature, she does not sentimentalize or romanticize it. She recognizes its brutality and humans' vulnerability to its forces.

Jewett enjoyed great popularity in her own time and exerted considerable influence on other American writers such as Willa Cather and Edith Wharton. She continues to be read widely. The growing interest in "nature writing" and environmentalists' growing appreciation of the interdisciplinary nature of their field have brought ever-increasing attention to her work.

LAURA COWAN

Cary, Richard. *Sarah Orne Jewett.* New York, 1962.

Donovan, Josephine. *Sarah Orne Jewett.* New York, 1980.

Silverthorne, Elizabeth. *Sarah Orne Jewett: A Writer's Life.* Woodstock, NY, 1993.

Johnson, Josephine Winslow (June 20, 1910–February 27, 1990). A writer and social activist whose efforts contributed to environmental awareness, Johnson was also influential in the labor movement of the 1930s and the antiwar cause during the Vietnam era. For Johnson, human abuse of nature was intimately tied to human injustice toward other humans. As her literary executor John Fleischman has written, "Her stories and novels always had a duality—a lyrical sense of the majesty of natural processes crossed with a savage indignation at the outrages of man."

Born to Benjamin and Ethel Franklin Johnson in Kirkwood, Missouri, Josephine was the second in a family of four girls. In her memoir *Seven*

Houses, she described finding her "niche in life" when, at the end of World War I, she wrote her first poem celebrating the triumph of "Liberty, Justice and Humanity." During much of her childhood and adolescence, Johnson lived in rural settings in the Kirkwood area, including the farm known as Hillbrook, where her family moved in 1922, and the country cottage where she spent summers with her aunts. This proximity to nature had a profound effect on Johnson: "I have always loved the country and nature more than anything else" (*Twentieth Century Authors*).

Johnson went on to study English and art at Washington University and the St. Louis Art School, although she abandoned her formal studies in 1932 and returned to the family farm to write. It was during the next decade that her deep commitment to social justice, perhaps influenced by her mother's Quaker background and pacifist stance during World War I, manifested itself: she wrote for the *New Masses*, engaged in pro-union activism (she was particularly active in support of rural laborers), and served as president of the Consumers' Cooperative organization in 1938.

In 1940, Johnson was quoted as saying, "It seems to me that the author must be participant as well as observer in many things and today writing is so bound up with actual living that it is ridiculous to quibble over the separation line" (*Contemporary American Authors*). Her writing career, which also blossomed during the 1930s, did indeed reflect the concerns that occupied her in the other areas of her life. Her first novel, *Now in November* (1934), which won the Pulitzer Prize in 1935, tells the story of the Haldemarnes, a Missouri farm family who struggled to survive the ravages of drought and debt during the Depression. Told from the point of view of Marget, one of the family's three daughters, the novel reflects the major concerns that would characterize Johnson's writing for the rest of her career: social criticism and protest, the effects of gender on identity, and observation of and reflection on nature. Johnson's portrayal of nature in the novel is particularly noteworthy: while she uses natural events and images to represent and explain human behavior, she also manifests a deep interest in the natural environment in its own right. She characterizes the powerful forces and cycles of nature as potential sources of sustenance and healing for the characters, in contrast to the problematical world of human society, but she never romanticizes nature as an Edenic realm of beauty and refuge. Instead, she sees in nature, as in human society, both beauty and ugliness, good and evil—a theme that persists in her later work.

Despite Johnson's lifelong love for the natural world, it would be her compassion for the struggles of human beings that would dominate her works during the 1930s and 1940s. These works, which took the form of both poetry and fiction, include *The Winter Orchard, and Other Stories* (1935); *The Unwilling Gypsy* (1936); *Jordanstown* (1937); *Year's End* (1937); *Paulina* (1939); and *Wildwood* (1945). *Jordanstown* in particular exemplifies the tradition of the proletarian novel, focusing on class struggle in a small, industrial midwestern town during the Depression. During this period, Johnson also taught at the University of Iowa (1942–1945) and at the Breadloaf Writer's Conference in Middlebury, Vermont.

At the same time that Johnson was establishing this astonishing record as an activist and writer, she met Grant Cannon, who was then a field examiner for the National Labor Relations Board and would go on to edit the *Farm Quarterly*, and they married in 1942. Johnson felt she had found an ideal companion in Cannon, whose optimism complemented her tendency to despair; in *Seven Houses*, she characterizes the initiation of their relationship as the major turning point of her life: "I seemed to be waiting to begin to live and not all the beauty, all the intensity of the words on paper, all the desperate search for reform and change . . . seemed to be the reality of living that I wanted to find. And then I met Grant Cannon and the waiting-to-live was over and the real life began."

By 1947, Johnson and Cannon had three children—Terence, Annie, and Carol—and they relocated to Newtown, Ohio, just outside Cincinnati. In 1956, they moved to a thirty-seven-acre farm, also in southwestern Ohio, which they decided to turn into a nature preserve by letting the "ecology develop." While Johnson continued to write throughout her time in Ohio, it was not until the early 1960s that she once again began to publish in earnest; in this phase of her career, she turned more to nonfiction essays and memoirs rather than to the poetry and prose she had favored in earlier years. Beginning with *The Dark Traveler* (1963), she went on to produce *The Sorcerer's Son, and Other Stories* (1965), *The Inland Island* (1969), *Seven Houses: A Memoir of Time and Places* (1973), and *The Circle of Seasons* (1974), as well as pieces in *Country Journal. McCall's*, and *Ohio Magazine.*

In these later works, as Nancy Hoffman puts it, "the naturalist in Johnson reappears in print." While *The Circle of Seasons* (a book of nature photographs for which Johnson wrote the text) and many of her later essays in *Ohio Magazine* demonstrate a keen eye for nature and deep concern about its destruction, *Inland Island* is particularly noteworthy for its environmental theme as well as its vehement protest against the Vietnam War. Reviewed favorably by Granville Hicks and Edward Abbey, *Inland Island* is a chronicle of life on Johnson's farm, the "island" of the title. Johnson's detailed descriptions of her natural surroundings and her narrative movement from January through December place her in the tradition of nature writers such as Henry David Thoreau and Aldo Leopold. Also like these literary predecessors, she feels a "sense of urgency" about the environmental destruction that threatens the land she loves and "a need to record and cherish, and to share this love before it is too late." And, most remarkably, Johnson is ultimately unable to see her farm as an island or natural refuge from the troubles of human society, and she extends her concern for the land and the people who live on it to Vietnam and the war being waged there. Characteristically, Johnson's life mirrored her writing; her son Terry was a conscientious objector against the war in Vietnam, and Johnson herself published an outspoken letter to the editor of the *New York Times* in May 1969, charging that the country was really being "uprooted" by the politicians, scientists, and businessmen responsible for the devastation of war, environmental destruction, and racism rather than by the college students, hippies, blacks, and others who were working for social change.

Although Johnson was greatly saddened by her husband's death from cancer in 1969, she continued to live on her farm; in an essay published in *Ohio Magazine* in 1990, she wrote, "I live alone in the country. I like it." An active writer and observer of nature and human society until the end, Johnson died of pneumonia. Perhaps her most enduring legacy is the manner in which she envisioned the causes of social justice and environmental preservation to be inseparable from one another.

KARLA ARMBRUSTER

Fleischman, John. Editorial on Josephine Johnson on the occasion of her death. *Ohio Magazine* (May 1990).

———. *"News from the Inland Island."* Photo. Robert Flischel. *Audubon* (March 1986).

Hoffman, Nancy. Afterward to *Now in November*, by Josephine Winslow Johnson. New York, 1991.

Kunitz, Stanley J., and Howard Haycraft, eds. "Johnson, Josephine Winslow." *Twentieth Century Authors: A Biographical Dictionary of Modern Literature*. 1942.

Millett, Fred B. "Josephine Winslow Johnson." *Contemporary American Authors: A Critical Survey and 219 Bio-Bibliographies*. 1940.

Johnson, Robert Underwood (January 12, 1853–October 14, 1937). An editor, poet, critic, lobbyist, and diplomat, Johnson was born in Washington, DC, but grew up on a homestead in the little village of Centreville, Indiana. From his parents, Catherine Coyle Underwood and Nimrod Hoge Johnson, he learned to appreciate the natural surroundings of his boyhood home. In his memoirs Johnson affectionately recalls seeing sky-darkening flocks of passenger pigeons, taking barefoot fishing trips to nearby creeks, and making regular visits to the local swimming hole. "Wherever we roamed," remembered Johnson, "we found a beautiful landscape."

After graduating from Earlham College in Richmond, Indiana, in 1871, Johnson accepted a clerkship in Chicago. In 1873 he began working at the *Century Magazine* (then *Scribner's Monthly*) in New York, where he remained for forty years, succeeding Richard Watson Guilder as editor in 1909. In the late nineteenth and early twentieth century, progressive social reform movements were often carried forward by magazines such as *Century, Harper's,* or *Atlantic.* Johnson and the other *Century* editors were public-spirited men who took seriously their publication's considerable influence on the formation of character and the amelioration of social evils in America. It was beneath the banner of *Century Magazine* that Johnson crusaded for causes such as international copyright law, the National Institute of Arts and Letters, and national parks and forest conservation.

Johnson's genuine commitment to conservation dates from 1889, when he first met explorer, preservationist, and writer John Muir in what Muir called the "confounded artificial canyons" of San Francisco's Palace Hotel. Having gone to California to arrange for a *Century* series on the gold rush, Johnson would leave persuaded of the urgent need for forest conservation. Characteristically, Muir had insisted that Johnson accompany him into the wilderness of the Sierras. It was by the campfire at Soda Springs that Muir explained the urgent need to protect Yosemite Valley and the surrounding areas from the encroachments of the timber and grazing interests that were already destroying the range. In immediate response to Muir's concern, Johnson proposed a Yosemite National Park, to be modeled after Yellowstone, which had been designated the first national park in 1872.

Muir agreed to write two pieces on the proposed Yosemite park for *Century,* and Johnson agreed to use his influence and that of his magazine to advocate the proposal back East. Returning to New York, Johnson published Muir's essays and used them to make his case before the public and before members of Congress. He was able to build an unlikely coalition of supporters including Southern Pacific Railroad and the Hearst family, and he had considerable pull with various members of the Committee on Public Lands. It is thanks to his efforts that Yosemite National Park was designated on October 1, 1890. Adopting Muir's tone of reverence for the natural world, Johnson would later refer to his efforts on behalf of forest conservation as "spiritual lobbying."

Although the battle for Yosemite National Park had been won, a succession of threats to the park continued for the next quarter century. In response to the earliest of these threats, Johnson proposed the formation of a Yosemite defense league. His suggestion resulted in the Sierra Club, which was formed in 1892 with Muir as its first president. Johnson was later elected honorary vice president of the club. Throughout the years of threat to Yosemite, Johnson was a tireless and effective advocate for preservation. For example, although the park had been established in 1890, the spectacular valley at its heart had been under state control since 1864. The scandalous mismanagement of the valley prompted conservationists to call for its retrocession to the federal government. After more than a decade of fighting side by side with Muir, the Sierra Club, William E. Colby, Frederick Law Olmsted, and even railroad magnate Edward H. Harriman, Johnson finally helped win retrocession of the valley in 1906. On several occasions, he helped kill bills that would have contracted park boundaries or otherwise threatened the wilderness it had been established to preserve. He served as chairman of the National Committee for the Preservation of Yosemite National Park as late as 1913.

After reading *Nature as Modified by Human Action* (1864; also called *Man and Nature*), George Perkins Marsh's pioneering ecological study of forests and the consequences of their destruction, Johnson rededicated himself to the cause of conservation. In the early 1890s he successfully encouraged John Noble, President Benjamin Harrison's secretary of the interior, to set aside a great many reserves in the Sierras and elsewhere; it is in this way that the Grand Canyon of Arizona came under federal protection. In 1895 he was decisive in the enactment of legislation to protect the Adirondack Forest Preserve, which had been established under the leadership of Charles S. Sargent. In 1896 Johnson persuaded the Honorable Hoke Smith, President Grover Cleveland's secretary of the interior, to create the

Forest Commission of the National Academy of Sciences, an influential body headed by Charles Sargent and including as members such men as Alexander Agassiz, Gifford Pinchot, and—in an unofficial capacity—John Muir. After examining the state of the nation's forests, the Sargent Commission recommended the creation of thirteen forest reserves containing a total of 21 million acres. President Cleveland created the proposed reserves immediately before leaving office in 1897 and averred that they were among the most important accomplishments of his administration.

Johnson continued to work for conservation even into the Roosevelt and Wilson administrations. Indeed, President Theodore Roosevelt's White House Conference on Conservation of 1908 was initially inspired by Johnson's assertion of the need for a forest summit. From 1906 until 1913 he lent his efforts to the battle to protect Yosemite's Hetch Hetchy Valley from damming as a water reservoir for the city of San Francisco. In what was to become one of America's most famous conservation battles, President Roosevelt and his chief forester Gifford Pinchot finally defeated the preservationist interests of Muir, Johnson, and the Sierra Club. The bill granting the pristine valley to the city was signed into law by President Woodrow Wilson on December 6, 1913.

Throughout his years as an editor and lobbyist, Johnson was a practicing writer of considerable merit. He edited the Century War Series and extended it by four volumes covering the Civil War. He was also a poet of distinction in his own day; his poetic works (mostly occasional poems in the romantic tradition) appeared in many volumes and were collected in *Poems of Fifty Years* (1931). He was one of the first fifty people to be elected into the American Academy of Arts and Letters (which he was instrumental in forming and which he served as secretary from 1903 to 1909), and he was friends with many literary luminaries of the Gilded Age, including James Russell Lowell, Mark Twain, Walt Whitman, John Burroughs, Joel Chandler Harris, and William Dean Howells.

After his retirement from *Century* Johnson traveled, wrote, and remained active in civic affairs. He served as director of the Hall of Fame of New York University after 1919, and he succeeded Thomas Nelson Page as U.S. ambassador to Italy in 1920–1921. In his career as a diplomat Johnson was decorated by France, Italy, Belgium, Poland, and Yugoslavia.

Although Johnson's work for forest preservation often took place behind the scenes—through anonymous editorials in *Century* or through "spiritual lobbying" of congressmen, senators, and secretaries of the interior—

his legacy to American conservation is considerable. He helped create Yosemite National Park, secure retrocession of the valley to the federal government, and protect the park against numerous assaults over many years. He was a primary force in the creation of the Sierra Club. And in addition to his own successful lobbying for forest preservation in the Sierras, the Adirondacks, and elsewhere, he was primarily responsible for the formation of the Sargent Commission (1896–1897), which indirectly resulted in the preservation of over 20 million acres of forest.

In 1876 Johnson married Katharine McMahon, who died in 1924. They had a son, Owen (who, like his father, became an author), and a daughter, Agnes (Mrs. Frank H. Holden). Johnson died at his home in New York City.

MICHAEL BRANCH

Cohen, Michael P. *The Pathless Way: John Muir and American Wilderness.* Madison, WI, 1984.

Johnson, Robert Underwood. *Remembered Yesterdays.* Boston, MA, 1923.

Jones, Holway. *John Muir and the Sierra Club: The Battle for Yosemite.* San Francisco, CA, 1965.

Turner, Frederick. *Rediscovering America: John Muir in His Time and Ours.* San Francisco, CA, 1985.

Jordan, David Starr (January 19, 1851–September 19, 1931). A naturalist, educator, author, and peace advocate, Jordan was born on a farm in Gainesville, New York, the fourth of five children of Hiram Jordan and his wife Huldah Lake Hawley. (During his early years he was educated at home and in the local ungraded school.) At fourteen he was sent to Castile Academy, a few miles from his home; but he protested to his parents that what he was being taught there he had already learned. Accordingly, and because he was such a promising student, he entered Gainesville Female Seminary, one of only two boys allowed to study at a school of education designed for women.

At the seminary, he excelled in algebra, geometry, and French, which he learned to read as readily as English. One of his teachers, an amateur geologist, would often take him and other students on field trips among the crystalline boulders scattered over western New York that had been brought down from Canada by glacial ice. Occasionally, they would come across

Devonian fossils and other objects that served to whet young Jordan's curiosity about nature and the makeup of the earth. On his own, he developed an interest in botany, the country around his home being rich in wildflowers. With the dedication of a budding scientist, he adorned the white walls of his bedroom with the names of all the plants he identified.

The beauty of the flowers he collected appealed to his aesthetic sense. He pondered their origins, structural resemblances, and ecological relationships. Concurrent with his passion for flowers was an admiration for the various trees that grew so abundantly about his home. And though he had no training in drawing, he began to paint pictures of his favorite trees and flowers. He dreamed of becoming a botanist, but having a love for animals, he also thought about becoming a sheep breeder. Being a superior student and intellectually motivated, however, he gave considerable thought to obtaining higher education.

In 1869, he won a competitive scholarship to Cornell, which had recently been founded. His academic ability was soon realized, and in his junior year he was appointed an instructor in botany. He delighted in teaching his favorite subject, but still interested in animal breeding, he researched and wrote his first paper on "Hoof-rot in Sheep," which he submitted to the *Prairie Farmer* in 1871. Three years later he prepared, with Balfour Van Vleck, *A Popular Key to the Birds, Reptiles, Batrachians and Fishes of the Northern United States, East of the Mississippi River*, the forerunner of his *Manual of the Vertebrates of the Northern United States* (1876).

In the fall of 1874, Jordan went to Indianapolis to teach science at its high school. He also pursued courses at Indiana Medical College, more to learn physiology than to prepare to practice medicine. In 1875 he was awarded an M.D. degree and went on to Butler College to teach natural history, which he did for four years before moving on to Indiana University. At Indiana he earned a reputation for being an excellent teacher, one who had an extraordinary knowledge of botany and zoology gathered from his own research and published papers. He also drew material for instruction from his own personal collection of almost 10,000 books and pamphlets on zoology that he had accumulated through the years.

On January 1, 1885, he was appointed president of Indiana University; in such a position he was able to initiate significant changes in curriculum, emphasize the teaching of science, and stimulate the faculty to research. He proved such an able administrator that in 1891 he was selected to become the first president of a new and promising institution of higher learning being established in California—Stanford University. After twenty-two years as president of Stanford, during which period the university underwent tremendous growth, he was relieved of administrative duties and elevated to chancellor, a title he held from 1916 until his death some fifteen years later.

Always diligent and effective in carrying out his executive duties, Jordan still found time for scientific research, especially that requested of him by the federal government. Between 1877 and 1891, for example, he served on the U.S. Fish Commission; then again between 1894 and 1909. He reported on fisheries from the Atlantic to the Pacific, from Alaska to the Hawaiian Islands. At the time, few individuals knew more about ichthyology than did Jordan, proof of which can be found in his *Guide to the Study of Fishes* (1905), *The Fishes of North and Middle America* (with B. W. Evermann; 4 vols., 1896–1900), *American Food and Game Fishes* (1902), and several other works. Between 1908 and 1910 he served as commissioner in charge of fur, seal, and salmon study set up to determine the principles governing the geographic distribution of animals and plants and to learn of their physical and biological environment.

At the turn of the century, Jordan became a peace activist. International arbitration, he advocated, should be the means of adjudicating difficulties between countries. Although there was considerable opposition to most of his recommendations to establish peace throughout the world, he continually crusaded to promote his ideals. He spoke in virtually every state and in many foreign countries. In his public addresses and published essays, he always stressed the deleterious biological effects of armed conflicts.

A prolific and versatile writer, Jordan wrote over 1,000 essays and books on scientific, educational, and social topics; among the most significant are such titles as *The Care and Culture of Men* (1896), *Evolution and Animal Life* (1907), *The Human Harvest* (1907), and *A Plan of Education to Develop International Justice and Friendship* (1925). His best book, perhaps, is his two-volume autobiography, *Days of a Man: Being Memories of a Naturalist, Teacher, and Minor Prophet of Democracy* (1922). In it, Jordan writes that he followed a twofold career of naturalist and teacher out of sheer love and that he assumed the role of an enthusiastic advocate for world peace from a sense of duty. The first volume of his autobiography (1851–1899) focuses mainly on his early life, teaching, and dedication to research; the second (1900–1906), on his travels and efforts to improve international relations.

Jordan's work merited worldwide recognition. In 1912, he was elected president of the American Association for the Advancement of Science. In 1922, he was welcomed as an Honorary Associate in Zoology by the Smithsonian Institute. He was either an active or honorary member of the American Philosophical Society, the Zoological Society of London, the Royal Academy of Science of Sweden, the Linnaean Society of New South Wales, the Naturalists' Club of Sydney, and the International Commission of Zoological Nomenclature. In 1886, he was awarded an honorary Doctorate of Laws degree by Cornell University. Other honorary degrees followed from Johns Hopkins in 1902, Illinois College in 1903, Indiana University in 1909, Western Reserve in 1915, and the University of California in 1916.

Jordan married Susan Bowes in 1875, with whom he had two daughters and one son; she died in 1885. Two years later, he married Jessie L. Knight, with whom he had one daughter and two sons.

GEORGE A. CEVASCO

"David Starr Jordan" [obit.]. *New York Times*, September 20, 1931.

"David Starr Jordan" [obit.]. *San Francisco Chronicle*, September 21, 1931.

"David Starr Jordan." *World's Works*, April 1914.

Kellogg, V. L. Review of *Days of a Man*, by David Starr Jordan. *Science*, March 23, 1923.

Just, Ernest Everett (August 14, 1883–October 27, 1941). Born in Charleston, South Carolina, to Charles Fraser Just (a builder of wharves and son of a former slave) and Mary Mathews Cooper (a teacher), Just was a marine biologist, zoologist, physiologist, and research scientist. As his father died when Just was four years old, his mother directed her son's early education.

With the death of Charles, Mary decided to dedicate herself to educating local African American youths. With this in mind, she taught Sunday school on James Island (a neighboring sea island), unknowingly kindling in her young son a lifelong fascination and love for marine life and nature.

In 1896, Just's mother enrolled him in the "Colored Normal, Industrial, Agricultural and Mechanical College" in Orangeburg, South Carolina. After completing a three-year course of studies designed specifically to train teachers, Just headed north to further his education. On concluding three years of college preparatory studies at the Kimball Union Academy in New Hampshire, he enrolled in Dartmouth College.

Although Just entered college expecting an atmosphere of intellectual stimulation, he soon discovered that life at Dartmouth focused more on sports than on academics. After completing a frustrating and disappointing freshman year, Just briefly considered leaving Dartmouth. In his sophomore year, however, he took his first biology course. His love for nature was soon rejuvenated, and Just became fascinated by the development of the animal egg. He graduated in 1907 with high honors, earning a degree in zoology. From Dartmouth, Just proceeded directly to Howard University in Washington, DC, where his first assignment was to teach literature and rhetoric.

Anxious to foster a career in biology, however, he began (in 1908) to investigate the possibility of graduate training. Because of his African descent, he was repeatedly advised to pursue medicine, which was less racially prejudiced than biology. Nevertheless, Just persisted in his preference and was put in touch with Dr. Frank Rattray Lillie, head of the Department of Zoology at the University of Chicago and director of the prestigious Marine Biological Laboratory (MBL) at Woods Hole, Massachusetts. Dr. Lillie was to become one of Just's most valued colleagues and friends, and from 1909 until 1930, Just would spend every summer (except one) at Woods Hole, there performing his most important research.

In 1909, Just commenced his graduate studies at Woods Hole, simultaneously working as Lillie's research assistant. During these summers, Just performed outstandingly well in assisting Lillie with his studies in the fertilization and breeding of the sandworm (*Nereis*) and the sea urchin (*Arbacia*). At this time, Just also began to ascend the professional hierarchy at Howard University, being appointed assistant professor in 1910, associate professor in 1911, and full professor in 1912. During the year of 1912, Just was also appointed head of the Zoology Department (a position he held until his death in 1941), as well as head of the Physiology Department of Howard University's Medical School (a position he held until 1920). As head of the Physiology Department, he greatly contributed to the improvement of medical education for African American students.

For significant research in biology, in 1915 Just was awarded the first Spingarn Medal (an award that goes to "the man or woman of African descent and American citizenship who shall have made the highest achievement during the preceding year or years in any honorable field of human endeavor"). In 1916, Just completed his dissertation, titled "Studies of Fertilization in *Platynereis megalops*," and was awarded the Ph.D. in zoology and physiology from the University of Chicago.

Over the next twenty-five years, Just was to complete groundbreaking research on the cellular processes of marine organisms, exploring in particular the mysteries of fertilization and embryology. Some of Just's most important research concerns the field of experimental parthenogenesis (the laboratory fertilization of an egg *without* the addition of sperm) and the role that ectoplasm (the material lying between the outside edge of the cell and the cell membrane) plays in fertilization and cellular water transport. Additional studies performed by Just concern cell division, hydration and dehydration in living cells, and the effect of ultraviolet rays on the number of chromosomes. From his studies, Just generated over sixty articles and two major documents: *Basic Methods for Experiments on Eggs of Marine Animals* (1939; revised and updated in 1957) and *The Biology of the Cell Surface* (1939).

For his exemplary work in the field of biology. Just became the recipient of numerous grants and awards. He was elected vice president of the American Society of Zoologists and was appointed editor of the journal *Protoplasm*. Despite his obvious brilliance and outstanding scientific achievements, Just was to remain plagued by the social stigma of racial prejudice throughout the entirety of his professional life. Unable to obtain appointment in any of the large American universities or research facilities, Just left the United States in 1930 for Europe, where he received unconditional welcome. He spent most of his last eleven years at the major European research facilities, such as the Kaiser Wilhelm Institute for Biology in Germany, the Naples Zoological Station in Italy, and the Sorbonne and Marine Station in France.

While in Europe, Just married a young German woman named Hedwig Schnetzler and settled with her in Paris. However, the subsequent German occupation of France forced a hasty departure, and Just returned to the United States in September 1940. Stricken by increasing health problems, Just was hospitalized and subsequently diagnosed with pancreatic cancer. Despite great attempts to resume teaching at Howard University in the fall of 1941, he became increasingly weak and died in Washington, DC.

In February 1996, Dr. Just was commemorated by the U.S. Post Office in the Black Heritage Stamp series.

TIMOTHY J. HORAN

"Dr. Ernest E. Just Honored on New Black Heritage Stamp." *Jet* (February 1996): 19.

Hayden, Robert C. *Seven African American Scientists.* New York, 1992.

Manning, Kenneth R. *Black Apollo of Science: The Life of Ernest Everett Just.* New York, 1983.

Sammons, Vivian O. *Blacks in Science and Medicine.* New York, 1990.

Krutch, Joseph Wood (November 25, 1893–May 22, 1970). An important figure in environmentalism for many reasons, chief among them the wide influence he exerted during the 1950s and 1960s, Krutch was an educator, drama critic, and naturalist. With his books, essays, lectures, and television productions, he awakened thousands to some of the most pressing ecological problems of our times. A prolific literary and drama critic most of his life, in his final decades he received national acclaim for his views on nature, conservationism, and the environment.

In his autobiography *More Lives Than One* (1962), he writes of his early years in Knoxville, Tennessee, as those of a shy and unaggressive adolescent, a good student not especially fond of school. What he failed to learn in the classroom, he eagerly taught himself through reading widely and being a keen observer of nature. Of later importance in his life, he notes, was his maternal grandmother's contribution to his fascination with animals and plant life.

In 1911, Krutch entered the University of Tennessee, a seat of learning he found "sleepily conventional," its faculty equally made up of genuine scholars and unenthusiastic pedagogues. He began as a mathematics major, but with the taking of a few English courses, he switched over to literature. After graduating with honors in 1915, he decided on graduate work in the humanities at Columbia University. Two years later with a master's degree in hand and course work for the doctorate completed, he was appointed an instructor of freshman English at Columbia. At the end of the year, he enlisted in the U.S. Medical Corps. His military career, as he remembered it, was "uneventful and inglorious."

In 1921 he accepted a position at Brooklyn Polytechnic Institute in its English Department. In the same year that he completed his doctorate he became drama critic for the *Nation*, a liberal weekly journal, and resigned from Polytech. He remained outside the academic world until 1937, the year he received an appointment as professor of English at Columbia. His appointment was a consequence of several factors, chief among them his reputation as a drama critic and the publication of a controversial book, *Edgar Allan Poe: A Study in Genius* (1926). In writing of Poe, he employed a

psychoanalytical approach. He was to go on to write two more well-received biographies, *Samuel Johnson* (1944) and *Henry David Thoreau* (1948), both of which avoided the psychoanalytical excesses that, as he later admitted, marred much of his work on Poe.

Like Johnson, whom Krutch greatly admired, he held humankind's rationality to be unique and sublime. Without degrading intuition or the emotions, Krutch posited that the freedom to make moral choices was contingent on rational decisions. Like Thoreau, Krutch came to hold that each individual had an obligation to perfect himself or herself. Like both Johnson and Thoreau, Krutch had a longing for an older, more contemplative way of life and a distaste for many of the inventions of the twentieth century. At times he went as far as to suggest in some of his essays that scientific reality was incompatible with the human spirit. "Man's humanity," he lamented, "is threatened by the almost exclusionary technological approach in social, political and philosophical thought."

In 1952 he gave up his position as drama critic at the *Nation* and his professorship at Columbia. He left New York to settle permanently in Arizona, where his home was an adobe brick house in the middle of the desert. "I didn't go West," he explained, "for its future, or its industry, its growth, its opportunity. I went for three reasons: to get away from New York and the crowds, to get air I could breathe, and for the natural beauty of the desert and its wildlife."

The nature of humankind and conservation of external nature, he decided, were essential to an understanding of the unity of being. He became a spokesman for the ecological movement long before many even knew what the term *ecology* implied. A string of memorable books on nature and conservation followed. In particular, he was fascinated by the desert flora and fauna and their survival in a region of scant rainfall. Some plants, he discovered, sank very deep roots in the soil to obtain the moisture they required, and he observed that certain animals, especially rodents, survived without ever drinking water.

Krutch wrote one and sometimes two books a year on the relationship between living organisms and their environment. Foremost among his most significant books on natural history and environmentalism are *The Desert Year* (1952), *The Voice of the Desert* (1955), and *The Great Chain of Life* (1956). In all three books he exhorts the reader to a participation in the mysteries of the natural world, beginning with rational understanding. "It is not ignorance," he maintained, "but knowledge which is the mother of wonder."

Krutch continually emphasized the disastrous spiritual outcome of natural imbalance. Though an advocate of the rational, his coming into communion with the environment of the West led him to reflect on the transcendental aspects of external nature. Analyzing its mysteries brought him close to an intuitive appreciation of its spiritual wholeness. Living side by side with Native Americans in Arizona, he came to accept their concepts of the land and the obligations of its inhabitants. In his nature essays, from *The Twelve Seasons* (1949) to *The Forgotten Peninsular* (1961), he explored the fragility of nature and humanity.

Humanity may have the ability to destroy the world, but Krutch hoped a way would be found to use wisdom and to cherish all life. Materialism, he warned, was at the root of ecological blindness. All too often humans lack a sense of the sacrality of the land they inhabit. In his books, Krutch describes manifestations of nature with the eye of a botanist. "Maybe the most I can claim," he once quipped, "is that I know more about botany than any other drama critic, and more about the theater than any other botanist."

In 1963 he received wide acclaim for a television show in which he starred and helped produce: an hour-long tour of the Sonora Desert near Tucson. Two years later he produced *The Grand Canyon*, in which he captured the grandeur and mystery of America's most awesome national park.

Krutch held membership in the American Academy of Arts and Letters, the American Academy of Arts and Sciences, the American Philosophical Society, and the Century Club of New York. He was honored with the John Burroughs Medal for Nature writing in 1954. In the same year, the Rockefeller Institute bestowed on him its Ettinger Award for Science Writing. He was also the recipient of several honorary degrees: a Doctor of Literature from Columbia University (1954) and Doctorates in Humane Letters from Northwestern University (1957), Arizona University (1960), and the University of New Mexico (1961). The many honors and awards Krutch received are testimony to his perception of, and articulation on, the grandeur and mysteries of nature.

An eloquent environmentalist in the tradition of Thoreau, John Muir, and Aldo Leopold, Krutch maintained that through waste, greed, and egocentricity, humankind had upset the balance of nature to such an extent that our very survival is endangered. Unless we share "this terrestrial globe with creatures other than ourselves," he insisted, "we shall not be

able to live on it for long." Foremost among the many lessons he taught in his forty or so books and edited volumes is the simple truth, as he put it, that "life is everywhere precarious, men everywhere small."

Krutch's importance, as the eminent botanist and ecologist Paul B. Sers once succinctly stated, "lies in the fact that with his literary craftsmanship and great respect for scientific accuracy, he exerted a wide influence over the country."

GEORGE A. CEVASCO

"Joseph Wood Krutch, Naturalist, Dies" [obit.]. *New York Times*, May 23, 1970.

Lehman, Anthony L. "Joseph Wood Krutch: A Selected Annotated Bibliography of Primary Sources." *Bulletin of Bibliography* 41 (1984): 74–80.

Margolis, John D. *Joseph Wood Krutch: A Writer's Life*. Knoxville, TN, 1980.

Pavich, Paul N. *Joseph Wood Krutch*. Boise, ID, 1989.

———. "Joseph Wood Krutch: Persistent Champion of Man and Nature." *Western American Literature* 13 (1978): 151–158.

Lacey, John Fletcher (May 30, 1840–September 29, 1913). As chairman of the House of Representatives Committee on Public Lands from 1895 until 1907, Lacey was one of the first outspoken conservationists to serve in the U.S. Congress. During his legislative career, Lacey introduced a number of important bills, such as the Park Protection Act, the Wichita Bison Preserve, the Antiquities Act, the Forest Service Act, and the Lacey Act. Lacey's vision of conservation stretched beyond the strictly utilitarian views of Gifford Pinchot to embrace the cause of wildlife and places of intrinsic natural or historical beauty.

Lacey was born in New Martinsville, Virginia, now part of West Virginia, to John M. and Eleanor (Patten) Lacey. In 1853, the fourteen-year-old John joined his family in a migration to Oskaloosa, Iowa. As an adolescent, Lacey worked on the family farm and learned the stonemasonry trade of his father. When Abraham Lincoln called for 75,000 volunteers in April 1861 to suppress the rebellion, Lacey enlisted in the Third Regiment of Iowa Volunteers.

After being captured by Confederate forces in November 1861, Lacey was paroled back to Iowa on the promise of not taking up arms again. He profitably used the time to read law under the tutelage of Samuel Rice, Iowa's attorney general. In the summer of 1862 a prisoner exchange allowed Lacey to reenlist. He joined the Thirty-third Regiment of Iowa volunteers and received a commission as first lieutenant. With Rice commanding the regiment, Lacey served as his adjutant. Lacey later acted as adjutant for General Frederick Steele. For the remainder of the war, he served in Arkansas, Louisiana, and Alabama. In 1865 he was mustered out with the rank of brevet major.

The Civil War was a critical experience for Lacey as it was for countless others in his generation. It imbued him with a sense of public duty and invested him with the future of the nation in a very personal way. He was active in influential veteran's organizations, such as the Grand Army of the Republic, and a spokesperson for their issues. For the rest of his life, he was fondly known as "Major" by all who knew him.

After the war, Lacey returned to the law. From 1865 until 1889 he put in long hours at his legal practice and established a reputation for his sound knowledge of railroad law. He published several legal works including the acclaimed *Digest of Railway Decisions* in 1875. He also held several minor political offices for short periods of time, such as city solicitor of Oskaloosa (1869), congressman in the Iowa House of Representatives (1870), and Oskaloosa city councilman (1880–1883).

In 1888 Lacey was elected to the U.S. House of Representatives to represent Iowa's Sixth Congressional District. He lost his bid for reelection in 1890 during a Democratic landslide but won the seat back in 1892. He held his seat in Congress until 1907. From 1895 until 1907 Lacey served as the chairman of the House Public Lands Committee. During his time in Congress Lacey was an Old Guard Republican who supported pensions to veterans, protective tariffs, and subsidies to industry, especially the railroads. He also advocated restrictions on immigration, expanded rights to Native Americans, and mine safety. By far Lacey's strongest contribution was in the area of conservation, especially wildlife protection. During the presidency of Theodore Roosevelt, Lacey was the administration's floor manager in the House conservation legislation.

Lacey's interest in conservation is still something of a mystery. Although he was a member of the Boone and Crockett Club, Lacey admittedly possessed little interest in outdoor activities. He did not hunt, fish, or take extended camping trips. Lacey's work as chairman of the House Public Lands Committee often led him to travel to places of natural beauty. While these areas may have influenced him, he probably recognized the need for a robust conservation program more from his nationalistic economic viewpoints than from a discernible love of nature. Ironically, Lacey's attention to conservation may have cost him his seat. Voters in Iowa's Sixth District had little to no interest in the subject their congressman devoted so much of his energy to.

Although the United States had created Yellowstone National Park in 1875, the wildlife within the boundaries remained unprotected. After the massive buffalo slaughter ended in the late 1880s, protection of the animal became a cause to conservationists. Although a couple of bills had been debated before Lacey joined Congress in 1889, nothing had been done to protect the buffalo. In March 1894 several buffalo were killed inside Yellowstone Park. Using national attention as a leverage, Lacey followed immediately by introducing the Park Protection Act. Although this law protected all wildlife in Yellowstone, it was designed primarily to provide refuge to the remnants of the wild buffalo herd. The act authorized the U.S. Army to patrol Yellowstone and enforce the law. It was the first time the federal government had engaged in wildlife protection.

In 1900, after eight years of fighting, Lacey gained approval of the wildlife protection law that bears his name. The Lacey Act made it a federal crime to traffic game meat that was killed in violation of state laws. Conservationists widely believed that the market for game meat led to local extinctions of waterfowl, deer, turkey, and upland birds. Game animals all over the country supplied the markets to the nation's largest cities. The Lacey Act forbade this trafficking if it was against the law of either state. It was a major victory for wildlife conservationists, but the onus remained on the states to forbid the sale of game meat and to establish hunting regulations, such as closed seasons, licenses, and bag limits. The full effect of the Lacey Act was felt in 1911 and 1912 when California, Massachusetts, and New York all banned the sale of game. In addition to utilizing federal power to regulate interstate commerce, the Lacey Act empowered the secretary of agriculture to introduce species and establish game regulations on federal lands. These last two measures took the established practices of the federal Bureau of Fisheries and applied them to game.

In 1902 President Theodore Roosevelt signed another piece of Lacey legislation, the Alaska Game Act. This law created hunting rules and regulations for the territory of Alaska. Lacey introduced and lobbied for this bill, which had the strong support of the Boone and Crockett Club and other eastern sportsmen. The Alaska Game Act protected brown bears, which were becoming threatened with extinction, and regulated the sale of game meat.

In 1905 Lacey introduced legislation creating the Wichita Bison Preserve in Oklahoma. As where the Park Protection Act sought to protect what few buffalo remained on the Plains, the Wichita Preserve Act reintroduced the animal into a part of its former range. The federal government supplied the land and an appropriation of $14,000 to enclose the Wichita Preserve in fence; the New York Zoological Society provided the fifteen animals that made up the initial herd. Currently known as the Wichita Mountains National Wildlife Refuge, the buffalo herd is maintained at about 600 animals.

Lacey's most important legislative achievement was the Antiquities Act of 1906. This act allowed the president to withdraw lands from those owned by the government to protect their historic value and increased presidential power at the expense of the legislature. As where the creation of a national park required a congressional act, the Antiquities Act allowed the president to do essentially the same thing through executive order. President Roosevelt wasted little time using this new power. Perhaps Roosevelt used the power more aggressively than Lacey had intended. Before the end of the year the president had created four new national monuments: Devil's Tower in Wyoming, El Morro in New Mexico, Montezuma Castle in Arizona, and the Petrified Forest in Arizona. This last one was something of a reward to Lacey, who had introduced a bill in 1904 to create a national park in the area, but it failed to gain the support of his colleagues. In 1908 Roosevelt created the Grand Canyon National Monument. Since 1906 over one hundred national monuments have been created using the act that Lacey authored.

In addition to preserving monuments, the Antiquities Act established a permit system for archaeological digs and made wanton relic hunting illegal. Lacey had seen for himself the damage that this practice posed to serious historic sites when he visited the Southwest four years before passage of the Antiquities Act. This last provision protected the interests of Native Americans, which were always something close to Lacey's heart.

Lacey warmly supported national parks and the forest reserves through several pieces of legislation he introduced. He assisted in writing the Forest Reserve Act of 1891, which allowed the president to withdraw land from national forests. In 1905 the U.S. Forest Service was created, and administration of the forest reserves was transferred from the Interior to the Agriculture Department. Lacey justified this on the grounds that forest reserves served as important watersheds that protected American agriculture. The following year, he crafted legislation to treat the forest reserves as game refuges, but he could not persuade his colleagues to support it. He also sponsored the bills that created Wind Cave National Park (1903) and adjusted the boundary of Yosemite National Park (1905). He suggested several other national parks that failed to become law, such as Pajarito and the Petrified Forest.

Although Lacey supported the popular President Roosevelt with conservation policies, he was losing touch with party loyalists in Iowa. In 1906 reformers led by progressive Republican Governor Albert Cummins successfully challenged Lacey in the primary. After completing his term in 1907, Lacey returned to Oskaloosa and resumed his legal practice. In 1908 he ran against Cummins in the Republican primary to replace recently deceased U.S. Senator William Allison, but he lost. In retirement many conservationists consulted him about navigating legislation through Congress. He continued to write legal papers and works of local history. Lacey served the interests of wildlife protection as a member of the League of American Sportsmen committee on conservation. After the Weeks-McLean Bill was passed in 1913 to protect migratory birds, Lacey was named to the advisory board established by the legislation to assist the secretary of agriculture in drawing up regulations. His death that year prevented him from making any serious contribution.

Lacey will always be remembered as one of the earliest champions of conservation in Congress and one of the strongest supporters of wildlife protection in American history.

GREGORY J. DEHLER

Conrad, Rebecca. "John F. Lacey: Conservation's Public Servant." *Antiquities Act: A Century of American Archaeology, Historic Preservation, and Nature's Conservation*, ed. David Harmon, Francis P. McManamon, and Dwight Picaithley. Tucson, AZ, 2006.

Gallagher, Mary Annette. "Citizen of the Nation: John Fletcher Lacey: Conservationist." *Annals of Iowa* 46 (Summer 1981).

Lacey, John F. Papers. State Historical Society, Ames, IA.

Trefethen, James B. *An American Crusade for Wildlife*. New York, 1975.

Leopold, Aldo (January 11, 1887–April 21, 1948). Acknowledged as the father of the profession of wildlife management in America, Leopold wrote the first textbook on game management and is one of the founders of the Wilderness Society. Leopold also initiated the first forest wilderness area in the United States, which is now the Gila National Forest. Furthermore, he was the first chair of Game Management at the University of Wisconsin, a position created for him in 1933. But Leopold's most important contribution to the history of the American environmental movement is the development of his biotic land ethic.

The evolution of Leopold's thinking about wildlife management is a study in the evolution of ecological ethics in the United States. During his early years, working for the Forest Service in the Southwest, Leopold campaigned to eradicate wolves, mountain lions, and other large predators to protect deer and cattle in Arizona and New Mexico. Later he came to understand the vital role these predators play in maintaining a healthy, balanced ecological system.

The development of Leopold's land ethic, which he describes in terms of a pyramid, came about gradually after years of observation and careful study. The theory presents a concrete image of a biotic pyramid Leopold describes as "a fountain of energy flowing through a circuit of soils, plants and animals," which depends on a series of ecological interrelationships within an evolutionary context. Susan Flader credits Leopold with playing a key role in transforming ecology from a mere descriptive science into a "functional approach to the total environment—a concern with processes and relationships, with causes and effects." In fact, she goes on to state that "Leopold may be said to have been thinking ecologically, in the functional or holistic sense, before ecological science had evolved a conceptual framework capable of supporting such thought."

Leopold's interest in wildlife management can be traced to his boyhood home in Burlington, Iowa, which overlooks the Mississippi River, where he spent his youth hunting and exploring. His interests in school included field ornithology and natural history. Beginning his studies at Yale in 1906 and working toward a career in forestry, Leopold graduated in 1909 with a master's degree and took his first job with the Forest Service in the Arizona and New Mexico territories' Southwestern District (District 3). He was promoted in 1912 and married Estella Bergere, settling into a pleasant life as supervisor of the Carson National Forest in northern New Mexico. But his satisfying lifestyle was soon to change. Due to a misdiagnosis, Leopold suffered acute nephritis after being caught in a blizzard and having to sleep in a wet bedroll. He took eighteen months to recuperate. In part, this time was spent reading, and it pinpoints the beginning of Leopold's shifting view of ecology.

Following this extended period of recuperation, Leopold devoted his efforts to game management. By 1915, he was working nearly full-time in organizing game and fish work in the Southwestern District, preparing a mimeographed work titled "Game and Fish Handbook," which defined the duties and powers of forest officers and drew attention to his organizational and administrative skills. Thus began Leopold's ongoing involvement with organizing wildlife protection. Flader points out how "extraordinarily persuasive in personal contact" Leopold proved to be. Declining position after position, Leopold focused his efforts on wildlife preservation, placing more and more emphasis on the importance of research. He became the secretary of the New Mexico Game Protective Association, where he was responsible for editing the *Pine Cone*, a quarterly newspaper he founded in December 1915. Through this mouthpiece, Leopold called for nonpolitical state conservation commissions. This work earned him the gold medal of W. T. Hornaday's Permanent Wildlife Protection Fund in 1917 and a special commendation from Theodore Roosevelt.

In 1924, as an administrator with the Forest Service, Leopold was instrumental in the designation of over a half-million acres in the Gila National Forest as wilderness, "setting the pattern," according to Flader, "for the system of roadless wilderness areas which was given force of law in the National Wilderness Preservation Act of 1964." Also in 1924 Leopold left the Forest Service to take a position with the U.S. Forest Products Laboratory in Madison, Wisconsin, a regrettable move, because of the burdensome administrative demands it involved, although it eventually led him into becoming more involved in conservation politics. Leopold was instrumental in the campaign leading to Wisconsin's Conservation Act of 1927, although this too was a bitter experience when the governor failed to appoint him commissioner of the new conservation agency.

In 1928, Leopold defined a new profession for himself in wildlife management. Working off a grant, he began traveling through several states, reporting, charting, mapping, and eventually publishing *Report on a Game Survey of the North Central States* (1931). During this same time, he presented a course of lectures on game management at the University of Wisconsin, establishing himself as the foremost authority on native game in the United States. He also became chair of a committee organized to develop an American game policy (eventually adopted by the Seventeenth American Game Conference in 1930), which, according to Flader, "signalled a new approach to wildlife conservation in the United States."

The stock market crash of 1929 provided Leopold, who now had five children to support, time to draft *Game Management*, his 1933 textbook, which is still highly regarded in the field. In this work he expressed a plea for ecological understanding, for the extension of ethics from the realm of human social relations to the whole land community of which the individual was an interdependent member.

In 1933, a chair of Game Management was created for Leopold at the University of Wisconsin, a position he held until his death. As a professor, Leopold emphasized research and academic excellence in a broad base of courses, although he could have been churning out graduates to fill administrative positions opening in many of the government's depression projects. However, nearly all his students went on to hold leadership positions in the fields of wildlife and natural resource management.

The mid-1930s were a time of ferment in Leopold's thinking. Three events explain the growth and maturity of his intellect at this time. First, in 1935, Leopold, among others, founded the Wilderness Society, a national organization that was designed to protect and extend wilderness regions. This organization strove to protect endangered species, including predators, which Leopold realized were vital to the health and proper functioning of the wilderness. Second, he took ownership of a worn-out farm, which became the setting for his *Sand County Almanac, and Sketches Here and There* (1949). Third, Leopold traveled to Germany, where he observed methods of forestry and wildlife management, leading him to reevaluate his former thinking.

Following this period, we see a decided shift in Leopold's concept of wildlife management. His earliest comprehensive statement of the new ecological viewpoint appears in "A Biotic View of Land," a paper read in June 1939 to a joint meeting of the Society of American Foresters and the Ecological Society of America. This was Leopold's first presentation of his biotic pyramid, later the central focus in what Flader calls his most important essay, "The Land Ethic," included in *A Sand County Almanac* (published posthumously).

Leopold died of a heart attack while fighting a grass fire threatening his farm. In "Thinking Like a Mountain," also published in *A Sand County Almanac*, Leopold compresses his lifetime evolution of thought into one dramatic scene as he witnesses the death of a wolf he has shot. He had been a young man when he killed the wolf, but he eventually realized what the dying green light in the wolf's eyes represented. He came to understand what it was to "think like a mountain" instead of just a human being. Leopold's former concept of wildlife management—a management meant to protect—had devastating effects on the land. By destroying predators like the wolf, the deer population mushroomed until it became unmanageable.

Because of his self-conscious ability to observe, reflect on, and evolve his ideas about ecology, Leopold stands as a key figure in the development of a vital biotic land ethic in the United States.

An extremely helpful listing of Leopold's publications can be found in *The River of the Mother of God, and Other Essays by Aldo Leopold* (1991), edited by Susan Flader and J. Baird Callicott. Leopold's journals have been collected and edited by Luna Leopold into a volume titled *Round River* (1953).

SHELLEY ALEY

Flader, Susan L. *Thinking Like a Mountain: Aldo Leopold and the Evolution of an Ecological Attitude toward Deer, Wolves, and Forests.* Columbia, MO, 1974.

Meine, Curt. *Aldo Leopold: His Life and Work.* Madison, WI, 1988.

Nash, Roderick F. *The Rights of Nature: A History of Environmental Ethics.* Madison, WI, 1989.

Rolston, Holmes, III. *Environmental Ethics: Duties to and Values in the Natural World.* Philadelphia, PA, 1988.

Leopold, A. Starker (October 22, 1913–August 23, 1983). A wildlife biologist, ecologist, educator, and author, Leopold was born in Burlington, Iowa, into a household filled with fervent enthusiasm for nature and science. His father, Aldo Leopold, pioneer of wildlife management and author of *A Sand County Almanac* (1949), instilled in him and his four younger siblings—Luna, Carl, Nina, and Estella ("Jr.")—an early appreciation for the natural world. When Leopold was four months old, he moved with his mother, Estella (Bergere), and father to Santa Fe, where Aldo was employed by the U.S. Forest Service. At three years of age, Leopold began to accompany his father on hunting and fishing trips and various excursions into the wilds. Early on, he displayed a talent for identifying wild animals, and as he grew older, he discussed wildlife ecology with his father, who taught him to formulate and test theories based on the close observation of natural phenomena.

Throughout Leopold's formative years and thereafter, the closeness of the family was largely built around the enjoyment of nature, hunting in particular. His mother was an excellent archer and would be Wisconsin's women's champion for five years running. Leopold and his sister Luna learned much about woodsmanship and the ways of wildlife through their participation in the Tome Gun Club, near Albuquerque, which their father joined in 1921. By the time Leopold turned sixteen and Luna fourteen, they were old enough to hunt alone, but the two spent as much time as possible outdoors in the company of their father. Naturally, sibling conflicts arose. Once, when only one of the children was permitted to accompany their father on a hunting trip, Leopold and Luna drew straws to determine who would go. When Luna won, Leopold exchanged his finest pencil for the privilege. As Curt Meine writes, "Luna soon owned all of Starker's pencils."

In 1924, the Forest Service sent Aldo and his family to Madison, Wisconsin, where Leopold's parents would reside for the rest of their lives. Leopold entered the University of Wisconsin in 1929, but having discovered liquor, coeds, and fraternities, he flunked out, much to the displeasure of his parents. In the summer of 1932, at the urging of his father, Leopold relocated to Poynette, Wisconsin, where he was employed by the U.S. Department of Agriculture Soil Erosion Service as a junior biologist. After two years of work (which would ultimately prove to be of great value as he began his teaching career), he persuaded University of Wisconsin officials to give him another chance. Having matured personally, Leopold became a serious student and earned a B.S. in agriculture in 1936.

In 1935 Leopold helped to build "the shack," the now-famous setting for *A Sand County Almanac* and Leopold family weekend outings, near the Wisconsin River in Sauk County. By the late 1930s, as the events in his life began to accelerate, Leopold saw his parents only a few times a year. In 1936, as his father had done three decades earlier, he entered the Yale School of Forestry, where he studied for a year. In 1938, he married Elizabeth Weiskotten, a marriage that would produce two children, Frederick Starker Leopold and Sarah Pendleton Leopold Klock. Completing his doctoral course work at the University of California at Berkeley, Leopold took a job as a wildlife field biologist for the State Conservation Commission of Missouri. This work provided him with the field setting to write his dissertation—written mostly at his parents' home in Madison—which is a discussion of the nature of heritable wildness in turkeys in the Missouri Ozarks. The dissertation, accepted by the university in 1944, influenced the Missouri state game department to modify its means of increasing the state's wild turkey population. Leopold's recommendations proved to be so successful that other state game departments subsequently adopted them.

With the war in full swing overseas, Leopold fully expected to be drafted into military service, but no notice would arrive. Waiting for the call, he served as a director of field research for the Conservation Section of the Pan-American Union in Mexico (1944–1946). The information he gathered from his fieldwork there would later be utilized in his book *Wildlife of Mexico: The Game Birds and Mammals* (1959). In 1946, Leopold joined the staff of the Museum of Vertebrate Zoology at the University of California at Berkeley. He became an associate professor in 1952, a professor of zoology in 1957, a professor of zoology and forestry in 1968, and professor emeritus on his retirement in 1978. Students of Leopold lauded the professor for his thorough knowledge and enthusiastic teaching style. One of his students said that Leopold "stood out from everybody else. He just had that ability to capture the whole spirit of a program and make it feel alive." Outside the classroom, Leopold spent time in the wilds, not only to conduct research but also to hunt and camp.

Leopold's output as an author on a variety of ecological subjects is impressive, the result of travel and study all over the United States and Mexico. In addition to writing numerous published articles, he wrote and co-wrote several books, including *Game Birds and Mammals of California* (1951), *Wildlife in Alaska: An Ecological Reconnaissance* (1953), *The Desert* (with the editors of *Life*; (1961), and *The California Quail* (1977).

Most of Leopold's published work was well accepted by the scientific community, but an article appearing in the *Sierra Club Bulletin* in 1953 titled "Too Many Deer," which proposed that hunters be allowed to shoot does to restore deer's natural food supply, stirred controversy among many conservationists. As one writer commented on some readers' reception of the article, "The notion that there could be too many deer was for many people as unthinkable as the idea that there could be too many redwoods." Leopold's essay proved to be prophetic: so many deer thrived in the 1940s and 1950s that herds began to dissipate from starvation.

Leopold was active in numerous public and private ecological agencies. In 1962, he was appointed by Secretary of the Interior Stewart Udall to serve as chairman of a special Advisory Board on Wildlife Management. The resulting "Leopold Committee Report," an exemplary study in the discipline, influenced the return of the national parks to their natural states and modified the animal control activities of the Bureau of Sports Fisheries and Wildlife. Among his many other activities, Leopold served as president of the Wildlife Society (1957–1958), received a presidential appointment to the Marine Mammal Commission (1972–1975), was a member of the Department of the Interior's Advisory Board on National Parks, served as vice president and a member of the board of directors of the Sierra Club, was a member of the board of governors for the Nature Conservancy, was involved with the Wilderness Society, and was associated with several scientific institutions. Several honors were bestowed on Leopold throughout his illustrious career, including his election to the National Academy of Sciences and the National Academy of Arts and Sciences.

Leopold has been described as "a big, friendly man." As Meine writes, he had a "ready sense of humor," was "easygoing and cheerful," and inherited his "mother's handsome Spanish features" and his "father's sharpness of mind and love of the outdoors." Leopold's gift for diplomacy without compromising his principles is often cited. As a friend remarked, quoted in the *Annual Obituary* of 1983, "He could tell that arguments between conservationists like him, and the preservationists who want to stop hunting, are stupid. . . . He was able to get the two sides talking to each other." His writings carry a unique understanding of the differing perspectives toward the wilds—those of hunters, government wildlife managers, conservationists, and scholars—but, writes one observer of Leopold's work, he always maintained a focus on the means for the "optimum survival of whatever species he chose to study, and with achieving a balance of resources and uses that will stabilize and preserve the natural state." After retiring, Leopold continued to research and write. He died in Berkeley of heart failure at the age of sixty-nine.

BRYAN L. MOORE

Gilliam, Ann, ed. *Voices for the Earth: A Treasury of* The Sierra Club Bulletin *1893–1977*. San Francisco, CA, 1979.
"Leopold, A. Starker." *Annual Obituary 1983*, ed. Elizabeth Devine. Chicago, IL, 1984.
Meine, Curt. *Aldo Leopold: His Life and Work*. Madison, WI, 1988.

Lindbergh, Anne Morrow (June 22, 1906–February 7, 2001). A writer and poet, Lindbergh was born in Englewood, New Jersey, the second of four children. Dwight Morrow, her father, was a lawyer, served as President Calvin Coolidge's ambassador to Mexico from 1927 to 1930, and was elected as a Republican to the U.S. Senate from New Jersey in 1930, having campaigned to repeal the Eighteenth Amendment. The Morrow children enjoyed a privileged upbringing that afforded them travel, good educations, and associations with people at the highest levels of society. While ambassador to Mexico, Morrow invited his friend Charles Lindbergh to fly to Mexico City. In December 1927, Lindbergh flew nonstop from Washington, DC, to Mexico City, where he met Anne, who was visiting her father.

Anne's marriage to Charles Lindbergh in 1929, a year after her graduation from Smith College, allowed her to see even more of the world. With her husband, she flew in single-engine planes across North America and over the South Atlantic. In *North to the Orient* (1935), *Listen! the Wind* (1938), and *The Wave of the Future* (1940), she wrote about her travels, determined to understand fully the sights she and her husband witnessed. Nature images, especially wind and water, are important in these books. The wind is often represented as an antagonistic force that threatens to prevent the successful completion of a journey. Water, however, is associated with changes and continuities in the flow of life. Years later, Lindbergh wrote that these journeys gave her a sense of the earth as a planet and, over time, a concern for the planet and the life on it.

Years before she became an active conservationist, Lindbergh was thinking and writing about the negative changes she saw happening to the planet. As early as the summer of 1941, Lindbergh warned her readers in "Reaffirmation," an essay published in the *Atlantic Monthly*, that technological developments were having detrimental effects on humankind: "The revolution that will have to take place over the world before it can again begin its march forward seems to me not alone the conquest of the machine by man, but much more deeply the conquest of spirit over matter." Here Lindbergh urges her readers to consider the importance of regaining control over the machine. She cautions against accepting, without questioning, the ever-increasing effects of technology, and her view eventually grew into a concern for preserving the natural life of the planet.

Lindbergh believed that matter's control of spirit was a primary cause of humans' alienation from their environment, and for her, this "alienation" was most acute for women. In *Gift from the Sea* (1955), she discusses a woman's feeling of aloneness in the modern world and ways to ameliorate this feeling. Her purpose is to explore the ever-encroaching problem of trying to maintain an inner serenity in the midst of life's distractions, of remaining balanced, no matter what forces tend to pull one off center. She speaks not only for herself but for the plight of all modern women who struggle to balance the duties of wife and mother: "I want to give and take from my children and husband, to share with friends and community, to carry out my obligation to man and to the world as a woman, as an artist, as a citizen. But I want first of all, in fact as an end to all these other desires, to be at peace with myself." The background for this quest is a two weeks' vacation at the seashore, on a small island. Alone on the beach, she finds and gathers shells that symbolize to her the various aspects of her life, on which she then reflects: the conflicts, the pressures, the incredible multiplicity. The lesson she learns from nature is the need to be alone. Every woman needs to be alone sometime during the year and some part of each week and each day. Moreover, the ocean teaches her that personal change is a possible and necessary process.

In 1956 Lindbergh published *The Unicorn and Other Poems*. Many of these poems focus on the liberating effects of contact with nature. In the title poem, the unicorn, a symbol of creativity, is fettered by the demands of daily life, just as the speaker is in *Gift from the Sea*. Freedom is found in nature. The six poems that follow, grouped in a section called "Open Sky," look to the sky as an inspiration for the release of creative activity. In "Flight of Birds," she writes that the effect of "Watching the patterns of . . . birds in flight" is that we see ourselves "Transmuted into bird or cloud or tree,/ Familiar fragments, here arranged in form." And in "Back to the Islands," she writes of the place of freedom, where she can feel her "pattern . . . change, / Living among the islands; all is strange—A bay, a cove, a sudden turn of tide, / An unexpected channel never tried / Before. Time and the mainland stand aside / Once more—Once more, life shatters open wide!"

Thus, in the 1950s the natural world was Lindbergh's teacher. By the 1960s, both Charles and Anne became active in the conservation movement. In fact, it was Lindbergh, concerned for her husband's depression, who encouraged him to become involved in the movement, in which he remained active until his death in 1974. For her part, Lindbergh, increasingly aware that her teacher was sick, published three essays and one book between 1966 and 1970 to address the need for a cure for earth's environmental ills. In *Earth Shine* (1969), Lindbergh explores the disturbing effects of humans' activity on the natural environment. The book combines two essays she originally published in *Life*, "Immersion in Life; a Brief Safari Back to Innocence" and "The Heron and the Astronaut." Originally published in 1966, "Immersion in Life" is based on the Lindbergh family's visit to Kenya and Tanzania and examines the relationship between animals and people. Her conclusion is:

> Animals are necessary to man, although man, insulated by his civilization, is often dulled to the need. . . . If the connection one senses with the animals is an authentic bond, perhaps the renewal one feels in their presence has a deeper significance than we realize. Immersion in wilderness life, like immersion in the sea, may return civilized man to a basic element from which he sprang and with which he has now lost contact.

"The Heron and the Astronaut" is based on the experience of watching the launch of the Apollo 8 moon-orbiting mission. The title indicates the central contrast, the natural world of the heron and the technological world of the astronaut. Both creatures fly, but one does so naturally, while the other is propelled by technology. Lindbergh explains the link between the two essays in the preface to *Earth Shine*: "Is there a bond between these two essays that they should be linked together in a book? Africa, a last stand of primitive life, and Cape Kennedy, a summit of a scientific civilization—what have they in common? At first, nothing but extremes. But today, even extremes are interrelated. . . . The sense of the earth as a whole, as a planet, is with us inescapably." The overall message of *Earth Shine* is one of hope. While she is quick to point out that our world does have problems, she believes that the new studies of environment and the new exploration of space can combine to give us "a terrestrial view" of ourselves. As a result of this view, Lindbergh believes that humans will strive to maintain their links with the environment as well as go forward technologically.

In February 1970, Lindbergh returned to Smith College to deliver a rare public address—"Harmony with the Life around Us"—at a conference on environmental pollution. While this is Lindbergh's last major statement on environmental issues, the essay crystallizes a major theme she explores in *Gift from the Sea*, *The Unicorn* and *Earth Shine*. This theme is that the individual is "alienated" from nature; this "artificial split between man and nature," caused in part by advances in technology, is responsible for "many of the physical, social and economic troubles of our time." She suggests that the violent reaction against the war in Vietnam resulted from a "deep instinctive protest against the growing dehumanization of our world—against industrialized, mechanized civilization." But Lindbergh did not blame all of humankind's problems on technology: "Science can only inform us about our world. It can bring us knowledge but it cannot set up values. Values are created by individuals." She goes on to express her hope in the ability of the new sciences in the fields of ecology and environment to "heal the breach" by creating an awareness of the "interrelatedness of man and life on earth."

Lindbergh took great interest in environmental and conservation issues. Her major contribution was her "feminist essay" *Gift from the Sea*, published eight years before Betty Friedan's *Feminine Mystique*. The meditation on seashells conveyed a revolutionary message about women's role and need for emotional independence. Ironically, her real satisfaction seems to have come from the knowledge that she helped her husband latch onto a cause that absorbed his attention and that all five of her children grew up to be active environmentalists.

DAVID BOOCKER

Herrmann, Dorothy. *Anne Morrow Lindbergh: A Gift for Life*. New York, 1993.
McBride, Mary Margaret. *The Story of Dwight W. Morrow*. New York, 1930.
Milton, Joyce. *Loss of Eden: A Biography of Charles and Anne Morrow Lindbergh*. New York, 1993.
Vaughan, David Kirk. *Anne Morrow Lindbergh*. Boston, MA, 1988.
Whitman, Alden. "Mrs. Lindbergh Sees a Hope for Man." *New York Times*, October 12, 1969.

Lindbergh, Charles Augustus (February 4, 1902–August 26, 1974). A pioneer aviator, combat pilot, Pulitzer Prize winner, and conservationist, Lindbergh was born in Detroit, Michigan, the only child of Congressman Charles August "C. A." Lindbergh and Evangeline Lodge (Land) Lindbergh. The son of a Swedish immigrant frontier-homesteader, C. A. held strong convictions as a father and as a politician, serving Minnesota in Congress from 1906 to 1917. He supported early legislation to conserve natural resources, addressing the House of Representatives on March 1, 1909, to explain that "no one can be more enthusiastic for the preservation of our natural resources than I." C. A. taught his son independence of mind, courage in adversity, and respect for the land.

"Charley" (as Lindbergh was called by his father) loved both science and the outdoors. Indeed, late in his life Lindbergh reflected that the two most important influences on his life were his maternal grandfather's lessons about Charles Darwin and his father's stories about Minnesota frontier days. A dentist in Detroit, his Grandfather Land taught him about the dawning of a new scientific age. His shelves and tables were stocked with magazines and books—*Scientific American*, *National Geographic*, a Webster's dictionary, a volume on pathology—all filled with pictures. At his grandfather's dinner table, he listened to discussions about the inventions of Thomas Edison and Guglielmo Marconi and Darwin's theories and learned of the potentiality in flight from the story of the Wright brothers. C. A. told his son stories about his pioneer days in Minnesota—stories of fishing in lakes and rivers, of hunting in forests, of break plowing virgin land. Lindbergh learned from his father that by the time he was born in 1902, the frontier was gone. Indians were placed on reservations, oxen were replaced by horses, forests were disappearing, and railroads connected an increasing number of towns in Minnesota. Still, Lindbergh loved nature, and when C. A. was not in Washington, father and son's favorite activities together were hunting, fishing, and camping.

While his father was in Washington, Lindbergh was in charge of their 120 acres. Only fifteen, he was fascinated with the idea of farming because it enabled him to combine his love of earth and animals with his interest in machinery. Moreover, he believed that farming allowed him to make a significant contribution to the war effort. Shortly after World War I ended, Lindbergh enrolled as a student in the College of Engineering at the University of Wisconsin at Madison. But faced with certain expulsion because of poor grades, he left the university and entered a civil flying school, where he learned to wing-walk, to parachute jump, and eventually to pilot an airplane. The life of an aviator seemed ideal to him because it involved skill and commanded adventure. He was actually glad to have failed his college courses because he perceived that mechanical engineers were fettered to factories and drafting boards, while pilots had the freedom to explore mountain peaks and clouds. It was while making solitary flights from Peoria to Chicago as an airmail pilot that Lindbergh dreamed of establishing aviation as a common means of transport and came up with the idea of competing for the Orteig Prize, $25,000 for the first nonstop flight between New York and Paris. With his historic transatlantic flight in 1927, he became the major symbol of scientific progress.

It was at the outset of his goodwill tour of Latin America in December 1927 that Lindbergh met Anne Morrow, daughter of the ambassador to Mexico, Dwight Morrow. They married in May 1929, and together they flew in single-engine planes across North America and over the South Atlantic. Already, they began to witness the negative changes to the planet that caused both to contemplate the detrimental effects of technological developments on life and that led them, by the 1960s, into becoming active conservationists. Humans needed to regain control over machines, both Charles and Anne agreed, and they became zealous about the preservation of life on planet earth.

Lindbergh, drawing on the boyhood lessons he learned from his grandfather and father, worked his whole life to maintain a balance between his love of science and nature. In *Of Flight and Life* (1948), Lindbergh discussed his life as a combat pilot in the Pacific and made a personal confession of faith in which he describes his own personal conversion from a worshipper of science to a worshipper of the "eternal truths of God." From this point in his life on, Lindbergh was truly concerned about humankind's future, fearing that because of the "idol worship" of scientific materialism and the atomic threat from Russia, the survival of civilization as humans had come to know it was threatened. Still, there was hope for salvation, which lay in giving science a moral direction, in controlling "the arm of western science by the mind of a western philosophy guided by the eternal truths of God." In 1949, Lindbergh received the Wright Brothers Memorial Trophy for "significant public service of enduring value to aviation and the United States." He proudly accepted the award and in his acceptance speech stressed the need for balancing the achievements of scientific progress with the quality of life, stressing that to survive each individual must place his or her character "above the value of his products. If we are to be finally successful, we must measure scientific accomplishments by their effect on man himself."

Lindbergh's activities and writings in the 1950s reflect a desire to maintain a balanced view of life. He supported Douglas MacArthur in the Korean conflict. He helped Wernher von Braun and his colleagues develop American rocketry. In 1954 he wrote an essay for the *Saturday Evening Post* in which he argued that America's survival depended on a buildup of offensive strategic weapons to ensure peace. These engines of destruction were a necessary evil, but he saw manned spaceflight and progress in commercial aviation as ways of avoiding conflict through international travel and cooperation. In 1953, when he published *The Spirit of St. Louis*, he further revealed a contemplative side in which he described his relationship to nature and the world beyond.

In the 1960s, both Charles and Anne became active in the conservation movement. Lindbergh was so disillusioned by America's glorification of scientific materialism that he found it difficult to balance his love for science and nature because the two were in conflict and increasingly difficult to reconcile. Although he remained interested in technological issues—writing forewords to Milton Lomask's tribute to rocketry pioneer Robert Goddard and to a history of the American satellite program, and supporting the Apollo program—the sixty-year-old Lindbergh sought a natural life apart from civilization and a cause to give his life meaning. In 1964, Lindbergh read in the *New York Times* about the formation of an American chapter of the Geneva-based World Wildlife Fund (WWF) and phoned chairman Ira Gabrielson to offer his services, shifting his major attention away from science and technology to become an active conservationist. According to some WWF activists, Lindbergh's involvement in the organization was owing to his wife's encouragement to get involved in the saving of wild animals as therapy for his depression.

Lindbergh's first public statement for his new cause was an essay that appeared in the July 1964 *Reader's Digest*. Titled "Is Civilization Progress?" the essay explores the effects of technological advances on life on earth. Lindbergh juxtaposed two experiences, his participation in a supersonic transport convention in New York with a trip to Kenya to look at land being considered for a national park. He acknowledges the "sense of magic" he experienced during his transatlantic flight, as well as the role he played in twentieth-century progress. But the trip to Kenya enabled him to understand firsthand how Africa's wild animals are endangered by the advance of civilization. Most significant, he revealed social Darwinist beliefs that he must have learned as a boy at his grandfather's knee: "Lying under an acacia tree . . . I realized more clearly facts that man should never overlook: that the construction of an airplane, for instance, is simple when compared to the evolutionary achievement of a bird; that airplanes depend upon an advanced civilization; and that if I had to choose, I would rather have birds than airplanes." At this point in the essay (and his life), Lindbergh began to question the definition he had assigned to the term *progress*, even going so far as to question the value to humanity of his flight to Paris, and he concluded that there is more progress in the history of the giraffe than in any of humankind's creations.

"Is Civilization Progress?" is accompanied by a personal appeal from Lindbergh to contribute to the World Wildlife Fund, and Lindbergh was active in this organization and others (the International Union for the Conservation of Nature, the Nature Conservancy, the President's Citizens' Committee on Environmental Quality) until his death. In 1968 Lindbergh was awarded, along with Senator Henry M. Jackson, the Bernard M. Baruch Conservation Prize. In July 1969, after witnessing the moon launch at Cape Kennedy, he participated in a WWF-sponsored dedication of a 900-acre bird preserve. He wrote several more essays for *Reader's Digest* and *Life* in which he appealed again and again for humans to live not by science alone, to be sensitive to what he called the "wisdom of wildness," a life of greater self-awareness found only in close contact with the natural world. He worked to save several endangered species: blue and humpback whales off the coast of Peru and the deerlike tamarau and monkey-eating eagle of the Philippines.

Lindbergh's work in the Philippines brought him into contact with another organization, PANAMIN, or the Private Association for National Minorities, which was dedicated to saving the primitive Tasadays from the spread of civilization. The Tasadays were a Stone Age tribe who lived on Mindanao Island. They had no history, no religion, and no contact with the outside world until the 1960s. Lindbergh was particularly amazed because they had no words for "war," "enemy," "murder," or "moral badness." He believed he had a moral obligation to protect the tribe from exploitation from the outside world. He worked with Manuel Elizalde Jr. to provide a proclamation from President Ferdinand Marcos to preserve more than 46,000 acres of Tasaday country.

In his *Autobiography of Values*, written in the early 1970s and published posthumously in 1977, Lindbergh reflected that he got involved in environmental causes because he recognized that "a catastrophic war was probably not the greatest danger confronting modern man. Civil technology vied with military in breaking down human heredity and the natural environment. Every day, increasing numbers of bulldozers and trucks tore into mountains, slashed through forests, leaving greater scars on the earth's surface than those created by bombs." He wrote about his many travels around the world. His unique perspective, flying above the earth, allowed him to see the changes that had taken place on the earth's surface: the disappearance of trees, changes in rivers due to erosion, and the spread of cities. *Autobiography of Values* represents Lindbergh's mature reflections about his life—about the self-reliant values he learned as a boy in Minnesota and how they carried him through his life in a world dominated by science and technology. Indeed, Lindbergh's apparent obsession with social Darwinism is visible throughout the book and has troubled some readers.

Lindbergh took great interest in environmental and conservation issues. While Lindbergh is certainly best remembered for his famous flight and as the victim of a tragic kidnapping, he reflected late in life that if he were entering adulthood in the late 1960s, he "would choose a career that kept [him] in contact with nature more than science." At his death, Lindbergh was hailed as a champion of conservationism. Indeed, his contributions to conservation in the 1960s were significant. His essays, which appeared in popular magazines such as *Reader's Digest* and *Life*, introduced millions of people to the conservation cause and were an integral part of a rapidly developing environmental movement in the 1960s. More important was Lindbergh's message warning his fellow human beings that they were destroying the environment in the name of technological progress and his appeal to lead a life less complicated by technology.

DAVID BOOCKER

Berg, A. Scott. *Lindbergh.* New York, 1998.

"Conservation Prize Going to Lindbergh and Senator." *New York Times,* January 13, 1969.

Hunter, T. Willard. *The Spirit of Charles Lindbergh.* Lanham, MD, 1993.

Luckett, Perry D. *Charles A. Lindbergh: A Bio-Bibliography.* New York, 1986.

Milton, Joyce. *Loss of Eden: A Biography of Charles and Anne Morrow Lindbergh.* New York, 1993.

Lopez, Barry Holstun (January 6, 1945–). Highly regarded as a nature writer, author of a dozen books of fiction and nonfiction as well as numerous essays, Lopez is the recipient of many prestigious awards and honors.

The elder of two children, Lopez was born to John Edward Brennan and Mary Frances Holstun Brennan in Port Chester, New York. (His brother Dennis was born in 1948, the year his family moved to California's San Fernando Valley, which was then largely undeveloped.) After his parents divorced in 1950, Lopez never again saw his father. For the next six years Lopez stayed in southern California with his brother and mother, who worked to support the family and encouraged Lopez's love of nature, taking him and his brother on trips to nearby deserts, lakes, mountains, and coastal towns. Lopez recounts that as a boy he used *Hammond's Illustrated Library World Atlas* to plan journeys that he hoped someday to take both in the United States and abroad. When Lopez was eleven, the family moved to New York City when his mother married Adrian Bernard Lopez, who legally adopted the boys and provided them with a quality education. Lopez attended Loyola, a Jesuit prep school in Manhattan, where he played varsity sports and became the president of his senior class. After graduating from high school, Lopez and a group of friends spent the summer traveling around Europe. He studied at New York University and the University of Notre Dame, where he played varsity soccer. Lopez traveled throughout the United States, having visited all but three states by the time he graduated from college. As a student he began keeping a journal, thus laying the framework for his future as a writer who is largely inspired by his travels.

In "A Literature of Place" Lopez wrote of the influence his childhood in rural California had on his development as a writer:

> My imagination was shaped by the exotic nature of water in a dry southern California valley; by the sound of wind in the crowns of eucalyptus trees; by the tactile sensation of sheened earth, turned in furrows by a gang plow; by banks of saffron, mahogany and scarlet clouds piled over a field of alfalfa at dusk; by encountering the musk from orange blossoms at the edge of an orchard; by the aftermath of a Pacific storm crashing a hot, flat beach.

As a boy, Lopez found both animals and language to be profoundly magical. The value and importance the writer places on physical observation is underscored in his comment that "animals regularly respond to what we, even at our most attentive, cannot discern." Lopez recounts that his earliest encounter with language "seemed as glorious and mysterious as a swift flock of tumbler pigeons exploiting the invisible wind."

Lopez received his B.A. and his M.A. from Notre Dame and moved to Oregon in 1967, where he enrolled in the graduate program at the University of Oregon. Since 1968 he has lived on the McKenzie River in rural western Oregon. Lopez, who intended to become an aeronautical engineer, considered becoming a Trappist monk, a sect known for their simplicity and hard work. Once a landscape and nature photographer, Lopez has been a full-time writer since 1970.

Lopez is perhaps best known as the author of *Arctic Dreams*, for which he was awarded the National Book Award for nonfiction in 1986. His most influential work, *Arctic Dreams* focuses both on the scientific and the cultural and portrays the exploitation of the Arctic as analogous to that of the entire North American continent. Another nonfiction work, *Of Wolves and Men* (1978), which studies the wolf's place in natural history, literature, folklore, and human culture, was a National Book Award finalist and won the John Burroughs Medal as the year's best book of natural history. Lopez has received numerous other awards and prizes, including the Literature Award from the American Academy of Arts and Letters; a PEN (Poets, Playwrights, Editors, Essayists, and Novelists) Syndicated Fiction Award; Pushcart Prizes for both fiction and nonfiction; the Distinguished Recognition Award in fiction from Friends of American Writers; the H. L. Davis Award for short fiction; a Lannan Foundation Award; as well as Guggenheim and National Science Foundation fellowships. Lopez's audio edition of *About This Life* was one of three finalists for the 1998 Audi Award for Best Abridged Nonfiction recording, and in 2002 he received the Orion Society's John Hay Award for significant contributions to the literature of nature.

A loosely connected trilogy of short stories written over a period of nearly twenty years, *Desert Notes* (1976), *River Notes* (1979), and *Field Notes* (1994) intertwine landscape and culture. *Desert Notes*, Lopez's first published book, is a collection of twelve short fictional pieces that share as a subject the desert. In all of Lopez's works, he focuses on the reader, gives voice to the landscape, and creates an atmosphere in which wisdom reveals itself.

> I know you are tired. I am tired too. Will you walk along the edge of the desert with me? I would like to show you what lies before us. All my life I have wanted to trick blood from a rock. I have dreamed about raising the devil and cutting him in half. I have thought too about never being afraid of anything at all. This is where you come to do these things. (*Desert Notes*)

More recent books include *Light Action in the Caribbean* (2000), a collection of short stories; *Resistance* (2004), nine interrelated stories that are framed as letters and constitute a criticism of the government's post-9/11 surveillance and repression; and *Home Ground: Language for an American Landscape* (2006), a dictionary of regional landscape terms that he coedited. Lopez was recently named a contributing writer to *National Geographic* magazine, which in December 2007 published "Coldscapes." Accompanied by Bernhard Edmaier's photographs, the essay addresses the problem of melting permafrost in the Arctic.

While doing graduate work at the University of Oregon, Lopez studied folklore, myth, and legends, and his research into Native American stories involving the coyote led him to write *Giving Birth to Thunder, Sleeping with His Daughter: Coyote Builds North America* (1977). In this collection of American Indian myths in which Lopez retells tales of the coyote as trickster and sage, he explores Native American tradition, culture, ethics, and folklore. Written for a youth audience, *Crow and Weasel* (1990) is a retelling of another Native American folktale.

Lopez, whose work reflects tremendous geographical scope, is drawn to cultures that recognize the divine in both humanity and nature. A self-described "writer who travels," Lopez often sojourns to remote places in Africa, Asia, Australia, the Arctic, and the Antarctic, where he relies on the insights of indigenous peoples, who, he finds, are keenly aware of and pay careful attention to nuance in the physical world; possess a deep combined tribal and personal history in a place; and live in ethical unity with the land—all of which Lopez views as a defense against loneliness. Lopez sees a loss of intimacy with the land as a loss of intimacy with the world, which contributes to loneliness—something he sees as a hallmark of modern culture. In "Remembering Orchards" he writes:

> He stood gazing at the stars. A woman lay at his side by his feet, turned away, perhaps asleep. The trees in that moment seemed not to exist, to be a field of indifferent posts. As the crows strode diagonally through orchard rows I thought of a single broken branch hanging down, and of Alejandro's ineffable solitude, and I saw the trees like all life—incandescent, pervasive.

Lopez is a graceful writer whose meticulous prose is poetic: "There will be a clarity in his description such that it will seem he is laying slivers of clear glass on black velvet in the afternoon sun" ("Directions"). He is powerfully persuasive: "I could make you believe you heard sandpipers walking in the darkness at the edge of a spent wave, or a sound that would make you cry at the thought of what had slipped through your fingers" (introduction to *River Notes*). His linguistic precision is hauntingly descriptive: "He came in quiet as air sitting in a canyon" ("The Falls"), and his work is replete with memorable imagery: "We camped up in those aspen and that was good water. It was sweet like a woman's lips when you are in love and hold back" ("The Falls").

Much of Lopez's work is about relationships: the relationship between human culture and nature; between landscape and culture; between landscape and literature. Commenting on the importance of fidelity to place, the author explains that the incorporation of landscape—and by that he meant not just line and color and contour and texture, but weather and the movement of landscape through time, in other words the flow of rivers and all that kind of material, everything occurring in the so-called non-human world—was not incidental to literature. To him it was integral.

Lopez, who was profoundly influenced by the Roman Catholic tradition in which he was raised, observes the spiritual nature both of animals' lives and of the landscape itself. One notes this spirituality in the reverence with which he details nature. In "The Language of Animals" Lopez wrote:

> I grew up in a forming valley in Southern California in the 1950's, around sheep, dogs, horses, and chickens. The first wild animals I encountered—coyotes, rattlesnakes, mountain lion, deer and bear—I came upon in the surrounding mountains and deserts. These creatures seemed more vital than domestic animals. They seemed to tremble in the aura of their own light. (I caught a shadow of that magic occasionally in a certain dog, a particular horse, like a residue.) From such a distance it's impossible to recall precisely what riveted my imagination in these encounters, though I might guess. Wild animals are lean. They have no burden of possessions, no need for extra clothing, eating utensils, elaborate dwellings. They are so much more integrated into the landscape than human beings are, swooping in contours and bolting down its pathways with bewildering speed. They travel unerringly through the dark. Holding their gaze, I saw the intensity and clarity I associated with the presence of a soul.

We see this same sanctity for nonhuman life in "Apologia," in which a long-distance driver delays his journey by stopping along rural roads to pull off the carcasses of animals that have been struck. For the author, who stresses the importance of a moral and compassionate existence, this is an act of respect, an awareness of our connection to the natural world. For Lopez, "Your work is your prayer."

Lopez expresses his concern over global warming; the loss of biodiversity; the desire for material wealth; technological progress; and the social damage of industrialization, colonialism, and materialism. Critical of business, government, the pressures of capitalism in America, and the resultant effects of consumerism, waste, and greed, Lopez believes that we have a social justice responsibility to both human society and the nonhuman world. He stresses the importance of leading a worthy life, a just life, a life led with moral intention, devoted to spiritual rather than material wealth.

As an act of resistance to the nation's current political climate, Lopez has become actively involved in higher education. In 2000 he collaborated with Harvard biologist E. O. Wilson to design for Texas Tech University's Honors College a curriculum that combines the sciences and the humanities in a new undergraduate major, the B.A. in natural history and the humanities. In 2001 Texas Tech acquired Lopez's manuscripts to be archived in the newly established James E. Sowell Family Collection in Literature, Community, and the Natural World, to which the papers of other writers have been added. In 2003 Lopez was appointed Texas Tech's first Visiting Distinguished Scholar. In 2004 he helped establish the endowed Formby Lectures in Social Justice and is currently working with the university's administration on curriculum development in ethics, Southwest cultural geography, and the conservation of water. Additionally, Lopez, who has demonstrated a tolerance and respect for cultural differences through an appreciation of each culture's unique stories and customs, has served as a liaison between Texas Tech University and the Comanche nation.

Deeply concerned about the fate of society, Lopez uses natural history as a point of departure for discussing broad-ranging questions and the larger issues to which he is drawn. In response to the clear-cut Cascade foothills:

> "It's difficult to see this because it fills me with grief," he said, indicating the cutover valley that provides habitat for beaver, elk, deer, mountain lion, lynx, skunk, bobcat, salmon, as well as human beings." I have a constant sense of grief about the relationships destroyed by all this. Ecology is not just about endangered species. It's about community. It's about relationships. You can't have a consumer-based culture unless you have an immoral relationship with non-human species."

Several years ago, Lopez purchased thirty-two acres of old-growth forest next to his home. He placed the land in a trust so that it could never be logged or developed. Lopez, who believes we must have a moral rather than an exploitive relationship with the land, takes frequent walks through this forest, whose stewardship he embraces.

A voracious reader, the author recognizes that he has been influenced by generations of writers, though, having read it three times before attending college, the linchpin for Lopez was Herman Melville's *Moby-Dick* (1851), where the seascape is indispensable to the story.

By combining attentive fieldwork with meticulous research, Lopez gains an intimacy with the environment. His intricate use of language, which invokes all the reader's senses, awakens one's appreciation of nature and landscape, as issues of moral significance merge lyrically with his observations on the natural world.

As broad ranging as his travels are the themes of Lopez's work, which include integrity; dignity; intimacy; responsibility; ethics; identity; spirituality; nature; culture; indigenous peoples; myths; science; history; geography; sociology; exploration; social justice; the spiritual nature of the landscape; the co-relation between the lives of plants, animals, places, and humans; our moral responsibility to place; and the relationship between natural history and cultural history.

Lopez's respect for language is apparent in the exquisite majesty of his prose, which is linguistically and stylistically perfect. Author of fiction, nonfiction, essays, short stories, travelogues, memoirs, nature writing, and spirituality writing, Lopez contributes regularly to a number of literary magazines, and his work is included in dozens of anthologies. The nation's premier writer about nature, landscape, and wilderness, Lopez examines community from various perspectives, including the relationship of the individual to both the state and society, as well as the connection between human culture and place. For Lopez, nature writing "assumes that the fate of humanity and nature are inseparable"; nature writing, which he believes may be our nation's greatest contribution to world literature, should continue to undermine complacency and induce hope. Lopez, who writes brilliantly and elegantly, delights the reader with his portrayal of the "magic and transcendence within landscapes, within story, within language itself." His love of nature intersects with his love of language and his love of God.

Lopez explains, "I think if someone were to read what I have written through the years, it would be clear that the issue I'm interested in is human dignity, and the nature of tolerance."

RONNA S. FEIT

"An Interview with Barry Lopez." *Baybury Review* (Ephraim, WI) 2 (1998).

Lang, Anson. "A Conversation with Barry Lopez." *Bold Type Magazine* 4.7 (November 2000).

Lopez, Barry. *About This Life*. New York, 1998.

Margolis, Kenneth. "Paying Attention: An Interview with Barry Lopez." Introduction by Stephen Trimble. *Orion* 9 (Summer 1990).

Martin, Christian. "On *Resistance*: An Interview with Barry Lopez." *Georgia Review* 60 (Spring 2006).

MacKaye, Benton (March 6, 1879–December 11, 1975). A forester, a regional planner, the father of the Appalachian Trail, a founder of the Wilderness Society, and an ecological humanist, MacKaye was born in Stamford, Connecticut, sixth of the seven children of playwright-producer-actor Steele MacKaye. His was an ardent, active, distinguished family. His grandfather, James Morrison McKay [*sic*], was a wealthy businessman; a founder of Wells Fargo Express and of one of the companies that became Western Union; a link in the Underground Railroad, moving escaped slaves to safety in Canada; and a member of the commission whose report to President Abraham Lincoln resulted in the Emancipation Proclamation. His father Steele was a theatrical innovator whose inventions included better lighting, better ventilation, and the folding theater seat. His brother James was an industrial chemist and democratic socialist philosopher who later taught philosophy at Dartmouth. His brother Percy was a noted playwright whose folk play *The Scarecrow* is still performed.

MacKaye's early childhood and schooling were in and around New York City, but in the summer of 1888 the family moved to what was to become its long-term base in the hill village of Shirley Center, Massachusetts. There young MacKaye found a well-integrated, indigenous community and woods, streams, and hills for a boy to explore. This "exploration" was focused and systematized when, during the summer and spring of 1890–1891, the family lived in Washington, DC. MacKaye haunted the Smithsonian Institution, becoming such a familiar that he was taken into the laboratories by the scientists working there. He also met and heard lecture the redoubtable John Wesley Powell and a young naval lieutenant named Robert Peary, who was preparing for Arctic exploration. He began to read Alexander Humboldt, who became a model for his own approach to the landscape.

On October 1, 1896, MacKaye enrolled at Harvard College. During those years at Harvard MacKaye was exposed to such faculty as Charles Eliot Norton, William James (a longtime friend of his father), George Santayana, Josiah Royce, Hugo Munsterberg, George Lyman Kittredge, Barrett Wendell, Fred N. Robinson, George Pierce Baker, Irving Babbitt, Albert Bushnell Hart, and Nathaniel Southgate Shaler, but the influence he most clearly remembered was that of geologist/physiographer William Morris Davis. Davis said his subject was "the earth as a habitable globe." That became a touchstone for MacKaye throughout his long career.

When MacKaye graduated in June 1900, he still did not have a career in mind, but a few weeks earlier his brother James gave him, on his twenty-first birthday, a copy of Thomas Henry Huxley's book *Physiography* (1877) in which Huxley presents his subject in terms of regions. The influences were accumulating. The winters of 1900–1903 MacKaye spent at various jobs in New York City, but the summers were used for mountain expeditions in Vermont and New Hampshire. In October 1903 he enrolled in the just-opened graduate program in forestry at Harvard.

MacKaye received his A.M. in forestry in June 1905, less than four months after the establishment of the U.S. Forest Service on March 3. He became a newly fledged professional, scientific forester just in time to join Gifford Pinchot's group of idealistic, energetic young men bent on conserving America's remaining forest lands. For the next few years, he divided his time between teaching forestry at Harvard and working in New England for the Forest Service. He helped survey and organize the Harvard Forest and made a pioneering forest survey that helped lead to the establishment of the White Mountain National Forest. It was the first national forest created on land purchased from private owners, an important precedent for eastern forests.

Still in Washington, in 1913 MacKaye helped form an informal group of social activists called the Hell Raisers. These government workers, journalists, congressional staffers, and political activists aimed at raising public awareness about various social and political issues. If they could create some newsworthy event—get someone in Congress to introduce a bill, even one impossible to pass, or to create a demonstration or protest—the journalists could report it, and the public would learn of the issue. Public awareness might make it possible to eventually create the needed reform. Issues ranged from corporal punishment in Maryland to exploitation of Alaskan public lands to women's suffrage. Besides MacKaye the group included William Leavitt Stoddard (Washington correspondent for the Boston *Transcript*), Judson King, Harry Slattery, Art Young, Stuart Chase, and a number of others. Lincoln Steffens was a frequent visitor to the meetings.

In 1914 professional concern for forest conservation joined with social issues in MacKaye's government assignment when he was sent to the Lake States to develop plans for use of cutover lands in northern Minnesota,

Wisconsin, and Michigan. The resulting plan proposed agricultural colonization in organized communities assisted by the federal government. The premise was that each settler facing the task of development alone, like Robinson Crusoe on a desert island, was less likely to succeed than groups working together. Each settler would have an individual farm, but the work that could be done more efficiently with big machinery—removing stumps and building roads and other community infrastructure—should be done collectively, with government assistance. From the Lake States in 1916 MacKaye went on to the Puget Sound region, where he proposed the establishment of stable logging communities based on sustained yield. In 1916 he also produced his first published article, in the *Journal of the New York State Forestry Association*, proposing recreation as a forest value, along with lumber and grazing, a forerunner of the later "multiple-use" doctrine of the Forest Service.

In 1915, MacKaye married Jessie Harding Stubbs, a prominent activist in the suffrage and world peace movements. She suffered an apparent mental breakdown in 1921 and committed suicide. There were no children from the marriage.

MacKaye had left government service after the Armistice, when, as he said in later years, "Washington came down like a circus tent." While visiting Charles Harris Whitaker's home in New Jersey, MacKaye broached his long-held idea for a long-distance hiking trail. Whitaker, editor of the *Journal of the American Institute of Architects*, promised to publish the idea if MacKaye would write it up. The result was "An Appalachian Trail, a Project in Regional Planning" in the October 1921 issue of the journal. The original proposal was for a footpath from Mount Washington in New Hampshire to Mount Mitchell in Georgia. More than just a hiking way, it was to create a belt of wilderness to stem the "iron flow" of megalopolis from the seaboard into the interior. Along this belt were to be recreation, logging, and farming camps where city dwellers could escape the urban at least for brief periods. The length of the trail was to provide an "infinite footpath" to which city dwellers could return year after year and find new experiences. Development of the trail was to be an excercise in community effort, with each segment the responsibility of a local trail club.

Much of MacKaye's time and energy over the next decade or more was devoted to organizing and promoting the development of the trail. With assistance from city planner Clarence Stein (brother of Gertrude Stein), landscape architect Allen Chamberlain, Dan Beard of the Boy Scouts, naturalist and writer Raymond H. Torrey, and many others, MacKaye spread the idea to various hiking and trail clubs and conferences up and down the Appalachians. In 1925 these groups organized a federation called the Appalachian Trail Conference. In the 1930s the northern terminus was moved up to Mount Katahdin in Maine, and the southern end to Mount Oglethorpe in Georgia. The final section was completed on August 15, 1937, and the full length opened to hikers. Hurricanes, road builders, and other disasters have closed sections of the trail from time to time, but the idea has been resilient, and the Appalachian Trail now has federal protection for its right of way.

Also during the 1920s MacKaye helped to found the Regional Planning Association of America (RPAA), a group that included Charles Whitaker, Lewis Mumford, Clarence Stein, Henry Wright, Stuart Chase, Robert Bruere, Edith Elmer Wood, Catherine Bauer, and Alexander Bing, among others. Pioneering in developing and disseminating the concepts of regional planning, the group was never very formal and ceased to exist in the 1930s. During those years, however, MacKaye wrote and in 1928 published his one book-length work, *The New Exploration: A Philosophy of Regional Planning*, which contains most of MacKaye's central ideas on the subject. Basically it proposes ways to achieve balance among the three functional human environments—urban, rural, and primeval—and calls for containment of the megalopolitan, which is none of the three but which threatens to engulf and destroy all human community. Beginning with the Spenglerian concept of the rise (spring), dominance (summer), decline (autumn), and death (winter) of civilizations, MacKaye saw Western civilization as having reached late autumn and approaching death. By stemming the tide of uncontrolled urbanization through new emphasis on indigenous culture and organic community, MacKaye believed we could short-circuit the Spenglerian cycle and keep our true civilization vital indefinitely.

In 1930 he published "The Townless Highway," demonstrating that through automobile traffic should not go into towns, the way horse and buggy roads traditionally did, but should bypass them, with restricted access along the route and outlets into each town it passes near. This was, in effect, the plan of the interstate highway system finally developed by highway and traffic engineers more than twenty years later.

Enactment of the Tennessee Valley Authority (TVA) under Franklin Roosevelt seemed a realization of RPAA ideas, and MacKaye went eagerly to work for the TVA in Nashville in 1934. He soon found, however, that as a planner he had no authority to implement the plans he made. After two years, he left TVA and returned to Shirley Center.

During those two years, however, he participated in the founding of the Wilderness Society. In October 1934, MacKaye, Harvey Broome, Bernard Frank, and Bob Marshall met, in response to earlier writings by Aldo Leopold and Bob Marshall, and founded the society. Harold C. Anderson, Aldo Leopold, Ernest C. Oberholtzer, and Robert Sterling Yard soon thereafter joined the group, and the wilderness movement in this country had a rallying point. MacKaye was the first to serve as vice president; in 1945 he became president; and in 1950 he was made honorary president, a title that endured for the last twenty-five years of his life.

During World War II, after working again for the Forest Service in the late 1930s, MacKaye worked as a planner for the Rural Electrification Administration out of St. Louis, on various projects including electrification along the route of the Alcan Highway. After the war, back again in Shirley Center, MacKaye began to advocate world federalism, writing a series of newspaper articles signed "Federalist." The Korean conflict, however, convinced him that world federalism would not soon be possible, however desirable.

In 1950–1951, MacKaye published a series of seven essays in the *Survey* under the general title "Geography to Geotechnics." He called these articles "a summary of my life work." In them he describes the development of the field of *geotechnics* (a term given him by Scottish planner Patrick Geddes) and reviews how the process of making the earth more habitable for humans had developed—and been set back—in the history of the United States.

MacKaye spent most of his retirement years alternating between summers in the MacKaye Cottage in Shirley Center, Massachusetts, and winters in the Cosmos Club in Washington, DC. He continued to work on what he called his "Opus," tentatively titled "The Geotechnics of North America," which was not completed. But MacKaye was concerned not just for his own country or his own culture. To paraphrase Lewis Mumford, if one part of MacKaye was attached to the rehabilitation of indigenous American values, he was equally concerned to develop a global strategy favorable to the maintenance of similar native values in other habitats and cultures.

MacKaye was not a technocrat. For all his technical expertise, which was considerable, he believed that the technical planner must make a community aware of the possibilities and then pursue the choices made by the community. The planner should never impose a plan. With this approach, writing was a key activity. The planner must write clearly, persuasively, accurately. Effective communication was the core of the planner's work. Accordingly, MacKaye's published writings fairly represent the whole of his work and his thinking. A good overview can be found in his book *The New Exploration*, published in 1928 and reprinted, with an introduction by Lewis Mumford, in 1962, and in *From Geography to Geotechnics*, a selection of his essays published in 1968. In 1969 the Wilderness Society published *Expedition Nine: A Return to a Region*, a look at the roots of MacKaye's thinking in his Shirley Center experience.

PAUL T. BRYANT

Anderson, Larry. "Benton MacKaye and the Art of Roving." *Appalachia* 15 (December 1987): 85–102.

Bryant, Paul T. "A Benton MacKaye Bibliography." *Living Wilderness* (January–March 1976): 33–34.

———. Introduction to *From Geography to Geotechnics*, by Benton MacKaye. Urbana, IL, 1968.

———. "MacKaye as Writer." *Living Wilderness* (January–March 1976): 31–34.

———. "The Quality of the Day: The Achievement of Benton MacKaye." Ph.D. diss., University of Illinois, 1965.

Chase, Stuart. "My Friend Benton." *Living Wilderness* (January–March 1976): 17–18.

MacKaye, Percy. *Epoch: The Life of Steele MacKaye by His Son*. 2 vols. New York, 1927.

Marshall, Robert (January 2, 1901–November 11, 1939). A forester, plant physiologist, public servant, and missionary for wilderness, Marshall was born in New York City, the son of constitutional lawyer and conservationist Louis Marshall and Florence (Lowenstein) Marshall. Few individuals, if any, have contributed so much to the American environmental movement in such a short period than Marshall.

Affectionately known as "Bob" to his professional contemporaries, he was like a meteor in the sky: brilliant, illuminating, and short-lived. The metaphor is apt not only for his ultimate contributions to American wilderness preservation but for his unquenching desire to explore and celebrate vast, undeveloped lands—a very special portion of the earth.

There is little doubt that the first twenty-one summers of his life spent at the family's summer home on lower Saranac Lake in the Adirondacks contributed extensively to his desire to learn about the natural world and spend his professional life in helping to manage it. Further, his father had been a delegate at the New York State Constitutional Convention of 1894, which placed in the state constitution the famous provision that the New York Forest Preserve shall be "kept forever wild." Additionally, extensive utilization of the family library introduced him to books and topographical surveys of the Adirondacks filled with beauty and the spirit of adventure. At the age of fourteen, he, along with his brother George and a guide, ascended a high Adirondack peak, thus cementing a lifelong love affair with wilderness exploration and celebration.

Marshall's higher education at three universities (B.S. in forestry, Syracuse; master's in forestry, Harvard; Ph.D. in plant physiology, Johns Hopkins) would serve him well as he developed a literary and resource management career. Independently wealthy, he nonetheless began his professional work after his master's degree, completing three years as a junior forester and silvicultural research assistant in Montana and Idaho for the Northern Rocky Mountain Forest Experiment Station–U.S. Forest Service. This western experience where he had the opportunity to observe large unbroken wilderness conditions served as the catalyst and foundation for his first major article on extensive natural landscapes. Titled "The Problem of the Wilderness" and appearing in the February 1930 issue of *Scientific Monthly*, it was a clarion call for setting aside and protecting large tracts of land in their natural and, to the extent possible, primeval condition. The article, called by Benton MacKaye (father of the Appalachian Trail) "A Magna Carta of Wilderness," elucidated four salient themes: (1) its great beauty and wildness with integrated aesthetic, mental, and physical values; (2) the rapid disappearance of wilderness; (3) the need to look beyond commodity value of resources in wilderness as the sole arbitrator of its value; and (4) the urgency to act for wilderness preservation.

Certainly his description of the beauty in wilderness is one of the most enduring aspects of the article. He described it as having several parameters, including its sheer size, not common to smaller manifestations of ocular beauty. Second, one looks at a painting or sculpture from the outside, but in wilderness you are surrounded or within it. Additionally, it is not static but evinces a dynamic beauty, as a tree has an ever-changing form from birth to death. Finally, wilderness gratifies all our senses, from sounds of singing birds to skin-piercing winds and fragrances of balsam drifting with the wind.

Soon after this prescient article was receiving widespread attention, he left for a return trip to Alaska, where he had previously studied trees in northern Alaska's timberline. His second visit would give him an opportunity to roam the immense and largely uncharted Brooks range and study native Alaskans, an increasing interest. This thirteen-month stay later culminated in his authorship of *Arctic Village* (1933), a comprehensive sociological work on whites and Eskimos of the upper Koyukuk region, and established Marshall as an accomplished observer of human experience. It became a bestseller and was reviewed in over one hundred publications.

After his Alaskan excursion ended in 1931, Marshall settled in Washington, DC, and immediately devoted his efforts to writing assignments. Collaborating with the U.S. Forest Service on *A National Plan for American Forestry* (1932), he contributed sections on national parks, wilderness, and recreation. A year later he published *The Peoples Forests*, in which he articulated the importance of conserving water, soil, and forests and again upheld forested areas in relation to human aesthetic needs and ascribed this as a pivotal, and perhaps most important, value to contemporary society.

In 1933 he was appointed director of Forestry, Office of Indian Affairs, where he helped develop sixteen wilderness reserves on Indian reservations and urged the secretary of the interior to keep roads out of undeveloped regions in his jurisdiction. Two years later, he was a leader of eight who founded the Wilderness Society "for the purpose of fighting off invasion of the wilderness and stimulating . . . appreciation of its multiform, emotional, intellectual, and scientific values." In 1937, he became chief of the new Forest Service Division of Recreation and Lands, a position for which he had infinite qualifications. Immediately he began moving official Forest Service policy in support of wilderness and drafted new administrative regulations relating to a Wilderness and Wild Area classification system. These new regulations sought to tighten protection by not allowing roads, lumbering, lodges, or permanent camps and would later become a benchmark for congressional legislation on a wilderness system. Approval came just months before an untimely death from heart failure at age thirty-eight.

In the 113-year struggle to develop a policy of wilderness preservation, from Henry David Thoreau's 1851 lecture in which he opened with "I wish to speak a word for nature for absolute freedom, and wildness" to passage of the Wilderness Act, creating a National Wilderness Preservation System, in 1964, few, if any, have advanced the cause more than Marshall. His efforts were tireless and his light brilliant. A light that continues to illuminate for all those who follow in his path—a journey that takes us not merely north or south, east or west, but into the myriad facets of wilderness, where humankind is a visitor who does not manipulate and does not remain.

Postscript: In 1964 as the Wilderness bill became law, the twentieth area to be named to the National Wilderness Preservation System was the Bob Marshall Wilderness Area in the Flathead and Lewis and Clark National Forests.

CHARLES MORTENSEN

Hendee, John, et al. *Wilderness Management.* Washington, DC, 1978.

Marshall, George. "On Bob Marshall's Landmark Articles." *Living Wilderness* (October–December 1976): 28–30. Following this article is a complete reprint of Marshall's "The Problem of Wilderness," originally published in *Scientific Monthly* (February 1930): 31–35.

Nash, Roderick F. *Wilderness and the American Mind.* New Haven, CT, 1982.

U.S. Department of Agriculture. Forest Service. "Bob Marshall Wilderness" [map]. Washington, DC, 1967.

Mather, Stephen Tyng (July 4, 1867–January 22, 1929). A wealthy borax manufacturer turned government servant, Mather helped establish the National Park Service and served as its first director. In collaboration with Horace Albright, he set a tone for the agency that has yet to be changed significantly.

Mather was born in California. His father, Joseph, was a bookkeeper for a mining company in San Francisco. His mother, Bertha, was a woman of poor health, and when Mather was only seven, she left California to return to her original home in the East. Mather remained in San Francisco, but following his graduation from the University of California at Berkeley, he traveled to New York to visit her. The stay became an extended one, and consequently he accepted a position with the *New York Sun.*

In 1893, Mather married Jane Thacker Floy. Because his wife's family disapproved of newspapermen, he left journalism to become publicist for the Pacific Coast Borax Company. In that position, he was credited with the development of the brand name "Twenty Mule Team" and the image that accompanied it. In 1903, however, Mather suffered his first nervous collapse. When he discovered that his employer, "Borax" Francis Smith, had failed to pay him during his period of recovery, he left the company and joined with friend Thomas Thorkildsen to form the independent and very successful Sterling Borax Company.

Mather, after entering the borax business, never lived in California again, but his annual vacations to the Sierra Nevadas remained his strongest contact with pristine backcountry. He was an active member of the Sierra Club, which in the early 1900s was as much a hiking club as it was an activist organization. Mather suffered a series of nervous breakdowns during his life, but the excursions into the mountains likely prevented them from occurring even more often than they did. On one trip in 1912, Mather encountered John Muir, who pleaded for help in confronting the damage caused by logging, mining, and the damming of rivers.

Mather's move to government service also was tied to his years in California. Both Franklin Lane, Woodrow Wilson's secretary of the interior, and Adolph Miller, Lane's assistant to the secretary, attended Berkeley at the same time as Mather. When President Wilson named Miller to the Federal Reserve Board, Secretary Lane asked Mather to fill the vacancy in Interior. Mather had written a letter to Lane, complaining of the misuse and poor management of Sequoia and Yosemite National Parks, and Lane's written response was, "Dear Steve, If you don't like the way the national parks are being run, come on down to Washington and run them yourself."

Mather was reluctant to take the position, but in January 1915, he agreed to two years of service on the condition that Horace Albright, a young Californian who had accompanied Adolph Miller to Washington, stay on to help him through the government red tape. The two-year commitment stretched into thirteen.

Mather's first responsibility as assistant to the secretary of the interior was to put life into the five-year-old attempt to create a national park bureau. Congress had been establishing national parks since the designation of Yellowstone in 1872 but had failed to authorize an agency to manage the parks as a unified system. Mather immediately put his expertise in public relations to work on the task. He personally took key legislators, publishers, and conservationists on a two-week camping trip in the Sierra Nevadas. He spearheaded a nationwide media blitz. With his own money, he hired

Robert Sterling Yard, a friend from his days with the *New York Sun*, as publicity director for the parks. He made sure that, prior to the congressional vote on the 1916 National Park Service bill, every member of Congress possessed a special national parks issue of *National Geographic* and a copy of Yard's illustrated book *National Parks Portfolio* (1916). The legislation passed, and Mather was named the National Park Service's first director.

During his tenure as director, thirteen monuments and seven parks were added to the system, among them Zion, Bryce Canyon, Mount McKinley, and Lafayette (now Acadia). Mather forwarded the notion that national parks should be not only in the remote West but also in the East where most of the people lived. He strongly supported Albright's call for historical and military parks, a direction that clearly differentiated the Park Service from the other resource management agencies of the time.

One area of controversy in Mather's administration was a tendency to favor recreation development over natural resource preservation. Mather espoused preservation as the first priority, but he also exploited the tourism potential of the parks. He courted railroad company executives, encouraging them to extend spur lines to the parks and construct fine hotels and restaurants. He improved, and at times extended, roads in the parks, believing the growing popularity of the automobile would open the national parks to a mass market. He even promoted the development of golf courses, tennis courts, and swimming pools inside park boundaries. In 1919, when a new law made it illegal for Mather to pay Yard's government salary, Yard resigned to found the National Parks Association, an autonomous group that would observe and comment on the management of the national parks. Mather personally financed this new association but then cut his support when Yard became one of Mather's strongest critics, accusing him of turning the parks over to commercial development.

Some contemporary historians also criticize Mather's entrepreneurial tendencies, but others defend them as a tool necessary for building a constituency of national park supporters. Robert Shankland, Mather's biographer, points out that Mather concentrated tourist amenities into small spaces and intentionally left the largest areas of the national parks undeveloped. Whatever Mather's motivations, he succeeded in keeping the young Park Service from being absorbed into the U.S. Forest Service, a genuine threat during the early years.

Late in 1928, Mather suffered a stroke. On January 12, 1929, he turned the directorship of the National Park Service over to Horace Albright. Ten days later, Mather died. Almost immediately, a private committee of business and government leaders proposed that a bronze plaque honoring Mather be placed in each national park and monument. Mather himself had opposed such commemorations in the parks, but in 1932, the dedications were made. The inscription on the plaque read, "He laid the foundations of the National Park Service, defining and establishing the policies under which its areas shall be developed and conserved unimpaired for future generations. There will never come an end to the good he has done."

STEVEN V. SIMPSON

Cahn, Robert. "Horace Albright Remembers the Origins." *National Parks* 59 (1985): 27–31.
Forestra, Ronald. *American National Parks and Their Keepers*. Washington, DC, 1984.
Shankland, Robert. *Steve Mather of the National Parks*. New York, 1951.
Swain, Donald. *Wilderness Defender: Horace M. Albright and Conservation*. Chicago, IL, 1970.

Matthiessen, Peter (March 22, 1927–). An author, Zen priest, and environmentalist, Matthiessen has written a number of books on natural history that reveal a profound anxiety about the chances for survival of leopards, tigers, and other animal species in the Far East, Africa, and the Americas.

Son of Erard A. Matthiessen, an eminent architect and the trustee of the National Audubon Society, and Elizabeth (Carey) Matthiessen, Peter was born in New York City into a socially prominent and comfortably fixed family. The Matthiessens owned a house on the Hudson River, a Fifth Avenue apartment, and a house on Fisher Island (where Matthiessen began his lifelong zeal for bird-watching). When his family moved to Connecticut, he started a collection of snakes (including several poisonous copperheads). His father also nurtured in him and his siblings an appreciation of outdoor activities, such as hunting and fishing.

He attended St. Bernard's School in New York City, then the chic Hotchkiss School in Connecticut. After graduating from Hotchkiss, he joined the navy and was assigned to Pearl Harbor in 1945 and 1946. After his discharge, he entered Yale University, pursuing an English major and writing sketches on fishing and hunting for the *Yale Daily News*. He also nourished his interest in nature by enrolling in courses in ornithology and zoology. Still, most of what he learned concerning wildlife and environmental science he learned through his own efforts. In 1986, he emphasized that he was not trained in any of the relevant disciplines but rather was a generalist.

After receiving a B.A. from Yale in 1950, Matthiessen published a few short stories in the *Atlantic*. In 1951, he married Patricia Southgate (but their union ended in divorce in 1958). He next moved to Paris to live a writer's life and joined a group of writers, some of whom, like William Styron, James Baldwin, and George Plimpton, were to gain future fame.

By the mid-1950s, Matthiessen had two children and by 1961 had published novels. Meanwhile, between 1954 and 1956 he had moved to Eastern Long Island and became a commercial fisherman. He enjoyed the work and the opportunity to participate in the experiences of working people. With time during the winter for writing, Matthiessen looked on these years as "the happiest time of my life," and "the life really suited me to a T.'

In 1956, traveling alone and by automobile, Matthiessen undertook a journey to survey wildlife refuges in the western United States. The trip, which resulted in the publication of *Wildlife in America* (1959), marked a turning point in Matthiessen's writing career. Hitherto, he had largely published fiction. Moreover, the journey introduced a theme in his career of travel and investigation of wildlife—he would journey to distant and often isolated locales around the world.

Wildlife in America is a careful and sober survey of North American animals from the inhabitants of the sea, rivers, and lakes to the fauna of land and the air above the land. He sets the tone of the book by describing the destruction of the last auk in 1844 on an island off the coast of Iceland. From that point Matthiessen moves across the continent region by region, detailing what one reviewer called an "appalling gallery of extinction or threatened species."

It was not that conservation efforts had been lacking in the 1890s and Progressive Era (1901–1916). During those years, Congress and the presidents established national parks (and in 1916 the National Park Service), national forests, and wildlife refuges, all of which marked a significant environmental effort. Still, as Matthiessen perceived, a certain element of the fauna fell victim to widespread animosity. The predatory animals, which were branded "criminals" and "vermin," were hunted, trapped, and poisoned. This class of fauna encompassed the Great Plains gray wolf, wild dogs, grizzly bears, cougars, and bobcats.

Regarding the role of conservation organizations, Matthiessen opined that their task is endless, and the sum of their efforts has doubtless saved and continues to shield the wildlife of North America that otherwise would have vanished long ago. Moreover, three years before Rachel Carson's *Silent Spring*, Matthiessen declared that the major concern of environmentalists was the dissemination (by the Department of Agriculture) of pesticides across America, which one zoologist denounced as the "greatest threat" North American wildlife has ever faced, perhaps more momentous than all the other threats combined!

Yet while recognizing how serious the problem of species extinction was, and why efforts must be made to halt this annihilation of American wildlife, Matthiessen did not resort to what one reviewer of *Wildlife in America* labeled "ranting and raving." Rather, continued this reviewer, Matthiessen told his "story as the poignant tale it is, and sad puzzling dilemma that we must solve at once or have on our conscience forever."

Indeed, in the years after the book's publication, an important shift occurred in the public's comprehension of, and support for, environmental legislation. And in the epilogue of the 1987 edition of *Wildlife in America*, Matthiessen noted the passage of the 1973 Endangered Species Act (as well as other environmental legislation). Moreover, he averred that there has been a "fundamental transformation in public understanding of the world and man's place in it and a significant redirection of conservation thought and practice." That having been duly recorded, Matthiessen redirected the reader's attention to the serious threat to biodiversity of American wildlife. "All of the six species of sea turtles that occur along our coast," he wrote, "are facing the threat of extinction." Nor was that all, as fish (like the Colorado River squaw fish), insects (for instance, the El Segundo blue butterfly), and some species of waterfowl confronted a similar fate.

The epilogue came to a close on a note of cautious optimism, muted, though, by a dilemma. "Our understanding of the magnitude and gravity of species extinction," wrote Matthiessen, "has grown enormously in recent years, without however a corresponding advance in the knowledge of how to remedy the problem." Nevertheless, Matthiessen insisted, important advances in conservationist thinking offered hope for averting at least some species losses that we might otherwise have had.

Still, Matthiessen observed, no one can doubt that the world confronts an "unprecedented impoverishment of the diversity of life." Unlike the widespread death of species that ended the age of dinosaurs, and a few other instances of species extinction that paleontologists have detected in the earth's distant past, Matthiessen asserted that "this one will be the exclusive work" of human beings.

Matthiessen concluded the epilogue with an observation and an admonition. He asserted that the variety of life in nature can be compared to a vast library of unread books, and the plundering of nature is comparable to the random discarding of whole volumes without having opened them and learned from them. "Our critical dependence on the great variety of nature for the progress we have already made," wrote Matthiessen, "has been amply documented." "Indifference to the loss of species," he stresses, is in effect "indifference to the future and therefore a shameful carelessness toward our children."

In the four decades or so following the publication of *Wildlife in America*, Matthiessen extended his itinerary to embrace Africa and the Far East. And a major concern of his was the damage he observed being inflicted on wildlife. In *The Tree Where Man Was Born* (1972), for example, he recounts his adventures in the Sudan in 1961 (on his way to join an anthropological expedition into New Guinea) and in East Africa in 1969. In the latter case, he focused on national parks. One such was the Serengeti Park (5,000 square miles in extent), but he found poaching to be a problem on park borders, especially among the poor who suffered from protein deficiency. The result was that an estimated 20,000 animals were killed annually in the Serengeti alone. Compounding the problem, the game departments were chronically bereft of funds and staff; rural Africans in the vicinity of game preserves and parks unfortunately—and also quite understandably—believed that the number of wild animals was inexhaustible. Moreover, hunting—as for Westerners on safari—was a ceremony and a sport with the gaining of prestige for the successful hunter.

Matthiessen's odyssey was far from over. Asked to join the tiger project, a joint U.S.-Russian venture to save the Siberian tiger, Matthiessen readily agreed. The outcome, at least for Matthiessen, was *Tigers in the Snow* (1949), which was almost two years in the making. As one reviewer observed, "[T]his beautiful book [a reference to the many photographs of tigers] is at once a celebration of the tiger, a lament for its decline, and a warning that without action, all tigers"—not just the Siberian species—"face extinction." Salvation of the tiger was (and remains) threatened by illegal poaching—whether for sport, for the skins, or for traditional medicines—and reduction of habitats (of the 160 habitats listed by Matthiessen, only 25 or so are large enough and with sufficient prey to remain viable long into the twenty-first century). As Matthiessen concluded:

> In arguing for heroic efforts on behalf of tigers, one could cite the critical importance of biodiversity as well as the interdependence of all life. Howard Quigley (co-director of the Siberian Tiger Project) reminds me, for example, how many attributes of its prey species—astonishing alertness and keen senses, speed, and strength of the deer and boar—might never have evolved without the tensions imposed on their ecology by this great predator. But finally these attractions seem less vital than our instinct that the aura of a creature as splendid as any on our earth, infusing man's life with myth and power and beauty could be struck from our experience of creation only at a dreadful cost. Life would be less without the tiger, Howard has observed, and I agree.

As the quote suggests, the "central theme" in Matthiessen's book is the "relationship between human beings and the natural world."

Indeed, as Matthiessen enters the eighth decade of his life, his concern for the status of the environment and wildlife remains unabated. Thus, in an article in the *New York Review of Books* (October 19, 2006), Matthiessen discussed congressional legislation to open up the Arctic National Wildlife Refuge for oil exploration. The very idea is abhorrent to him. As he wrote:

> Wild northern Alaska is one of the last places on earth where a human being can kneel down and drink from a wild stream without measurably more poisoned or polluted than before, its heart and essence is the Arctic National Wildlife Refuge (ANWR) in the remote northeast corner of the State, the earth's last sanctuary of the great Ice Age fauna that includes all three North American bears, gray wolves and wolverines, musk ox, moose, and in the summer, the Porcupine River herd of caribou, 12,000 strong. Everywhere fly sandhill cranes and seabirds, myriad waterfowl and shorebirds, eagles, hawks, owls, shrikes and larks and longspurs, as well as a sprinkling of far-flung birds that migrate to the Arctic slope to breed and nest from every continent on earth. Yet we Americans,

> its caretakers, are still debating whether or not to destroy this precious place by turning it over to the oil industry for development.

The reserve, originally established in 1968, was more than doubled in size by Congress in 1980. As Matthiessen noted, "Most of the 19.6 million acres permanently set aside for wildlife protection were steep rocky mountains uninhabitable by large creatures other than the white Dall mountain sheep. The one great wildlife region inside the refuge was the flat coastal plain between the Brooks Range foothills and the Beaufont Sea." Yet Matthiessen observed:

> Even if Congress should succeed today in bestowing the refuge on the corporations, the first leases could not be issued before 2008, after seismic exploration, test wells, permits, and truncated Environmental Impact Statement (EIS) required for the lease sale are completed. Next would come seven more years of construction of hundreds of miles of roads and pipelines and hundreds of acres of infrastructure, from flow stations to cesspools—all this to be done during eight or nine dark months of ice and blizzard, followed by a brief summer season when roads and installations sink and shift in the endless swamps of water-logged tundra.
>
> Not before 2015 could the oil extracted from the Wildlife Refuge affect energy supplies, and even then it would represent an inconsequential fraction of our gluttonous US consumption. (A Department of Energy report of September 2005 predicted that ANWR oil production, peaking in 2025, would slash the gas price at the pump by no more than one penny per gallon.) As most of our legislators know well, to flog this questionable source as a solution to our wasteful habits is not only dishonest but a long-term disservice to the nation.

It so happened that an exploratory well by Chevron was judged "a disappointment," and subsequently, Exxon, too, lost interest in the ANWR's limited potential. For Matthiessen this was good news indeed.

Matthiessen closed his *New York Review of Books* essay by quoting Republican senator Lincoln Chaffee (Rhode Island). "[That I have been to forty-nine of the fifty states and] this is the most beautiful place I have ever seen."

Matthiessen, a tall man, is "handsome in a craggy-featured way" and "looks like Hollywood's image of an intrepid author adventurer." He married three times (his second wife died of cancer) and had four children. He resides in Sagapowack, a farming community on eastern Long Island (and plans to remain there until farms surrender to summer residences).

Matthiessen, an environmental activist, has almost certainly won converts to the conservationist causes through his essays, novels, and especially nonfiction. In the latter case, his writing style, rooted in his manifold travels and observations, as well as questioning zoologists and other scientists, is clear, direct, and convincing. Matthiessen has alerted us to the importance of, and threats to, biodiversity and furthermore has warned that the loss of any species is an irredeemable environmental tragedy. Indeed, he considers the environmental cause the "defining issue of our times, no question."

RICHARD P. HARMOND

Dowie, William. *Peter Matthiessen.* Boston, MA, 1991.
"The Nature of Peter Matthiessen." *New York Times Magazine*, June 10, 1990.
Nicholas, D. *Peter Matthiessen, a Bibliography, 1951–1979.* Canoga Park, CA, 1979.
"Peter Matthiessen." *Contemporary Literary Criticism.* Vol. 32. 1991.
"Peter Matthiessen." *Dictionary of Literary Biography.* Vol. 6. 1980.

Mayr, Ernst (July 5, 1904–February 3, 2005). While often not fully honored as an important environmentalist, Mayr's work in systematics, biogeography, and evolution is of prime importance to a fuller understanding of the environment. His overall scientific contributions to environmentalism are such that in the judgment of many he was one of the foremost evolutionists of the twentieth century.

Mayr, the second son of three, was born in Kempten, Bavaria. From his earliest childhood on, his parents took their three boys on weekend outings with emphasis placed on observation of the local fauna and flora. As a consequence, he became an excellent bird-watcher, being able to identify all local birds by sound as well as by sight.

Mayr developed into a lifelong naturalist but published few papers on his field observations, his empirical research being almost exclusively based on museum specimens. After his father, a jurist, died of cancer in 1917, his mother took the family back to her home city of Dresden. Immediately after completing the Gymnasium in March 1923, Mayr's bird-watching interests led to the sighting of a pair of Red-crested Pochards (*Netta rufina*) in the neighborhood of Dresden, a species that had not been reported in Germany since 1846. This sighting resulted in an introduction to Professor Erwin Stresemann, curator of birds at the Natural History Museum in Berlin. Mayr visited Stresemann on his way to the University of Greifswald, where he started university studies in medicine, following a family tradition; he completed his basic medical sciences, obtaining his Candidate in Medicine, which would allow him to complete his clinical training, should his venture in zoology be unsuccessful. In March 1925 he switched to zoology at the University of Berlin, specializing in ornithology with Stresemann and completing his degree in 1926, just before his twenty-second birthday. Stresemann fulfilled his promise to Mayr for changing his studies to zoology by arranging an expedition for him to New Guinea in 1928, which was further extended to the Solomon Islands as part of the American Museum of Natural History's (AMNH) Whitney South Sea Expedition. After his return to Germany in 1920, Mayr received an invitation to come to New York as a visiting scientist and study the material collected during the South Sea Expedition. When the AMNH obtained the vast Rothschild ornithological collection in 1932, Mayr obtained a permanent position as the first and only Whitney-Rothschild curator and worked on this outstanding collection for the next two decades. The empirical research done during this period formed the foundation for his theoretical work that is so important for environmentalists.

Mayr first became widely known internationally for his 1942 book *Systematics and the Origin of Species* in which he analyzed individual and geographic variation, the species concept, and the process of speciation, all of which provided links between the newly developed field of population genetics and the well-established area of systematics. This volume is the second of the books forming the foundation of the evolutionary synthesis of 1937–1948 in which evolutionary theory was reformulated closer to that presented by Charles Darwin in 1859 than most approaches held during the intervening decades. In his book, Mayr presented his well-known biological species concept: "Species are groups of actually or potentially interbreeding natural populations, which are reproductively isolated from other such groups." He also distinguished between external isolating barriers, which were needed during the process of speciation, and intrinsic isolating mechanisms, which are attributes of the species and prevent gene flow from one species to others. He argued against the widespread idea of sympatric speciation, pointing out that difficulties existed in the proposed mechanisms of external barriers keeping apart two sympatric populations of the same species during the speciation process. Mayr made clear distinctions between the species concept and the species taxon; between the nondimensional species concept and the multidimensional species taxon; and between polytypic species, including those possessing diverse morphological types, and sibling (cryptic) species. All of these ideas are central to environmental studies because it is impossible to undertake such investigations without a clear understanding of the taxa involved.

In addition to his work on species and speciation, Mayr made a number of important contributions to evolutionary theory as well as to its history. Significantly he showed that in his 1859 book, Darwin did not propose one theory, as he always claimed and almost everyone believed, but a bundle of five interrelated, independent theories. These are: (1) evolution as such; (2) evolutionary change is gradual; (3) multiplication of species in addition to phyletic evolution; (4) the causes for evolutionary change (Darwin's natural selection); and (5) common descent (organisms have descended with modification from common ancestors). He showed that the acceptance of these several theories varied greatly, and most people were interested in that evolutionary change took place and in common descent, with considerably less concern about the actual mechanisms of this change. Mayr stressed that the process of phyletic evolution differed from that of speciation and that the first never resulted in the appearance of new species but in the modification of the species taxon over time. He also argued that phyletic evolutionary change results from the simultaneous action of two different causes, namely, an accidental one (origin of genetically based phenotypical attributes) and a design one (selective demands arising from the external environment). In 2001, his last work in biogeography was a major analysis of the distributional history of the birds of northern Melanesia, testing his ideas of island biogeography that he proposed first in 1933 and in more detail in 1940. Although Mayr never published a book-length treatise on biogeography, about 10 percent of his approximately 800 publications deal with this topic. The subjects of these papers range from his analysis of the spread of a species with stress placed on the characteristics of the borders of the species range, ideas on island biogeography (including mountaintop "islands in the sky"), the origin and composition of an entire fauna such as the North American avifauna, and a detailed inquiry into the preplate tectonic theory of continental drift.

Mayr's biogeographical studies have significance for much environmental and conservation work, such as whether to expend much effort to conserve a species at the very periphery of its range; the importance of the identification and conservation of biological hot spots; the importance of dispersal in avian distribution and conservation; and the understanding of island distribution, including mountain "islands." Unfortunately no one followed his lead in analyzing the spread of bird species (as he had done in his Ph.D. thesis published in 1926), though there are at least a dozen species that extended their range significantly to the north in the eastern half of North America during the twentieth century.

Mayr let it be known in late summer of 2003 that secondary cancer lesions were discovered in his liver; the primary site was never found. Medical problems developed later, and in early December 2004, he entered the nursing wing of Carleton-Willard Retirement Facilities (where he lived after the death of his wife Gretel in 1990, although he never actually retired from active scholarly work). This writer visited him there for the last time in mid-December, at which time he was still hoping to return to his apartment, and he was doing physical therapy to gain the needed strength. This never happened. With his two daughters at his bedside, he quietly slipped away. Over the July 4, 2005, weekend, Mayr's family gathered as usual at their summer place in New Hampshire—The Farm; this writer was invited to join them. On Sunday morning, everyone gathered on the glacial esker bordering Burton Pond and covered with lady slippers in the spring—Ernst and Gretel's favorite spot. His ashes were scattered along this low ridge to join hers.

Mayr was surely one of the greatest ornithologists and evolutionists of the twentieth century whose work is of major importance to all environmentalists. He was well known to scholars around the world and will be missed by all. For this writer, he had been a teacher, a mentor, and mainly a close friend for more than fifty years.

WALTER J. BOCK

Bock, Walter J., and M. Ross Lein, eds. *Ernst Mayr at 100*. Ornithological Monographs 58. Washington, DC, 2005. Includes a DVD, *A Taped Interview with Ernst Mayr.*

Greene, J., and M. Ruse, eds. "Ernst Mayr at Ninety." *Biology and Philosophy* 9 (1994).

Haffer, Jürgen. *Ornithology, Evolution, and Philosophy: The Life and Science of Ernst Mayr, 1904–2005*. New York, 2007.

McDonough, William A. (February 21, 1951–). As an architect of more than simply buildings, McDonough's writings and design ideas have overturned conventional assumptions about buildings and corporations to help define the modern green building and corporate social responsibility movements. McDonough is known for his high-profile projects designed for major corporations such as Nike, Ford, and Gap and, more recently, the Chinese government. McDonough coauthored a manifesto of sustainable design with his professional partner, German chemist Michael Braungart—*Cradle to Cradle: Remaking the Way We Make Things* (2002). McDonough is the winner of three U.S. presidential awards: the Presidential Award for Sustainable Development (1996), the National Design Award (2004); and the Presidential Green Chemistry Challenge Award (2003). In 1999, *Time* magazine recognized him as a "Hero for the Planet."

Perhaps McDonough's most remarkable contribution to the environmental movement is his ability to influence the minds of industrialists, builders, and corporate managers with his biocentric concepts. He renders a world where the environment and human commerce and civilization are not in conflict: "Imagine a world in which all the things we make, use, and consume provide nutrition for nature and industry—a world in which growth is good and human activity generates a delightful, restorative ecological footprint." As a designer, McDonough finds that corporations are not morally wrong in creating pollution but instead are simply using "outdated, unintelligent design." Though sometimes described as naive or utopian, McDonough's numerous built architectural and industrial projects include a ten-acre green roof on the rehabilitated Ford River Rouge factory and a building at Oberlin College that purifies its own water. His book *Cradle to Cradle* is printed on technical nutrient, a polymer that is endlessly recyclable.

McDonough was born into 1951 postwar Tokyo where his American father was working as a Seagrams executive. From a young age, he admired the craftsmanship and fine details of Japanese architecture. His grade school years were spent in Hong Kong, then an impoverished city rapidly swelling with refugees from Communist China. Until reaching teen years, his childhood experiences were divided between both the Eastern and Western hemispheres and could not have been more different from one another. McDonough spent his school months in Hong Kong and summered at his grandparents' log cabin in the Yukon. Living in old growth forest of the Yukon and learning about oyster farming, he relished the abundance of nature and interconnectedness of all species. He has said that his grandparents gave him a "sense that the world is full of gifts." As a child, he had familiarity with two global extremes: one of a developing country's harsh disparities and intensive human domination of the landscape, another of a fecund ecosystem where humans were barely a smudge on the time line of recent history. The duality of his childhood homes dealt him an early dosage of the forces of globalization converging in the later twentieth century and may have uniquely prepared him to be one of the early voices in the sustainable development movement.

McDonough's family moved to Westport, Connecticut, for his teenage years, where he brought an expatriate's awareness of wealthy New York suburbanites' ostentatious American lifestyles. He graduated from Dartmouth in 1973 and from Yale School of Architecture in 1976. In 1977, McDonough demonstrated his commitment to sustainable buildings by constructing one of the first solar houses in Ireland. Several years after graduating, he founded his own practice, and in 1985 he undertook his first notable commission, designing the New York headquarters for Environmental Defense without using unsafe chemicals or materials. The Environmental Defense Fund gave him one caveat, that they would sue him if any employees got sick from poor air quality. As McDonough set out to learn more about building material chemicals, he was shocked by the reality. At the time, environmentally conscious construction called for more energy-efficient buildings that were also more airtight. Inadvertently, this more efficiently captured and retained the harmful chemicals off-gased by many common building materials. As one of the first green office buildings in the United States, the Environmental Defense office set a new standard for building designers to protect inhabitants from harmful chemicals and be more integrated into the natural environment. McDonough would later be one of the charter members of the U.S. Green Building Council, which created a green building certification standard.

In 1991, McDonough met German chemist Michael Braungart at a party for the Environmental Protection and Encouragement Agency and immediately engaged in a heated exchange of ideas that led to the forging of a career-long partnership. McDonough and Braungart formalized their vision on sustainable development in *The Hannover Principles: Design for Sustainability* (1992), which was commissioned by the city of Hannover to be the official design guidelines for the 2000 World's Fair. In nine simple statements, this doctrine recognizes the interdependence of humans and nature and envisions sustaining growth through regenerative systems that have no waste, use natural energy flows, and respect long-term values. The city presented the *Hannover Principles* to the 1992 United Nations Earth Summit in Rio. The 1992 Earth Summit event marked one of the first international gatherings to address global environmental concerns where McDonough took up a role he would frequently fill in the future, offering hopeful and imaginative ideas to people concerned with environmental degradation.

McDonough had already established himself as more than a green architect by 1995, when he paired with Braungart to form McDonough Braungart Design Chemistry (MBDC), a polymathic consultancy with expertise spanning from the "design of molecules to products, to buildings, to communities, to cities, to regions." MBDC works with industrial, chemical, and manufacturing clients with the goal of engineering what the pair call "the next industrial revolution." The *Cradle to Cradle* paradigm outlined by McDonough and Braungart was a dramatic departure from the mainstream environmental views of the time. McDonough criticized the familiar mantra "reduce, reuse, recycle" as only slowing down the inevitable loading of landfills and famously posited that "being less bad is not good enough." In a similar rebuttal of efficiency strategies, McDonough demands that good intention and purpose must first be sound and brands the existing design conventions as a "strategy of tragedy." On this issue of growth, which neo-Malthusian environmentalists rallied against, McDonough offers another view: "When we follow nature's rules, growth is good. The question before us is not growth versus no growth. It is: what would good growth look like? And this is a question of intent, of design. What if we grow health instead of sickness, home ownership instead of indigence, education instead of ignorance?"

McDonough combines his visionary statements with business sense that appeals to corporations. Perhaps because of that, he was all the more controversial in environmental arenas. Critics questioned his conclusion that regulation was a design failure and found blaming industrial pollution on simply poor design as dismissive. Yet McDonough never flinched in carrying his biocentric ethics into corporate boardrooms:

> While this may seem like heresy to many in the world of sustainable development, the destructive qualities of today's cradle-to-grave industrial system can be seen as the result of a fundamental design problem, not the inevitable outcome of consumption and economic activity. Indeed, good design—principled design based on the laws of nature—can transform the making and consumption of things into a regenerative force.

McDonough's built projects are a testament to his aptitude for making green corporate conversions. McDonough designed The Gap headquarters in San Bruno, California, in 1998, complete with a native grass–covered roof. In 1993, he built a factory for Herman Miller called the "Greenhouse" that won *Business Week*'s first "Good Design Is Good Business" award for improving the environment and increasing productivity by 24 percent. Perhaps McDonough's most emblematic corporate design is the River Rouge Plant, a 600-acre site, once the world's largest factory complex, that had fallen into refuse and abandoned land for twenty-five years. McDonough's redesign will take twenty years and $2 billion to complete and includes a living roof, wetlands, and wildlife habitat. While some critics view the architectural projects as greenwashing, McDonough's buildings may have slowly seeped into new corporate ethos and led to social responsibility policies and changing business practices. As a sign of his corporate influence, McDonough acts on an advisory board for one of the benchmarks of corporate social responsibility, the Dow Jones Sustainability Index.

Recently, McDonough is taking his sustainability message to China, where he is designing six new cities and one village. His green designs include a sustainable water treatment system for a town with an open sewer, extensive solar power, and rooftop farmlands. McDonough brings his *Cradle to Cradle* logic to a global scale when discussing international trade between China and the United States:

> Currently, the two nations suffer from the commercial exchange of toxic products that damage the economic, social and environmental health of both nations. While China becomes the world's low-cost producer of toxic products, the U.S. brings those products to market with the world's

> most "efficient" distribution system, moving goods in a rapid, one-way trip from retailer to consumer to landfill. In many cases, the U.S. sends the most toxic products back to China, where lead and copper are unsafely recycled from computers and televisions. This is trade as mutually assured destruction.

In following with McDonough's pathologically optimistic tact, he points out the opportunities posed by China, such as the potential to lower the world price of solar panels simply through Chinese demand. As a designer, McDonough's work is helping to make this a reality.

McDonough served as the dean of the Architecture School at the University of Virginia, a position he held for five years before transferring to the commerce school to teach sustainability to business students. Today, McDonough lives in Charlottesville, Virginia, with his family. He is the principle of two design firms, founder of a nonprofit, and serves as an adviser or board member of numerous sustainable development institutes. McDonough teaches at several universities. He writes and speaks extensively and continues to vocally challenge inane conventions and contribute creative solutions to global environmental problems.

IRENE BOLAND

"The Monticello Dialogues." New Dimensions Radio. Part 1, 2002.
Rosenblatt, Roger. "A Whole New World." *Time*, February 22, 1999.
Shulman, Ken. "Think Green." *Metropolis*, August–September 2001.
Stubbs, Stephanie. "Being Less Bad Is Not Being Good." American Institute of Architects National Convention and Design Exposition, Los Angeles, CA, June 8–10, 2006.

Mellon, Paul (June 11, 1907–February 1, 1999). Although best known as a devotee of art, with major responsibility for completion of the National Gallery of Art in Washington and the founding of the Yale Center for British Art, Mellon was also concerned about the preservation of the environment and supplied funding for the creation of the Cape Hatteras National Seashore in North Carolina and the Sky Meadow State Park in Virginia, among other projects.

Mellon—philanthropist, art collector, horse breeder, and scion of one of the richest families in America—was born in Pittsburg to banker and later

Secretary of the Treasury Andrew W. Mellon and Nora (McMullen) Mellon, who came from Hertford in England. The reserved Andrew and his emotional wife had a rocky marriage. Mellon's parents separated when he was two years old and divorced when he was five. Mellon later wrote that for him and his older sister Ailsa "it was not much fun to be a divorced person or to be the offspring of divorced parents in Pittsburg in 1912." Though Mellon was afraid of his "dry and censorious" father, the divorce settlement called for the children to spend most of the year with him and four months with their mother, who had moved back to England. It was in the periods he spent with his mother that he began to develop his love of the English countryside.

Mellon began his schooling at the Shady Side Academy in Pittsburgh and later went to Choate in Connecticut. He had considered enrolling in Princeton, but at the last minute he decided instead to study English and history at Yale. At college he was on the board of the *Yale Daily News* and joined the secret society Scroll and Key. After earning an A.B. in 1929, Mellon continued his education at Clare College, Cambridge University in England, where he received a second A.B. degree in 1931. Already an avid horseman, during his time at Cambridge he also became interested in foxhunting. It was also during these years that he began collecting books.

After his graduation from Cambridge, Mellon stayed in England with his father, who was serving as the U.S. ambassador to the court of St. James from 1932 to 1933. Paul then returned to Pittsburg and, bowing to his father's wishes, began an internship at the Mellon bank and was named a director for a number of companies in which his family had an interest. In 1934, Mellon began his own collection of art when he purchased a painting of the racehorse Pumpkin by George Stubbs. The following year Mellon married his first wife, Mary Elizabeth Conover Brown, with whom he had two children.

In 1930, Mellon had been appointed to the board of the A. W. Mellon Educational and Charitable Trust, established to help complete the National Gallery of Art in Washington and arrange for the transfer of the elder Mellon's collection of art to the museum. This marked the beginning of Mellon's long relationship with the National Gallery. After Andrew Mellon's death in 1937, the trust took on the responsibility for the museum's construction and early operation.

Mellon soon found that his heart was not in the family business, and on November 29, 1936, he met with his father and told him that he did not want to be "an inadequate replica of yourself, or a counterfeit." Surprisingly, the elder Mellon agreed and said his son did not have to take an active role in business but should keep "vaguely in touch" with the top management.

In 1937, Mellon and his wife left for Europe, where he would complete an M.A. at Cambridge the following year and study in Zurich with noted psychiatrist Carl Jung. Back in the United States in 1940, Mellon enrolled in St. John's College in Maryland to fill in some "gaps" in his education, but developments on the world scene would change his plans. A few months after the official dedication of the National Gallery on March 17, 1941, with World War II on the horizon, Mellon consulted with General George C. Patton about enlisting in the armed forces. In June of that year, Mellon enlisted as a private in the U.S. Army cavalry. He was soon selected for training as an officer and was commissioned in March 1942, after which he spent a year as an instructor in horsemanship at Fort Riley, Kansas. By 1943, Mellon had joined the Office of Strategic Services (the precursor to the Central Intelligence Agency) in Europe. Despite trying to get into the action, Mellon spent much of the rest of the war mainly doing propaganda work, eventually reaching the rank of major. In his autobiography, Mellon describes his war years mostly as a series of sometimes humorous mishaps—including being inadvertently sent to a venereal disease ward for a short time while recovering from pneumonia and spending a chaste night in a brothel when a small unit he was with was temporarily housed there.

With the end of the war and his release from the army, Mellon returned to his Virginia home, to take up horse breeding at his Rokeby Farms and Stables, while also indulging in his love of foxhunting. In October 1946, however, his life was disrupted when his wife Mary died unexpectedly of causes related to asthma. In 1948, Mellon married Rachel Lloyd, nicknamed Bunny, a descendant of the Lambert family, who had developed Listerine.

During these postwar years Mellon expanded his philanthropic efforts. In 1945, he formed the Bollingen Foundation with the mission of encouraging the translation of foreign works (including those of Carl Jung) into English, as well as the publishing of books in the areas of art and philosophy, among other subjects. The foundation also created the Bollingen Prize, which became the nation's premier poetry honor.

Art continued to be Mellon's consuming passion. In 1945, he rejoined the board of the National Gallery, a position he would serve in for the next four decades. In these years, Mellon continued his generosity to the museum—being instrumental in setting up a series of lectures named for his father; initiating fellowships for the advanced training of museum professionals; donating works by such artists as Canaletto and Paul Cézanne; providing $200,000 to fund a major national art library; and culminating in providing major funding for a second building for the National Gallery. This East Building, designed by I. M. Pei, was officially dedicated in 1978.

The National Gallery was not the only museum to benefit from Mellon's philanthropy. At his alma mater, Mellon built the Yale Center for British Art, which opened in 1977, to house his substantial collection of works by British masters such as Thomas Gainsborough, William Blake, and J. M. W. Turner. Mellon also donated art to and served as a trustee for the Virginia Museum of Fine Arts.

In the 1950s, Mellon became interested in trying to help the environment. There does not seem to have been one dramatic instance that caused his interest in conservation issues. Rather, he said in his autobiography that "I think many of us feel unhappy that the requirements and pressures of modern civilization are causing a steady deterioration in our surroundings." Mellon added that he "always enjoyed visual awareness" and found himself "unconsciously depressed at the sight of a long stretch of hardtop road punctuated with billboards and concrete gas stations." Through the Avalon Foundation, created in 1940 by his sister Ailsa, and Paul's own Old Dominion Foundation, set up in 1941, the Mellons shared the cost of acquiring about 28,600 acres for inclusion in the Cape Hatteras National Seashore, dedicated in 1958. The two foundations also funded the National Parks Service publication *Our Vanishing Shoreline* (1956), a survey of Atlantic and Gulf coast areas in need of preservation.

Later, the Andrew W. Mellon Foundation—formed in 1969 by the consolidation of the Avalon and the Old Dominion Foundations—helped pay for the purchase of some 8,500 acres of the Cumberland Island National Seashore. On another occasion, Mellon, acting on his own, acquired 1,100 acres slated for development near his home in Upperville in Virginia. This land was turned into Sky Meadows State Park. Mellon wrote that supporting these environmental preservation projects gave him the "profoundest pleasure and the most heartwarming satisfaction." He also supported the work of such groups as the Conservation Foundation and the National Audubon Society, though he admitted that his efforts would "play a small part in trying to hold back this tide."

Mellon died in 1999. In *Reflections in a Silver Spoon* he wrote, "I have been an amateur in every phase of my life; an amateur poet, an amateur scholar, an amateur horseman, an amateur farmer, an amateur soldier, an amateur connoisseur of art, an amateur publisher, and an amateur museum executive. The root of the word 'amateur' is the Latin word for love, and I can honestly say that I've thoroughly enjoyed all the roles I have played." One of the more notable points about his philanthropic projects is how seldom they bare his name. Mellon did not believe that philanthropy should be used to gain power or preferential treatment. It was his belief that privacy was the "most valuable asset that money can buy."

Mellon's environmental legacy includes the projects at Cape Hatteras and Sky Meadows, but perhaps more important is the continuing work of the Andrew W. Mellon Foundation in supporting projects in conservation and the environment. In its early years, the foundation allocated much of its funding to the sort of land acquisition projects noted above. In the latter part of the 1970s, it began supporting research in energy, natural resources, and the environment, including the oceans. In more recent years, the Mellon Foundation has supported basic research on how natural ecosystems work. It also expanded its original focus on work done in the United States to now include Latin America and South Africa.

WILLIAM L. KEOGAN

Iacullo, Maria T. "Paul Mellon." *Scribner Encyclopedia of American Lives.* Vol. 5. 2002.

Mellon, Paul, and John Baskett. *Reflections in a Silver Spoon: A Memoir.* New York, 1992.

Paul Mellon's Legacy: A Passion for British Art. New Haven, CT, 2007.

Russell, John. "Paul Mellon, Patrician Champion of Art and National Gallery, Dies." *New York Times*, February 3, 1999.

Merriam, C. Hart (December 5, 1855–March 19, 1942). An ornithologist, physician/surgeon, mammalogist, anthropologist/ethnologist, prolific writer, and influential leader in his field, Merriam was born in New York City and raised near the Adirondack Mountains on the family homestead in Locust Grove, New York. Hart, as he was called, came from a distinguished lineage of well-educated and politically prominent citizens, including the family's first American settler, Joseph Merriam, an Englishman, who settled in Concord, Massachusetts, in 1638. His father, Clinton L. Merriam, owned a successful bond and brokerage business and was a member of the House of Representatives. His mother, Caroline (Hart) Merriam, whose father was a judge and member of the state assembly, was a college graduate. He had an older brother, Charles Collins Merriam, and a younger sister, later known as Florence Merriam Bailey, who also became a highly respected author and ornithologist.

Merriam's genius and talents manifested themselves at an early age. His curiosity about plants and animals, fostered by his proximity to the Adirondack Mountains and coupled with his ability to shoot accurately, led to a growing collection of birds that he learned to mount himself. This hobby previewed his later large-scale collections of birds and mammals. By seventeen years of age he had participated in the Hayden Survey, a government expedition into the western states that included many notable scientists of the day. This trip both shaped and reflected Merriam's destiny, by creating professional contacts and friendships that lasted a lifetime and by initiating him into the West, where much of his subsequent research would take place. The trip culminated in a fifty-page report by Merriam—"Report on the Mammals and Birds of the Expedition" (1872)—that was a forerunner of the extensive writings and edited collections he would produce in the future.

Merriam's professional energies seem to have been boundless. During the time he was a student attending and graduating from Yale (1877) and, later, the College of Physicians and Surgeons at Columbia University (where he received his M.D. in 1879), and while maintaining a large and successful practice as a country doctor for several years, Merriam continued as an active naturalist, conducting research and developing a growing interest in mammals as well as birds. Publications from that era include *A Review of the Birds of Connecticut* (1877) and *Remarks on Some of the Birds of Lewis County, Northern New York* (1878), as well as numerous short papers and his groundbreaking *The Vertebrates of the Adirondack Region, Northeastern New York*, an exhaustive study with detailed "life histories" of each animal, published in two volumes (1882–1884).

At that time, compared to ornithology, the study of mammalogy had been neglected. Mammalogists were virtually unknown, and collections of such animals were sparse. It was assumed that most, if not all, mammals had been discovered, that there was little left to learn about them, and that the principal task that remained was to put the known mammals into the proper taxonomic categories. Merriam, who by 1884 had a private collection of mammals containing approximately 7,000 specimens and who had described his first new species, a small shrew (*Atophyrax bendirii*), clearly took a leadership role as an innovator who would dramatically change society's understanding of and knowledge about North American mammals.

Merriam carried on his medical practice in Locust Grove while conducting a massive project of bird migration for the American Ornithologists' Union. In 1884 he represented America, along with Elite Coulees, on an important international committee involved in worldwide bird observations. The work on bird migration grew to such an extent that in 1885 Congress established a section of ornithology in the Entomologist Division of the Department of Agriculture. Merriam was asked to head the office, and he quit his medical practice to organize this new division and to assume the title of ornithologist. Within the first year his section had changed to the Division of Economic Ornithology and Mammalogy, and in 1905 it became known as the Bureau of Biological Survey, which Merriam headed up for twenty-five years. By 1939, long after Merriam had retired, the office came under the Department of the Interior and was called the Fish and Wildlife Service, its name today.

As head of the Biological Survey, Merriam rapidly established himself as a noted authority. He brought his enthusiasm, genius, and high standards to the task of discovering a growing number of new species and subspecies of American mammals and of developing more effective methods of gathering, preserving, analyzing, and classifying them. It should be noted that he was assisted in these impressive endeavors not only by his colleagues at the Biological Survey but also by a young assistant in the field, Vernon Bailey, and by a helpful invention, the Cyclone Trap. Bailey, a Minnesota farm boy, gathered specimens in the field for Merriam and eventually came to Washington and worked with him for many years. (Bailey also became Merriam's brother-in-law by marrying Florence.) In addition, the invention of the Cyclone Trap (1887) made possible the capture of previously elusive and totally unknown species, most notably many nocturnal animals.

Within the first two years of his tenure as chief of the Biological Survey, Merriam had described seventy-two new species and several genera of mammals, inaugurating a new era in mammalian discovery and classification. As a division head, he brought together naturalists and collectors and organized and carried out numerous expeditions in the western states. These included his well-known work in Arizona on the causes of plant and animal distribution.

Under his leadership his division carried out fieldwork in the San Francisco Peak area in Arizona. On the assumption that temperature was the most important factor limiting the northward and southward distribution of organisms, he formulated his Temperature Laws in 1894. He then attempted to ascertain which phase of temperature was the critical one relative to the biotic units, or "life zones," a concept then in its formative stage. Based on this research, he concluded that there were seven life zones. He later modified his theory to apply to all of North America, publishing his classification in its final form in 1898. Further research has subsequently subjected Merriam's laws of temperature control and life zones to substantial criticism. Ultimately, it was determined that his temperature control laws could not be accepted and that his "life zones" were also problematical. Criticism centered on his descriptions of the zones, his explanations of the causal factors, and the technical errors in his work. Fault was also found with Merriam's complacent attitude that all the problems had been solved and all the necessary laws formulated. Despite this, it is important to note that his field surveys represented accurate descriptions of the vegetation zones he examined, that his was the first significant attempt to interpret the distribution of North American plants and animals based on climatic data, and that his findings prompted a good deal of important future research.

Furthermore, in pursuit of the study of mammals and their distribution in North America, he organized a large and successful expedition to Death Valley for the purpose of studying the mountains and deserts of California and Nevada. He was also invited by President Benjamin Harrison to be one of the Bering Sea commissioners and to study the fur seals on the Pribilof Islands in Alaska, which he did in 1891. In addition, Edward H. Harriman requested that Merriam select and organize many noted scientists of the day for what became known as the Harriman Alaska Expedition. In March 1899 Merriam participated enthusiastically as the expedition secretary, then returned to give exceptional care in editing and publishing the fifteen-volume report of that experience.

A meticulous author and editor, during his leadership as the head of his division several highly acclaimed bulletins were produced by his staff. In

1889 he founded a series of monographs known as the *North American Fauna* that reported his division's efforts to map the geographical distribution and relations of the birds and mammals of North America. A perfectionist, he essentially wrote the first ten numbers. While these publications undoubtedly increased the visibility of Merriam's division and, in conjunction with his charismatic personality and boundless energy for the profession, helped establish him as a national figure, his control over the earliest issues apparently caused hard feelings among some of his subordinates. By the late 1890s, however, he was entrusting members of his staff to write issues while he continued to edit them.

Merriam revolutionized the study of mammalogy, and for many years his methodology and ideas were widely accepted. In later years, critics expressed concerns over the wide latitude permitted the observer in making what were sometimes felt to be hair-splitting distinctions in determining species. After the 1900s, scientific advances in biochemical and genetic research provided new approaches and turned the tides. Merriam, who clung tenaciously to his traditional methods, found himself increasingly in the minority. His work with North American grizzly and brown bears exemplified his extreme position when he identified eighty-six forms based for the most part on their skull characteristics, conclusions unacceptable to contemporary mammalogists.

He spent many summers in the West conducting fieldwork where he maintained a home in Lagunitas, California. He knew the paths and stories of the pioneers and established contacts and friendships with Native Americans, developing over the years an extensive collection of Indian basketry. Toward the end of Merriam's tenure at the Biological Survey, which was demanding more political astuteness for which he had little aptitude, friends successfully appealed to Harriman's widow to set up the Harriman Trust. The trust was administered by the Smithsonian Institution and provided generous funds to support Merriam's research for the rest of his life. Although colleagues and friends expected him to carry on his research with mammals, in fact Merriam shifted his focus to what he believed to be more important work, the study of vanishing Native American tribes and their culture. With the support of his wife he photographed dances and rituals and collected extensive information on the folklore, traditions, and languages of many Indian tribes. He published a number of articles based on this fieldwork, and he left at his death a large collection of data, including notes and unpublished manuscripts, now housed in the Smithsonian.

A true leader throughout his life, he founded and headed up many major organizations. He helped organize the Nuttall Ornithological Club (1871) and the Linnaean Society of New York (1878), where he was elected the first president. He was president of the Lewis County Medical Society and the Biological Society of Washington, DC (1891–1892). He was a cofounder of the American Ornithologists' Union (1900), where he was elected the first secretary and chairman of the Committee on Bird Migration. He was a founder of the National Geographic Society and an associate editor of the magazine and served on the board of directors for fifty-four years. He was a member of the Philosophical Society and helped to found the Washington Academy of Sciences, taking a leadership role in their publications. He was a cofounder and the first president of the American Society of Mammalogists (1919–1920) and president of the American Society of Naturalists (1924–1925), and he served as chairman of the U.S. Board on Geographic Names (1917–1925). He received many honors and awards, among them election to the National Academy of Sciences, the Linnaean Society Medal, and the Roosevelt Memorial Medal (1931) for "distinguished work in biology." He declined several offers to receive honorary doctorates.

Merriam's lifetime publications makes an impressive list. An early publication that gave him special pride is his exhaustive study "Monographic Revision of the Pocket Gophers" (1895). His wide-ranging expeditions and fieldwork led him to discover and describe approximately 600 new mammals. His extensive studies of the American brown and grizzly bears eventually resulted in his book *Review of the Grizzly and Big Brown Bears of North America* (1918). Because of his expertise, colleagues and friends long anticipated that he would write a major work on the mammals of North America. Theodore Roosevelt had encouraged him to write such a volume; on realizing it would not be forthcoming, he wrote to another naturalist about Merriam, "He was fitted to be a great architect. He has trundled wheelbarrows with bricks instead." Merriam himself appears to have been tormented later in life by his inability to complete the mammalogy book and other writing projects he had undertaken. While he may have had difficulty synthesizing and theorizing, it should be duly noted that he took infinite pains in his work, perfected new methods, and continually shared his expertise via his great willingness to train and assist others.

Merriam, who has been called the father of mammalogy, was an advocate of both laboratory and field studies. In each of his endeavors, Merriam developed exceptional expertise and dedication and drew many around him by his magnetic personality. A man of high standards and meticulous methods, he revolutionized the technique of making study specimens of small mammals, recognized the importance of using a uniform system of measurements, and insisted on maintaining accurate data about each animal's habitat. His approach was adopted by many researchers nationally and internationally. A brilliant, industrious, and complicated man, his accomplishments both as a naturalist and as an ethnologist had an outstanding effect during his lifetime and continue to reverberate.

He married Elizabeth Gosnell, of Martinsburg, West Virginia, who had been his secretary, on October 15, 1886. They had two daughters.

ANDREA W. HERRMANN

Camp, Charles L. "C. Hart Merriam." *California Historical Society Quarterly* (September 1942): 284–286.

Daubenmire, Rexford F. "Merriam's Life Zones of North America." *Quarterly Review of Biology* (September 1938): 327–332.

Kendeigh, Charles S. "A Study of Merriam's Temperature Laws." *Wilson Bulletin: A Quarterly Magazine of Ornithology.* 44 (1932): 129–143.

Palmer, T. S. "In Memoriam: Clinton Hart Merriam." *Auk.* 71 (1954): 130–136.

Sterling, Keir B. *Last of the Naturalists: The Career of C. Hart Merriam.* New York, 1974.

Storer, Tracy I. "Mammalogy and the American Society of Mammalogists, 1919–1969." *Journal of Mammalogy* 50 (1969): 785–793.

Talbot, Z. M., and M. W. Talbot. "C. Hart Merriam." *Science* 95 (1942): 545–546.

Merton, Thomas (January 31, 1915–December 10, 1968). Also known as Trappist monk Father Louis, Order of Cistercians of the Strict Observance, Merton was a spiritual writer; poet; critic of nuclear weapons, war, and racism; and advocate for social justice. His bestselling spiritual autobiography *The Seven Storey Mountain* (1948) details the first twenty-seven years of his life as a child in France, a rebel student at Oakham School in England, an undisciplined university student at Clare College, Cambridge, a student at Columbia University, and his first years in the monastery. During his time at Columbia under the mentorship of Professor Mark Van Doren, Merton earned his M.A. in English literature, writing his thesis on

William Blake's theory of art and nature. "I always felt at Columbia," wrote Merton,

> that people around me, half amused and perhaps at times half incredulous, were happy to let me be myself. (I add that I seldom felt this way at Cambridge.) The thing I always liked best about Columbia was the sense that the university was on the whole glad to turn me loose in its library, its classrooms, and among its distinguished faculty, and let me make what I liked out of it all. I did. And I ended up by being turned on like a pinball machine by Blake, Thomas Aquinas, Augustine, Eckhart, Coomaraswamy, Traherne, Hopkins, Maritain, and the sacraments of the Catholic Church. After which I came to the monastery in which (this is public knowledge) I have continued to be the same kind of maverick and have, in fact, ended as a hermit who is also fully identified with the peace movement, with Zen, with a group of Latin American hippie poets, etc., etc. (*Love and Living*, pub. post. 1979)

After his conversion to Roman Catholicism and subsequent entrance into the Trappist Monastery at Gethsemani, Kentucky, on December 10, 1941, Merton wrote prolifically on desert spirituality, prayer, and contemplation and published some of his reworked journals as *The Sign of Jonas* (1953), *The Secular Journal of Thomas Merton* (1959), and *Conjectures of a Guilty Bystander* (1966). His deep commitment to solitude and prayer did not alienate him from the human community but paradoxically enabled him to turn toward the world with compassion and prophetic wisdom. He published numerous essays on the evils of war, the value of nonviolence, and the dangers of technology. Merton was in correspondence with such luminaries and social activists as Joan Baez, Daniel Berrigan, Ernesto Cardenal, Lawrence Ferlinghetti, Thich Nhat Hahn, Coretta Scott King, Clare Booth Luce, Czslaw Milosz, Boris Pasternak, Evelyn Waugh, D. Z. Suzuki, and Louis Zukovsky. In his later years, Merton was exploring the connections between Eastern mysticism, especially Zen Buddhism, and Western contemplation. His effort to make these connections begins to be evident in *New Seeds of Contemplation* (1962) and to some extent in *Emblems of a Season of Fury* (1963), *The Way of Chuang Tzu* (1965), *Mystics and Zen Masters* (1967), and *Zen and the Birds of Appetite* (1968). From 1965 until 1968, he lived as a hermit on the grounds of the monastery until his untimely death at age fifty-three, by accidental electrocution, in Bangkok, Thailand, where

Merton had been an invited speaker for a meeting of Asian Benedictine and Cistercian monks.

Although Merton is noted primarily as a writer on spiritual and monastic themes and as a prophetic voice for social justice, there is recent convincing evidence that he was becoming interested in environmental concerns. His published writing and poems reveal a transformation from seeing nature as something separate from humankind to seeing human beings as part of nature, "though we are a very special part, that which is conscious of God" (*Conjectures of a Guilty Bystander*). During the course of his public life as a writer, Merton relinquished his dualistic view of nature for a more sacramental one in which he not only saw himself in ecological balance with the birds near his wooded hermitage (*Day of a Stranger*, 1965) but realized that everything he touched was prayer: "where the sky is my prayer, the birds are my prayers, the wind in the trees is my prayer, for God is all in all" (*Thoughts in Solitude*, 1958).

The almost 3,000 pages of Merton's recently published personal journals reveal his increasing view of and interaction with nature in several distinct ways: as *weather report*, which both set the tone for his day and deepened his sense of place; as *trigger for memories* of earlier times; as *analogy* to explain the conundrums of his life; as vehicle for his *poetic eye*; as *language to mediate* the ineffable experience of prayer; and as *healing influence* to provide Merton with a sense of coming home.

Merton's life experiences offer some explanation for this change of view. His father, Owen, a New Zealander, was a painter; his mother, Ruth, also an artist, recorded in Merton's baby book how the light, the trees, and small animals delighted him. At Columbia University, Merton was cartoonist and art editor for the *Columbia Jester*. After entering the monastery, he daily chanted biblical psalms praising the God of creation; he studied the fathers and doctors of the church and steeped himself in Benedictine reverence for the rhythm of work and prayer, a rhythm that acknowledges the power of the seed and its interdependence with the soil. Merton also read deeply and widely, not just theological treatises and spiritual mystics like Meister Eckhart, who described creation as an act of God's passion, but philosophy, literature, poetry, and anthropology. He paid attention to current events and to the stirrings of his own heart. Indeed, Merton's contemplative vocation was not an experience of separation or alienation *from* the world but an invitation to community *with* the world, that is, all things on this planet.

In 1949, Merton began spending some hours each day in the woods on the monastery grounds, where he allowed the sun, the trees, the birds, and the deer to deepen his spirituality. In the 1950s, as master of scholastics and later master of novices, he acquired from the government hundreds of loblolly pine seedlings to control soil erosion and creek runoff. In later years, three public actions indicate that Merton was becoming an advocate for the environment.

In 1963, after reading Rachel Carson's controversial *Silent Spring* (1962), Merton wrote to Carson, applauding her "diagnosis of the ills of our civilization," which Merton identified as our "awful irresponsibility with which we scorn the smallest values . . . and dare to use our titanic power in a way that threatens not only civilization but life itself." We "must make use of nature wisely, and understand [our] position, ultimately relating both [our]-self and visible nature to the invisible—in my terms, to the Creator, in any case, to the source and exemplar of all being and all life" (*Witness to Freedom: Letters in Times of Crisis*, 1994).

Four years later, Merton wrote a book review of George H. Williams's *Wilderness and Paradise in Christian Thought* (1962) and Ulrich W. Mauser's *Christ in the Wilderness* (1963). In addition to comments on the desert and wilderness as prerequisite for contemplation, Merton added that these texts help us understand our "criminal wastefulness with which commercial interests in the last two centuries have ravaged and despoiled the 'paradise-wilderness' of the North American mountains, forests and plains." Monks, wrote Merton, "would seem to be destined by God, in our time, to be not only dwellers in the wilderness but also its protectors" (*Cistercian Studies*, 1967).

In February 1968, Merton wrote a lengthy book review of Roderick Nash's *Wilderness and the American Mind* (1967) in which he criticized Nash for remaining in the historical mode and refusing to recognize the crucial issue of the 1960s, namely, that the savagery the Puritans projected "out there" onto the wilderness had turned out to be savagery within the human heart. Merton called on the reader to come to terms with a deep conflict imposed by our culture, namely, the tension between the wilderness mystique and the mystique of exploitation and power in the name of freedom and creativity. Merton concluded his review by asking all of us if Aldo Leopold's ecological conscience could become effective in America today.

Were Merton alive today, he would be in the vanguard of contemporary nature writers and environmentalists, not simply because creation is holy but because we humans have a moral obligation to be the voice for the voiceless.

MONICA WEIS, S.S.J.

Shannon,. William H. *Silent Lamp: The Thomas Merton Story.* New York, 1992.

———. *Something of a Rebel: Thomas Merton, His Life and Works. An Introduction.* Cincinnati, OH, 1997.

Weis, Monica, S.S.J. "Thomas Merton: Advance Man for New Age Thinking about the Environment." *ISLE: Interdisciplinary Studies in Literature and Environment* 5.2 (Summer 1998): 1–7.

Miles, Emma Bell (October 19, 1879–March 19, 1919). The startling contrasts in the life of an impoverished mountain wife helped make her into a widely published author and an environmentalist of some note. While rearing four children in a series of ramshackle cabins near Walden's Ridge, Tennessee, and enduring their sometimes sensitive, sometimes brutish father, Miles supported her family by writing two books as well as numerous poems and short stories; painting panels, miniatures, and postcards of birds; writing a newspaper column on the relationship of humankind and nature; and giving lectures on the birds, wildflowers, and general natural beauty of the Southern Appalachian Mountains.

Though her parents, Benjamin Franklin Bell and Martha Ann (Mirick) Bell, were college educated, Emma never attended school regularly. Particularly after her parents' move to Walden's Ridge in 1890, Emma roamed the woods, sketching plants and animals and learning her lessons from what she would later call the "Great Gray Mother." Her parents noted her artistic ability and sent her to the St. Louis, Missouri, School of Art in 1899, but she was miserable there, missing the beauty of the mountains and the company of Frank Miles, a young mountain man who shared her love of plants.

Her parents opposed her relationship with the illiterate Miles, but Frank and Emma were married in October 1901. In their mountain shack, the young couple read Henry David Thoreau's *Walden* (1854) together, Emma vowing, like Thoreau, to confront life simply and directly. She got her wish as the improvident Frank moved her from shack to shack over the next few years while she gave birth to twins Judith and Jean in 1902, Joe in 1905, Katherine in 1907, and Mark in 1909. Between bearing and raising these children, keeping house, and struggling to feed and clothe her growing family, Miles published *Spirit of the Mountain* in 1905, a sensitive portrait of the mountain people, their culture, and their environment.

In addition to her book, Miles's poetry and short fiction, all characterized by detailed and accurate descriptions of nature, were appearing in such national magazines as *Harper's Monthly*, *Century*, *Lippincott's*, *Putnam's*, and *Red Book*.

By the time *Spirit of the Mountain* appeared, Miles was a well-known figure in Chattanooga, selling nature portraits to the well-to-do city folk who summered on Walden's Ridge and giving lectures on nature at various forums in the city. On February 12, 1908, she gave a lecture titled "How to Be Happy on Walden's Ridge" in which she commented that most people spend too little time in nature. She advised her audience to get out early and often, to observe systematically, and to record their observations in a notebook and sketchbook. She concluded that rare plants should not be disturbed, only sketched. Her lecture was covered by all three Chattanooga newspapers.

Over the next few years, Miles would continue her close observation of nature, recording in the spring of 1909 twenty new species of birds on Walden's Ridge. In July 1913, Miles gave a talk at the Chattanooga Episcopal Church on the birds and flowers of the mountain summer; it was intended as a guide for those who summer in the woods. The *Chattanooga News* reported the talk as a sublime nature study as well as a study in psychology that revealed the correlation between nature and mind and illustrated how life among the mountains could bring out the sublime as well as the sordid.

In the spring of 1914, Miles began writing for the *Chattanooga News*. She introduced a column called "Fountain Square Conversations" that purported to record conversations between the statue of a fireman in the square, who represented the view of humans, and the birds who visited the statue, who represented the view of nature. The participants discussed such topics as the cruelty of slaughtering birds and small animals for personal adornment, the restrictiveness of city life in which nature is controlled by design, and the failure of human beings to understand their effects on nature. Miles also wrote a series of editorials on environmental topics—discussing such issues as the wanton destruction of wildflowers by careless summer visitors and advising her readers on which flowers could be picked safely and which ones were so rare they should be left undisturbed. By June her wildflower articles had borne fruit when the city government imposed fines for picking wildflowers and initiated an educational program on the importance of nature in the Chattanooga public schools.

In the midst of her growing success as a writer and lecturer on environmental issues, Miles left Chattanooga to return to the mountain with Frank. She was pregnant, ill, and in despair of reconciling her need to write and her need for art with her need for her family, for Frank, and for the solitude of the mountain.

Miles spent the next year on the mountain, observing birdlife and making careful notes. In the winter of 1915 she returned to Chattanooga to give a series of lectures on the necessity of preserving birdlife and the importance of nature study in the public schools.

In February 1915, she was diagnosed with tuberculosis and admitted to the Pine Breeze Sanitarium. In the quiet solitude of Pine Breeze, she continued her careful observation of birds, sketching them, making notes, and preparing careful observation notebooks.

When she was well enough to leave the sanitarium, Miles returned to an abandoned cabin on the mountain, spending whole days observing and sketching birds. Over the next two years she would support her family by selling colored drawings of birds and by giving lectures on a growing variety of topics such as ferns, the conservation of evergreens, and of course, birds.

The last months of Miles's life were spent in preparation of the text and illustrations of *Our Southern Birds*, a book intended for schoolchildren and to be published by the National Book Company. *Our Southern Birds*, attractively bound and illustrated with her sketches of the birds she described, appeared near March 1, 1919. She was pleased with the tangible evidence of her years of careful observations and hoped that the proceeds from the book would ensure her children's future. Miles died at the age of thirty-nine, a victim of tuberculosis and poverty.

Miles's introduction to *Our Southern Birds* advised her readers and would-be bird-watchers to do nothing that would startle the timid creatures they wished to observe. She further admonished her readers to refrain from touching nests or eggs, as many species' sense of smell is so keen they will not return to a nest that has been touched. Finally, she suggested some procedures for accurate bird observations, advising her readers to describe the bird's size, color, shape and color of bill, marks such as wing bars, notes, and song. The book concludes with a note to teachers that urges that their classes should include discussions of such questions as "Why birds should be protected" and "How we may protect birds."

How do we assess Miles's impact? No major legislation bears her name; she can only be credited with a Chattanooga, Tennessee, ordinance against picking wildflowers. No parks or natural areas bear her name, though the Chattanooga Chapter of the National Audubon Society is named for her. Her two books, *Spirit of the Mountain* and *Our Southern Birds*, are available today; but though admired by those who know them, they are not widely known. Perhaps her greatest legacy is the realization that simple, quiet people doing simple, quiet things finally may make the greatest contributions toward saving our environment. Thanks to Miles, certainly the people of Southern Appalachia are more aware of the fragile beauty of the world they inhabit.

J. KAREN RAY

Gaston, Kay Baker. *Emma Bell Miles.* Signal Mountain, TN, 1985.

———. "Emma Bell Miles and the 'Fountain Square Conversations.'" *Tennessee Historical Quarterly* 37 (1978): 416–429.

Rowell, Adelaide. "Emma Bell Miles, Artist, Author, and Poet of the Tennessee Mountains." *Tennessee Historical Quarterly* 25 (1966): 77–89.

Miller, Harriet Mann [Olive Thorne Miller] (June 25, 1831–December 26, 1918). Born in Auburn, New York, Miller was the eldest of four children of Seth Hunt Mann, a banker, and his wife Mary Field (Holbrook) Mann. When she was eleven, the Mann family moved to Ohio and subsequently to Wisconsin, then Illinois, and finally Missouri. Late in life, she related some of her childhood adventures in an autobiographical set of children's stories titled *What Happened to Barbara*, published in 1907. Educated in private schools, young Harriet's greatest pleasure was in reading and writing, and she shared her fondness for writing with some of her friends, several of whom, like her, were to become well-known authors.

At age twenty-three, she married Watts Todd Miller, a native of northern New York. The couple began married life in Chicago. Eventually they would produce four children: two daughters and two sons. Miller was an excessively shy and private person who expressed herself best in writing, but "it was not until my children were well out of the nursery," as she later wrote, that her career as an author began. A born reformer, she first tried essays on self-improvement, but these consistently met rejection. Then a friend suggested she turn to dealing with factual material.

Between 1870 and 1882, she produced five books based on personal research. These included *Little Folks in Feathers and Fur, and Others in Neither* (1879); *Queer Pets at Marcy's* (1880); and *Little People of Asia* (1883)—all published by E. P. Dutton in New York. These did so well that she continued with other works in the same vein. By 1885, over 370 of her articles on a wide range of subject matter had been published in children's magazines such as *Our Young Folks, St Nicholas,* and the *Youth's Companion.* Her work also appeared in newspapers such as the *Chicago Tribune*, as well as religious weeklies and general interest magazines, including *Harper's* and *Scribner's.* During this period, Miller adopted the nom de plume Olive Thorne, later changing this to Olive Thorne Miller.

In 1880, Miller and her husband moved from Chicago to Brooklyn, New York. Soon after their arrival she received a visit from Sara Hubbard, an enthusiastic birder friend whom she had known in Chicago. The two women began taking walks in a nearby park. During these outings, Hubbard introduced the bookish Miller to local bird species in the area. Miller had little firsthand experience of nature but became deeply intrigued. Quickly she acquired a working knowledge of birds, their life cycles, and habitats and then continued with extensive avian studies. Hubbard became an associate of the American Ornithologists' Union (AOU), founded in 1883. Miller soon followed suit. Women were not then admitted to full membership, so both women were listed as associates—among the first women to join this important professional group.

Writing in the *Auk* (the AOU journal) in 1893, William Brewster, one of the union's founders, took note of Miller's "rare powers of observation," saying that "[i]t is a pity that writers like Mrs Miller . . . cannot be induced to record at least the most important of their discoveries in some accredited scientific journal, instead of scattering them broadcast over the pages of popular magazines or newspapers, or ambushing them in books such as . . . her *Little Brothers of the Air.*"

In response to Brewster's "gentle admonition," Miller pointed out that she was "naturally of literary rather than scientific proclivities." By writing in "unscientific" publications, she felt she could better express her "great desire to bring into the lives of others the delights to be found in the study of Nature." "[N]ever having studied scientific ornithology . . . and, moreover, having an intense love of live birds, and an almost Buddhistic horror of having them killed," she would defer to others "who . . . spend their days killing, dissecting and classifying." She would devote herself to "the study of life" and to doing what she could to preserve "the tribes of the air from the utter extinction with which they are threatened."

In this riposte, Miller tacitly challenged the standard methodology of ornithologists, who collected specimens to advance the science. In using the word *killing*, she was issuing a chiding of her own—to scientists who reduced bird populations to study them, even in those cases when the species were rare and declining. It would be many more decades before a majority of bird experts utilized field glasses, cameras, and recording devices instead of guns.

Seven years before Brewster's comments appeared, Miller had started specializing in ornithological writing with *Bird Ways* (1885), written primarily for an adult audience, though she did not forsake writing books for children. Other adult bird books followed, including *In Nesting Time* (1887), *Little Brothers of the Air* (1892), *A Bird Lover in the West* (1894), *Upon the Tree Tops* (1897), *The First Book of Birds* (1899), *The Second Book of Birds* (1901), *True Bird Stories from My Notebooks* (1902), *With the Birds in Maine* (1903), *The Bird Our Brother* (1908), and finally, a children's version of the *First* and *Second Book of Birds* titled *The Children's Book of Birds* in 1915, which was her last book. Some books were illustrated with photographs, others with bird portraits by Ernest Thompson Seton (1860–1946) and Louis Agassiz Fuertes (1874–1927), the latter probably the most proficient American bird and animal artist of his day. Their willingness to illustrate her books speaks clearly to the esteem in which Miller was increasingly held by some of the leading bird experts of the era.

Some of Miller's writing about the family life of birds tended to anthropomorphism. For example, in describing the infancy of a bird in the *Children's Book of Birds*, she wrote, "While he is blind, naked and hungry, he must have a warm, snug cradle. So when the bird fathers and mothers come in the spring the first thing they do is to find good places and build nice cradles, for they are very fond of their little ones." Robert H. Welker has commented that Miller expressed some observations about birds "in terms of human life, and has noted that the 'manners and customs' [of the birds] were interpreted as though proceeding from human motives." It could be said, however, that she adjusted her writing to her youthful audience.

Arguably, children are more likely than adult readers to need human reference points. On the other hand, he adds, she did not indulge in "misguided sentimentality." In her defense, it must be noted that Miller was considered one of the four leading American women writing about birds at the turn of the century, the others being Florence Merriam Bailey, Neltje Blanchan, and Mabel Osgood Wright.

Miller pursued other interests during her career. For instance, she remained active in several New York metropolitan clubs, publishing in 1891 *The Woman's Club: A Practical Guide and Hand-Book.* But she was never far from bird interests. For many years she maintained a bird room in her Brooklyn home, where she could observe birds flying about freely indoors during the winter months. Besides attending annual AOU meetings in various eastern cities, Miller was active in the Linnaean Society in New York. Also a member of the Audubon Society, she lectured—despite her innate reserve—on birds and their preservation. She also took a strong, continuing interest in conservation issues.

During most summers from 1883 to 1903, Miller spent several months in some of the northeastern states, the South, the Midwest, and the West, keeping careful field notes of her observations, which she would then prepare for publication each autumn. Her last such publication appeared in 1902, after which she focused her attention on writing for children. By 1915, three years before her death, Miller had published a total of two dozen books, eleven of them about birds, a good many of them for children, and nearly 800 articles of one kind and another. Most of this output was completed after she reached her midfifties, while she and her husband lived in Brooklyn.

Miller began visiting several of her adult children in California in 1902. After her husband's death in 1904, she spent her remaining years in "El Nido," a house in a Los Angeles suburb overlooking a bird- and flower-filled arroyo. This place had been especially designed for her use with the imaginative help of her younger daughter. She died in Los Angeles at age eighty-seven.

KEIR STERLING

Bailey, Florence Merriam. "Mrs. Olive Thorne Miller." *Auk* 36 (April 1919).

Bonta, Marcia Myers. *Women in the Field: America's Pioneering Women Naturalists.* College Station, TX, 1991.

Clegg, Blanche Cox. "Harriet Mann Miller." *American National Biography*. 1999.
Welker, Robert Henry. *Birds and Men: American Birds in Science, Art, Literature, and Conservation, 1800–1900*. Cambridge, MA, 1955.

Mills, Enos Abijah (April 22, 1870–September 21, 1922). A naturalist, conservationist, nature guide, businessman, and author, Mills is most noted for being the "father of the Rocky Mountain National Park." He was born and raised near Pleasanton in southeastern Kansas. Little is known of his parents, Enos A. Mills Sr. and Ann (Lamb) Mills. It is known that they were accompanied by Elkanah Lamb to the gold fields of Colorado in 1860. The memories of this were passed on by the mother to the son, and this constituted Mills's single most vivid recollection of his childhood. He was a sickly child with a major digestive disorder, and doctors were unable to provide a cure. They thought that a change of climate might help, so at the young age of fourteen, Mills set out for Colorado on his own. After several years his health returned after a change in diet.

He arrived in Estes Park, a summer resort town in Colorado, in the winter of 1884–1885. The area was virtually an unspoiled and unexplored wilderness. He homesteaded on some land approximately nine miles south of Estes Park in what was called Longs Peak Valley. He built a twelve-by-fourteen-foot cabin located so that it had a magnificent view of Longs Peak. In the summers he worked for Elkanah Lamb in the operation of the Longs Peak House and provided guided tours for guests throughout the valley, including summit climbs of the peak. Mills familiarized himself with the terrain of the peak to such an extent that he was able to find his way to the top and back in all kinds of weather. Over the years, he logged a total of 297 summit climbs.

The necessity of employment during the winters, not available around Estes Park, took him to Butte, Montana, where he worked for the Anaconda Copper Company. A quick study, Mills worked his way up to plant engineer over the winters. As the winter of 1889 approached, a fire closed the copper mines, and Mills went to San Francisco. While there, he had a chance meeting with John Muir, and a lifelong friendship followed. Muir encouraged Mills to write about what he had seen in Colorado "in a manner to make other people believe they had seen it." According to Mills, "If it hadn't been for him [Muir], I would have been a mere gypsy," that is, a wanderer, not a writer. He did wander, however, for the next ten years all over the United States and Europe but finally settled and began to write.

Mills was largely self-educated and became an avid reader. His success in writing—sixteen books and several hundred articles and essays in a span of a little more than two decades—was due less to native talent than to determination and hard work. According to Mills, his success was based on an intense enthusiasm for his subjects, a deep interest in nature, and an unusual array of material about which to write. His articles of wilderness adventures and wild animals were printed in both large-circulation magazines, for example, the *Saturday Evening Post*, where he published fifty articles, and more specialized country, outdoor, and juvenile magazines, for example, *Country Gentleman, Country Life,* and *American Boy.* He wrote to entertain, educate, and help protect nature. His first book, *The Story of Estes Park and Guide Book*, was self-published in 1905 and later republished as *Early Estes Park* (1959, 1963) by his wife Esther B. Mills. The first commerically produced work, *Wild Life on the Rockies* (1909), was published by Houghton Mifflin, publisher of a number of his succeeding works, including *The Spell of the Rockies* (1911), *The Story of a Thousand-Year Pine* (1913), *In Beaver World* (1913), *The Rocky Mountain Wonderland* (1915), *The Story of Scotch* (1916), *Your National Parks* (1917), *Being Good to Bears and Other True Animal Stories* (1919), and *The Grizzly: Our Greatest Wild Animal* (1919). Later works were published by Doubleday Page, including *The Adventures of a Nature Guide* (1920), *Waiting in the Wilderness* (1921), *Watched by Wild Animals* (1922), *Wild Animal Homesteads* (1923), *The Rocky Mountain National Park* (1924), and *Romance of Geology* (1926). Selected titles have been reprinted by the University of Nebraska Press (1988–1991).

Stripped to their essentials, Mills's writings are a mix of scientific information, field observations, and personal anecdote written to provide nature guiding in a more refined, organized, and compact form. Mills was one of the first to bring a description of the Rockies to the rest of the nation and Europe. Often enhancing the writings were photographs taken by their author, who preferred traveling with a camera instead of a gun.

In the winter of 1901–1902, he purchased the Longs Peak House from the Lambs and renamed it Longs Peak Inn. It burned down in 1906, but he rebuilt it, and it became one of the better-known hostelries in the nation. With his business obligations and writings, he began to train others for the duties of being a nature guide. According to his daughter Edna, the training of nature guides is a unique and significant characteristic of Mills. Until this time no one had formally trained nature guides, and by doing so, Mills was able to spread further his love of the outdoors. One of the early guides, Esther Burnell, stayed on to become his secretary, and in August 1918, they were married. They had one child.

Mills gave occasional talks, which led to a career as a speaker. His reputation and influence as a writer provided him with the audiences he needed for lecturing on things he believed in. His career as a lecturer assumed a life and momentum of its own. He would talk about his unique life in the mountains, his devotion to the outdoors, and forestry and forest preservation. Teddy Roosevelt appointed him to the position of government lecturer on forestry, a position he held from 1907 to 1909. During that time, he logged 2,118 addresses. His talks awakened interest in trees, wildlife protection, and outdoor adventure. He also appealed to his audience to see America first, and with this he campaigned for government involvement in providing better roads into the areas he talked about. His overriding interest as businessman, nature guide, and outdoor enthusiast was to bring people to the wilderness areas he knew and loved.

In 1909, the Estes Park town fathers created the Estes National Park and Game Preserve as a way to attract more visitors. This was the germ of the national park idea, and Mills made more than 300 appearances on behalf of the park campaign. For Mills, the broad scope of his work to have more people enjoy the outdoors became crystallized in the national park idea. Though he used a variety of arguments, including keeping tourist dollars at home, there was a lot of opposition from grazing, timber, and water interests, as well as a local group who called themselves the Front Range Settlers Group.

In addition to campaigning for a national park, Mills insisted that the federal government create a separate department to administer the national parks. To Mills, the National Forest Service was not adept to run a national park. Mills argued that a national forest was a business involved with the sale of timber and grazing rights. A forester was a man with an axe. A national park was an open-air museum that was meant to be preserved from commercial development. Scenery was a noble resource. These differences created different demands on management, and they could not be met within one department. Mills won his arguments and was very influential in the resulting legislation. In January 1915, Congress created the Rocky Mountain National Park and a year later the National Park Service.

Ironically, Mills's final campaign was against the National Park Service. Shortly after its creation, the service began granting monopolies on concessions. Mills was particularly irate at the granting of a transportation monopoly within the boundaries of a national park. When irritated, Mills became a formidable foe with an uncompromising attitude. He argued that the roads should be free for all and that transportation companies overcharged and were allied to specific hotels and inns. The Park Service argued that controlled access to national parks was necessary for sound and efficient management. They considered Mills an old-time park booster with vested interests in the Longs Peak Inn. Mills brought a lawsuit, but the Park Service prevailed. It was not until approximately one month before Mills's death that the transportation monopoly was overturned.

In a strange twist of fate, Mills was in a subway accident while in New York and suffered two broken ribs and a punctured lung. This and his exhaustive schedule of travel and lectures led to his death at Longs Peak Inn when he was fifty-two.

Mills once said, "My chief aim in life is to arouse interest in the outdoors." In this, he was overwhelming successful. It has been said that his books reached thousands, his articles and essays, millions. In addition, he was a popular lecturer. In all, he demonstrated that the average individual could be interested in stories of trees, rock formations, the works of glaciers, and wildlife. Clearly, the September 15, 1915, celebration dedicating the Rocky Mountain National Park was the high point of Mills's public career, even though its approximate 360 square miles was about half the size Mills proposed. By the time of the dedication, states were aware of national parks, and there was hardly one that did not wish to secure such a park for itself. Moreover, from the success in the United States, the national park ides spread to other countries.

Mills was to the Rockies what Muir was to the Sierras. For Mills, nature was universal, and its preservation was necessary for the enjoyment of all. The Rocky Mountain National Park was predestined to become one of the most visited and most enjoyed of all scenic reservations of the federal government. Additional legacies of Mills found around the Longs Peak area are Enos A. Mills Grove, Mills Moraine, and Mills Glacier, as well as the Mills cabin, which survives and serves as a family-run museum and store.

BYRON ANDERSON

Chapman, Arthur. *Enos A. Mills, Author, Speaker, Nature Guide*. Longs Peak, CO, 1921.

Enos A. Mills Papers. Western History Collection. Denver Public Library, Denver, CO.

Hawthorne, Hildegarde, and Ester Burnell. *Enos Mills of the Rockies*. Boston, MA, 1935.

Pickering, James H. Introduction and Notes to *The Rocky Mountain Wonderland*, by Enos A. Mills; Lincoln, NE, 1991. Reprint of original text, 1915.

Mills, Stephanie (September 11, 1948–). With her remarkable and yet shocking speech as valedictorian for her college graduation in 1969 in Oakland, California, Mills burst onto the scene of the rising environmental movement of the 1960s and 1970s. The speech, titled "The Future Is a Cruel Hoax," called for urgent and personal action to combat the crises of overpopulation and environmental degradation. She pledged to her audience that she would never bear children, a statement of commitment to the cause of protecting the earth and improving the conditions of humanity. The speech was so radical that news of it swept the media in such publications as the *San Francisco Chronicle*, the *San Francisco Examiner*, the *Oakland Tribune*, as well as national publications like the *New York Times*, *Time* magazine, and *Look*. She is known as a bioregionalist, an avid lecturer on environmental issues, and an author, and she has served as editor of a number of publications. A self-described Luddite, she prefers the life experience of living simply, without the excessive technological input and pace of so-called modern life.

Mills, the only child of Robert and Edith Mills, was born in Berkeley, California, and raised in Phoenix, Arizona. Her mother (whose maiden name was Edith Garrison) was originally from Mississippi and worked in the Women's Army Corps before marrying and becoming a homemaker. Her father Robert was raised in mining towns throughout the intermountain West and was educated in mechanical engineering. He made a lengthy career working as a salesman of mining equipment. Her paternal grandparents were from Michigan. As an only child in an environment where her father respected her mother's intelligence and Christmas presents were nonsexist, including items traditionally given to both girls and boys, Mills developed a strong sense of self, not bound by internalized limitations of what women might grow up to be.

Mills went to a public grammar school, attended a private prep school for junior high, and then went to public high school. She expressed her interest in writing in high school as a columnist for the school paper—even then she was controversial. Mills attended college at a women's school,

Mills College, in the mid- to late 1960s, pursuing a liberal arts curriculum in history, literature, philosophy, and art. Ironically, she showed little interest or aptitude for biology throughout most of her college career. She continued her interest in journalism, again serving as columnist for the school paper and also edited the literary magazine for the college. Her great influences included works of Carl Jung, Joseph Campbell, James Joyce, and Thomas Mann and her professors Hunter Hannum and Diana O'hehir. Late in her senior year, Mills became interested in environmental issues as a result of what she refers to as her first "armchair wilderness experience." One night in Berkeley, Mills picked up a Sierra Club book while visiting a friend. The book was *The Place No One Knew: Glen Canyon on the Colorado*, a book of photographs and essays by Eliot Porter and David Brower (then executive director of the Sierra Club). She was instantly amazed and transformed by the book's images and Brower's "stirring prose to inspire wilderness defense," in his efforts to show the astonishing beauty and senseless destruction of Glen Canyon for the building of a dam. Around this time she also read biologist Paul Erlich's work on population, and Mills decided to organize a campus event on ecology. As a result of her rapidly growing knowledge of and interest in environmental issues, Mills developed a passionate stance toward addressing the growing population crisis at the time of her graduation and proclaimed in the infamous valedictorian speech: "I am terribly saddened by the fact that the most humane thing for me to do is to have no children at all."

After college, she worked for a year for Planned Parenthood as a campus organizer and began giving speeches routinely in this position. Her career as a lecturer and writer on environmentalism was launched simultaneously, as her innumerable lectures on population coincided with the environmental movement at the time, and she received countless invitations to speak on environmentalism. In 1970 and 1971 she received grants from the Point Foundation to host dinner parties for nongovernmental organizations participating in the United Nations Conference on the Human Environment in Stockholm, and in Berkeley. These "parties" were known as "salons" where people were able to exchange ideas in a setting conducive to trust and the act of "breaking bread" together. Mills wrote a piece on the salons for the *Co-Evolution Quarterly*, a publication of the Whole Earth Catalogue that was timelessly influential. Years later the editors of the magazine *Utne Reader* asked her to write another piece on the concept to help them launch the Utne salons that were aimed at fostering conversations among neighbors and exchange of ideas in hopes for a better world. Mills continued living in the San Francisco Bay Area, throughout the 1970s until the mid-1980s, working as a freelance writer and editor, traveling broadly to deliver speeches on population, environmentalism, and the state of Western society. She worked as editor for the *Earth Times*, as editor of *Not Man Apart* (the Friends of the Earth publication), and as editor of *Co-Evolution Quarterly*. She has written for many publications and has continued to write for the *Whole Earth Review* throughout her life.

Although *Whatever Happened to Ecology* (1989) has often been referred to as the work that gained her notoriety/reentry into the environmental movement of the 1970s, she coedited a volume with Robert Theobald in 1973, titled *The Failure of Success*, in which one of her early essays on the state of the earth, population, and human responsibility was published. It was titled "O and all the little babies in the Alameda gardens." The essay was originally printed in *Ecotactics*, the Sierra Club handbook (1970). In this essay, she predicted state control of population (prior to its implementation in China in the 1980s) if individuals did not take voluntary responsibility as she promoted. While liberals and conservatives have criticized Mills for her stance on the issue, in this essay she acknowledged, that "birth control is regarded as tampering with nature." Yet she also points out the contradictions in our society because death control (medical intervention for living longer and reducing disease) is not considered to be tampering with nature. Despite the gloomy picture of the future she described in her valedictorian speech a year earlier, in this piece Mills espoused her value orientation that has not wavered since; she called for "passion for humanity, love of the earth, joy of existence, and hope for the future."

In *Whatever Happened to Ecology*, Mills chronicles experiences that led her to become an environmentalist, beginning with childhood in Phoenix, where she felt disconnected from the land, and describes the artificial environment where she was raised—contrasting Phoenix, which could have been "Anywhere, USA," with the actual environment of the desert and her current home in Michigan. Mills noted that in Phoenix the ethic to dominate nature becomes exacerbated because "the desert is extreme country, far less forgiving and forgetting than was the great forest or the prairie." As a child, the outdoors, she writes "was the irrigated city, and the air conditioned mall."

In this book she also discloses some of her most profound realizations on her journey as a renowned environmentalist. One such example is the time when Mills and her friend Jamie Nelson were working for the Friends of the Earth's publication *Not Man Apart*. They diverged from classic environmental journalism to comment on the irony of the World Wilderness Congress to be held later that year in the fall of 1978 in Johannesburg. They were covering the Friends of the Earth Seventh International Meeting and visiting other European Friends in Europe when they learned of the upcoming wilderness conference. Seizing the opportunity to provide commentary on the political climate of South Africa's black majority, they wrote: "Certain ideals [should] follow other ideals." Attempting to explain their position regarding a "nontraditional approach to saving the earth," they wrote, "We wish to discuss human social ecologies." Mills was unsettled to realize it was naive to expect that others (including her boss, Brower) would react as strongly against holding the World Wilderness Conference in a land of apartheid. She had erroneously assumed some automatic connection among those who wished to save the environment with ideals of social justice. Yet in posing this notion, she was certainly ahead of her time, because it was not until a decade later that the discourse of environmental sustainability began to include talk of equity and justice for all humans.

In a seemingly impulsive shift, Mills left San Francisco in 1984 to move to Michigan and soon after marry Philip Thiel. They had met at the First North American Bioregional Conference, and she felt she had finally found a life partner with whom she could begin to live a life closer to the principles of bioregionalism: respect for and connection to the land of one's local environment (or commonly called "sense of place"), resources produced as close to home as possible, and a sense of community. Unfortunately, less than a year after their self-described "storybook marriage," they were in a tragic car accident in which she almost lost her leg and Phil was almost killed. After years of rehabilitation and surgeries, they both recovered and were able to build their own house. The marriage lasted five years, and Mills ended up with the property. Mills reveals that living the bioregional life she had imagined—one of simplicity, sustainability, and wood stoves free of city accoutrements—was not as easy as she had envisioned. Yet she made her home in Michigan, the Traverse City area, and learned the ways of the land and people there, becoming a family of friends with "grown-up hippies," as she called them. She continued to explore solutions for sustainable living such as her involvement with the New Alchemy Institute. By the time of the late 1980s, however, Mills was looking for a more spiritual self-realization. Starhawk, an author and practitioner of earth-based spirituality and a promoter of nonviolent social change, became a significant influence.

Mill's next book, *In Praise of Nature* (1990), was an edited collection of essays regarding the deeper concerns of environmentalism—that is, our relationship to nature. In this collection, she highlights the philosophical context that underlies personal connections to the five elements: earth, air, fire, water, spirit. Essays in each section speak to the gifts of the specific element, its threats, and hopes for attending to solutions to those threats. The organization of the collection is exceptional in that after the opening essay of each section, reviews of ten to twelve illuminating books on the issue are provided, including reviews of the authors' contributions, such as John Muir, Aldo Leopold, Rachel Carson, and Wendell Berry. In addition to providing reviews of works by the classic figures mentioned above, it also provides reviews of practical guides regarding topics like organic gardening, green cities, and wastewater treatment. The remarkable accomplishment of the book is that Mills manages to choose selections that discuss the personal and spiritual connections the various authors and essayists reviewed have with the earth.

Mills developed a keen interest in and activism for restoration ecology, as outlined in her book *In Service of the Wild* (1995). She took it upon herself to learn as much scientific background as possible on the subject, and she combines the notions of science, deep ecology, and service to humanity in this narrative work of her journey toward bioregionalism and restoration of her chosen homeland in Michigan. Once again, she skillfully provokes the reader to contemplate our lives and connections to wild places. Mills acknowledges that aboriginal peoples occupied lands we now consider wild, but she explains that because of the sheer numbers of people in our so-called civilized society, we have done more cumulative damage to the land than previous inhabitants. She writes, "In this troublous time, ecological restoration work represents a gallant gesture of sodality with our fellow beings here on Earth, and effort to counter the trend of extinctions, and the sincerest possible expression of concern for life's future."

Her next book, *Turning Away from Technology*, was published in 1997 and is a collection of condensed transcripts of talks at conferences on technology: the "Megatechnology Conference" in 1993 and 1994 and the 1980 conference "Technology: Over the Visible Line?" In this volume, Mills and others question whether our dependency on technology has become unhealthy and counterproductive. The book also confronts issue of globalization, women, and patriarchy as regards megatechnology.

In *Epicurean Simplicity* (2002), Mills develops a sophisticated literary style and draws on the philosophies of Epicurus to write about her quest for simplicity, living as close to nature as possible while avoiding technological and fast-pace annoyances of modern life. In an interview with *First Monday*, she explained: "It's not that I'm proud of being a Luddite, it's that I'm content with the pace, volume and style of communication I do enjoy." *Epicurean Simplicity* has received widespread attention because Mills describes the path and philosophies that led her to live the simpler life that many in today's society long for. Perhaps her greatest contribution has been through sharing this voyage to simplicity through her poignant, self-reflective, and often autobiographical writing. She inspires her audience to live in a way that reflects one's beliefs in environmental causes and the profound importance of developing a regional and community connection to both people and place.

Mills lives near Traverse City, Michigan, where she is president of Bay Bucks, an organization dedicated to supporting local economic resilience, community enterprise, and social capital. Through the use of locally produced currency, "Bay Bucks," residents of northwestern Lower Michigan are able to keep resources within their bioregion. She lives in a modest home on thirty-five acres of land. She continues to accept invitations to lecture and has delivered keynote speeches at prestigious conferences such as the E. F. Schumacher Society but prefers to stay within her home region and give talks for local organizations. Her most recent book is *Live Well, Live Wild* (2006) coedited with Bill McKibben.

REBECCA AUSTIN

Mills, Stephanie. *Tough Little Beauties: Selected Essays and Other Writings.* North Liberty, IA, 2007.

Mills, Stephanie, and Bill McKibben. *Live Well, Live Wild.* North Liberty, IA, 2006.

Mongillo, John, and Bibi Booth, eds. *Environmental Activists.* Westport, CT, 2001.

"Stephanie Mills." *First Monday Interviews* 7. 6 (2002). Conducted by Edward Valauskas. www.firstmonday.org/issues7_6/mills/index.html.

Miner, Jack (April 10, 1865–November 3, 1944). A naturalist, author, researcher, and lecturer, Miner was born in Dover Center, Ohio, as John Thomas Miner. He moved to Canada at the age of thirteen. For several years, he hunted geese, selling them to the markets, but abandoned that pursuit after he developed an interest in the life and study of birds. He became a prominent bird conservationist—indeed, a leading authority in the field.

Miner began his study of Canadian geese by creating artificial ponds by flooding excavations around a small factory on the shores of Lake Erie, where he manufactured draining tiles. There he fed the Canadian geese en route their migrations north and south. The birds came to recognize this feeding site and often landed there for rest. He tagged the legs of certain geese with slips containing a few different words from the Bible and was thus able to track their migratory habits because the birds would return to his sanctuary year after year. Hunters also sent him slips from legs of geese, noting the location information where they found the tagged birds.

In 1904, Miner established a bird sanctuary at Kingsville, Ontario, and gained a reputation as a foremost bird conservationist. At this sanctuary, he also established a hospital and first-aid treatment facility for sick and injured birds. This sanctuary became very popular and served as a major attraction for visitors, many of whom were converted into conservationists or became further entrenched as such. In fact, while Miner is known as "the man who made the wild geese tame," at his Kingsville sanctuary thousands of ducks and swans took their rest and refuge during their spring and fall migratory flights.

Miner became one of the earliest conservationists in action, preceding even John Muir, who in the early years of the twentieth century became renowned for his efforts to preserve the wilderness of the Sierra Nevada mountains. A pioneer in the field of conservation, Miner established a breakthrough as an activist studying, protecting, and widely educating and influencing the American and Canadian public about birds and nature.

His more renowned naturalist predecessors of the nineteenth century, Henry David Thoreau and Ralph Waldo Emerson, wrote about the human relationship to nature, but as observers; as transcendentalist philosophers, their focus was more on the process of humankind than on the process of nature. And in his *Journals*, Thoreau wrote about "the marriage of the soul with Nature that makes the intellect fruitful." Miner, on the other hand, led by example how to preserve nature, and he tirelessly lectured the public on this issue at a time when birds were unprotected as wildlife in Canada. A most active ornithologist, he lectured in many Canadian towns and cities and as a guest speaker at many U.S. gatherings. His prominence was such that he even shared the platform with President Herbert Hoover on one occasion.

Miner authored several books on his personal experiences, including *Jack Miner and the Birds* (1923), *Jack Miner on Current Topics* (1929), and *Jack Miner and His Religion and Life* (published posthumously, 1969). His publications made him a leading bird conservationist and authority on the migratory patterns of birds, a new scientific topic and issue at the turn of the century. Although Miner was not formally educated (he did not attend any college), in 1943 he was awarded the Order of the British Empire "for the greatest achievement in Conservation in the British Empire."

Active in conservation societies and associations, Miner was a member of the Dominion Bird Protective Association, the Hamilton Bird Society, and the American Forestry Association, holding the position of vice president at the latter. In 1931, his friends and followers established the Jack Miner Migratory Bird Foundation to aid his research and to ensure his legacy. He advanced his work further with this financial support until his death at seventy-nine.

CLAUDINE L. BOROS

"Jack Miner Dead; Bird Protector, 79; Canadian Naturalist Created Sanctuary for Geese and Checked Migrations" [obit.].

New York Times, November 4, 1944.

McCay, Mary. *Rachel Carson*. New York, 1993.

Who Was Who in America with World Notables. 1968.

Muir, John (April 21, 1838–December 24, 1914). A preservationist, naturalist, author, and outdoorsman, Muir was born in Dunbar, Scotland. His father, Daniel, an austere and dour Christian zealot for whom the Bible was the only approved reading material, ran a prosperous grocery business. His mother, Ann (Gilrye) Muir, was a pious, conservative, and affectionate person who was also fond of poetry and painting. At the age of eleven he immigrated to America with his family, where his father purchased a tract of eighty acres of virgin farmland in Marquette County, Wisconsin.

Muir's childhood was not an especially tranquil or happy one. Corporal punishment was the rule both in the Scottish primary school he attended and at home, where Daniel Muir used it to make certain that John learned the Bible by heart. One of the happy exceptions to this austere upbringing was the childhood walks he took with his grandfather to the ocean and fields of Dunbar. These excursions, and grammar school nature books, were John's first exposure to the delights of the natural world, concern for which would be his life's vocation.

Muir was an active outdoorsman and lover of the wilderness long before he became an outspoken crusader for the protection of wilderness areas. When he was a student at the University of Wisconsin, he first became interested in botany, a field of study he was to enjoy and become expert in during the years to follow. After an accident in which Muir nearly lost the vision in one eye, this interest in botany, plus his lifelong love of the outdoors, was put to good use. Muir decided on a "grand sabbath day three years long" in which he would botanize and enjoy the beauty of fields and flowers. This "thousand mile walk" from Kentucky to the Gulf of Mexico was later to lead him on to Cuba, to New York, and finally to California. It was here, in California, that Muir accomplished the work for which he is most remembered today.

Muir had become concerned about conditions in and around Yosemite, in central California. As time went on, he became convinced that only by making the Yosemite a national reserve—a national park—would the trees, flowers, grasses, and other scenic wonders of the area be protected.

There were several problems in the Yosemite region. The Yosemite Valley had been under California state control since 1864 and suffered at the hands of an unpaid and weak board of governors who met only twice a year. Business interests in the valley, mainly those of hotel and stable owners, took precedence over concerns for the land and the natural beauty of the area. This lack of control over the area by the State Commission led to meadows and fields being dug up and trees being cut down to supply the needs of these commercial enterprises. The highlands surrounding the valley were totally unprotected federal lands that were suffering from extensive sheep grazing ("hoofed locusts" as Muir called them), which had destroyed most of the herbaceous vegetation.

Muir's efforts to have the Yosemite made into a national park took form when he agreed to write two articles for the influential *Century* magazine at the request of its editor Robert Underwood Johnson. Muir's articles, written to rouse public opinion, extolled the natural beauty of the area and made a case for protecting the region from the ravages of sheep and from some types of human behavior. Muir's argument was that even under the government's protection the Yosemite was being destroyed; what was needed was to set aside the area as a reserve, a place where humans, exercising due concern for the land, could still enjoy its beauty. Muir's campaign also included fervent newspaper articles and interviews, all urging passage of the Yosemite National Park bill then in Congress. Thanks largely to this incessant literary barrage, Yosemite became a national park in October 1890, during the Harrison administration. Muir and Johnson's efforts were also crucial for the creation of the Sequoia and General Grant National Parks, also in 1890, formed to save the valuable stands of sequoia timber there.

In the years after 1890, Muir continued to lobby for federal protection of the Yosemite Valley itself through its incorporation into Yosemite National Park. In 1905, during the administration of Theodore Roosevelt, this became a reality when the valley was receded to the federal government and placed under the care of the National Park Service.

The friends Muir made during his national park campaign became a more permanent organization with the creation of the Sierra Club in 1892. Muir was its first president, a post he held until his death in 1914. Among the stated aims of the Sierra Club, which is active to this day, was that of encouraging the support and cooperation of people, both in and out of government, for the "preservation of the forests and other natural features of the Sierra Nevada Mountains."

The last great battle of Muir's life, in the years 1901 through 1914, was a great disappointment to him. The Hetch Hechy Valley, a protected area included in the National Park Act of 1890, closely resembled the better-known Yosemite Valley. Named by the Indians because of its grassy meadows, the beauty of Hetch Hetchy had drawn Muir's attention in the early 1870s. To the city of San Francisco, however, the Hetch Hetchy represented a potential for a cheaper supply water and better service than they were receiving from a private company. In 1901 a bill allowing for water conduits through national parks passed Congress. Muir at once began another writing and lobbying campaign to convince officials, the general public, and President Roosevelt as well that the integrity and beauty of the entire Yosemite area must not be sacrificed. Muir proposed a less damaging compromise, which would involve using Lake Elanor a few miles north of Hetch Hechy and within the park. Nevertheless, in 1908 San Francisco received a permit, pending congressional approval, to dam Hetch Hetchy. Amid much dispute between preservationists and those who advocated the tapping of park resources, the bill to dam Hetch Hetchy languished in Congress until passed in 1913; it was signed into law by Woodrow Wilson.

Muir, who had remained active in this dispute until the very last votes were counted, was both angry and resigned. His conviction that some "compensation" would come out of this "dam-dam-damnation" was realized in 1916, after his death, with the creation of the National Park Service. Largely as a result of the loss of the Hetch Hetchy Valley, a bill was passed by which sixteen national parks and twenty-one national monuments were put under a director within the Interior Department with the stated purpose "to conserve the scenery and the natural and historic objects and the wild life therein and to provide for the enjoyment of the same in such manner and by such means as will [leave] them unimpaired for the enjoyment of future generations." Notably, the first director of the new National Park Service was a Muir disciple and Sierra Club member, Stephen T. Mather. Mather organized the department according to the principles long advocated by Muir—that wilderness life and beauty must be preserved and made accessible to the public for recreation and inspiration.

Another of Muir's legacies is the literary work he left behind. His first book, *The Mountains of California*, was published in 1894 when Muir was fifty-six years old. Taken from his journals, the book revealed Muir's intimate knowledge of the Sierra Nevada mountains. A series of articles for the *Atlantic* were published in 1901 under the title *Our National Parks*. Six years and a dozen printings later, this book established Muir as "a major voice of the wilderness." The posthumously published *A Thousand Mile Walk to the Gulf* (1916), edited from the journals that accompanied him there, makes clear the extent to which this nature trek convinced Muir of what his life's vocation should be. Through these books—indeed, through all his writings—the majesty and wonder of the wilderness come alive to the reader.

Muir died of pneumonia. He had lived an exceptionally active life in the literary and political arenas as well as in the outdoors. There was no mistaking his first love, however, since he never tired of telling people they must experience the wilderness themselves rather than just reading about it. To the end, Muir's philosophy, which guided his every action, was that America must find a place for the rights of the wilderness.

Muir married Louie Wanda Strentzel of Martinez, California, in 1880, and they had two daughters, Wanda and Helen.

ROSE ZUZWORSKY

Austin, Richard Cartwright. *Baptized into Wilderness: A Christian Perspective on John Muir.* Atlanta, GA, 1987.

Fox, Stephen. *John Muir and His Legacy: The American Conservation Movement.* Boston, MA, 1981.
Wadden, Kathleen Ann. "John Muir and the Community of Nature." *Pacific Historian* 29 (1985): 94–102.
Wolfe, Linnie Marsh. *Son of the Wilderness: The Life of John Muir.* Madison, WI, 1973.

Murie, Olaus Johan (March 1, 1889–October 21, 1963). The son of Norwegian immigrants, Murie was born and reared in the frontier community of Moorhead, Minnesota, where he and his brothers spent many hours exploring the shores of Red River. During these jaunts, Murie often played the role of a Native American. He looked to the books of Ernest Thompson Seton for inspiration and was especially impressed by *Two Little Savages* (1903), which details the wilderness adventures of two youths near their rural home. Seton's personification of wild animals impressed on Murie the belief that wildlife and humanity were not separate entities but rather players in a larger natural system. He realized at an early age that quality interactions with nature were an important source of spirituality and inspiration; without such interaction, human existence was diminished. This belief formed the foundation for his future works in wildlife biology and wilderness conservation.

After graduating from high school, Murie pursued his wildlife interests at Fargo College in North Dakota and Pacific University in Oregon, where he studied zoology. Upon graduation in 1912, he spent two years as a conservation officer in Oregon. During this period, Murie became fascinated with the Arctic region. He was given the opportunity to pursue this interest in 1914 when he accompanied a Carnegie Museum expedition to Hudson Bay to collect specimens for the museum. It was here where Murie developed a love for high-latitude wilderness. His experiences with Eskimo and Cree cultures strengthened his convictions regarding the relationship between humans and nature. Murie believed that the quality of humans' moral fabric was driven in part by wilderness encounters and that the value of such experience was related directly to the effort one makes to discover that wilderness.

In 1920, Murie joined the U.S. Bureau of Biological Survey, where he was given the task of constructing a comprehensive study of the Alaskan caribou. It was during his stay in Alaska when he was introduced to his future wife, Margaret Gillette. The caribou project was part of a larger effort by the survey to examine the possibility of domesticating the large animal. The survey was under increasing pressure from the U.S. Congress to focus its efforts on projects with economic potential. Such projects included large-scale predator warfare against wildlife, like wolves and coyotes, that were deemed destructive by ranchers and farmers. This economic focus was at odds with Murie's view of nature. During an extensive study of the elk herds in Jackson Hole, Wyoming, Murie concluded that traditional approaches to wildlife management were destabilizing many natural systems. Murie was among the first biologists to recognize the concept of an ecosystem; he found that human attempts to understand and regulate natural systems were often overly simplistic and in many cases caused a decrease in the quality of the wildlife. As such, Murie disagreed with many of the survey's policies, especially those related to predator management.

These disagreements were representative of the ongoing debate over the use of natural resources that characterized the late nineteenth and early twentieth centuries. During this period, most individuals saw wilderness from a utilitarian viewpoint. For many, nature was a resource to be utilized by all for the benefit of all. This attitude resulted in the destruction or degradation of many wilderness settings. Murie agreed that all should enjoy nature, but only through quality natural encounters; wilderness must be conserved as a source of inspiration, moral goodness, and relief from the tensions of modern society. From this belief evolved Murie's democratic ethic of wilderness preservation, in which the opportunity to interact with wilderness was equated with freedom of choice. The destruction of wilderness limited this freedom of choice. Murie's democratic ethic matured eventually to include the rights of wildlife. As time passed, Murie became less concerned with his technical work for the survey and more involved with conservation efforts.

In 1937, Murie joined the Wilderness Society, where as a council member, director, and eventually president, he campaigned for the rights of wilderness and the importance of maintaining natural settings. His writings and testimonies provided the framework for many important decisions involving wilderness conservation, including the creation and defense of the Jackson Hole National Monument. After resigning from the Biological Survey in 1945, Murie focused his attention on the threat that hydroelectric power production posed to wilderness areas. Among these efforts was the successful campaign against the construction of the Glacier View Dam in

Montana. This dam would have flooded several thousand acres of Glacier National Park and set a dangerous precedent by allowing development in a national park.

Murie continued to engage in scientific work despite the large amounts of time he devoted to conservation efforts. In 1956, he accompanied a team of researchers to the eastern portion of the Brooks Range in Alaska to gather ecological data on the region. Murie hoped these data could be used to justify the creation of an Arctic wildlife refuge. This dream was realized in 1960, when the Arctic National Wildlife Range was established.

By the time of his death, Murie had left a lasting impression on wildlife conservation in the United States. His work earned him both the National Audubon Society's Audubon Medal and the Wilderness Society's Leopold Medal. His efforts stand today as a foundation for modern conservation programs.

ADAM W. BURNETT

Kendrick, Gregory D. "An Environmental Spokesman: Olaus J. Murie and a Democratic Defense of Wilderness." *Annals of Wyoming* 50 (1978): 213–302.
Vogt, Bill. "O. J. Murie." *National Wildlife* 30 (1992): 26.
Worster, Donald. *Nature's Economy: The Roots of Ecology*. San Francisco, CA, 1977.

Murphy, Robert Cushman (April 29, 1887–March 20, 1973). An ornithologist, expedition leader, museum curator, and writer, Murphy was an early and eloquent advocate of conservation credited with speaking about *ecology* well before the term came into popular usage. He led several scientific expeditions that helped build a world-class collection for the bird department of the American Museum of Natural History. His more than forty books range in scope and tone from the scholarly *Oceanic Birds of South America* (1936) to the elegant and highly praised adventure narrative *Logbook for Grace* (1947). Among Murphy's many honors were a Congressional Medal, awarded for his Antarctic research, and a John Burroughs Medal (1938). He was also named honorary president of the National Audubon Society.

Murphy was born in Brooklyn, New York, the son of Thomas Daniel Murphy and Augusta (Cushman) Murphy. As a small boy, he frequently visited the Brooklyn Museum, where he was befriended by museum director F. A. Lucas, whose stories about whaling expeditions in the polar regions helped awaken young Murphy's interest in the natural world. In 1894 his family moved to Long Island, where he became an enthusiastic and skilled bird-watcher. In 1912, less than a year after he graduated from Brown University, he accompanied an expedition sponsored by the American Museum to the South Georgia Islands. Over the next several decades, he led expeditions to such diverse locales as Baja California, the Peruvian coast, the western Mediterranean, New Zealand, and the Caribbean.

His field researches demonstrated that the migrations of sea birds are not chaotic but rather reflect a close relationship between the birds and their environment. In the course of his expeditions, he also made a variety of nonornithological discoveries: he had two mountains named after him, as well as several species of animals and a louse. (Of the latter he reputedly said, "As a scientist, I'd as soon have a louse named for me as a mountain.")

From 1911 to 1920, Murphy held various curatorships at the Brooklyn Museum. In 1921 he took a position as assistant curator of birds at the American Museum, then became curator in 1926. He officially retired in 1955 but would remain associated with the American Museum for the rest of his life. The many exhibits and dioramas he created for the museum, and the substantial educational role they played for the public, must be considered a substantial part of Murphy's legacy.

An early advocate for a variety of environmental causes, he served as a consultant on conservation problems for several nations, including Chile, Peru, Venezuela, and New Zealand. His 1964 book *Fish-Shape Paumanok: Nature and Man on Long Island* pleaded for the protection of that region's ecosystem. He spoke out on radio in the 1960s, "demolishing the slanderous attacks on Rachel Carson that followed the publication of *Silent Spring* [1962]" and later "threw himself into one of today's great crusades, the battle to save the whales."

In 1912 Murphy married Grace Emeline Barstow, herself a writer and the woman for whom he had kept the diary that became *Logbook for Grace*. The couple frequently worked together, whether constructing dioramas at the museum or collecting specimens in the field. The couple had three children, Amos C. Barstow Murphy, Robert Cushman Murphy Jr., and Alison Barstow (Mrs. Steven L. Conner).

DAVID MAZEL

Brooks, Paul. *Speaking for Nature*. Boston, MA, 1980.

Fox, Stephen. *The American Conservation Movement: John Muir and His Legacy.* Madison, WI, 1985.

"Robert Cushman Murphy, Noted Naturalist, Dies." *Washington Post,* March 21, 1973.

Whitman, Alden. "Robert Murphy, Sea-Bird Expert, Dies." *New York Times,* March 21, 1973.

Muskie, Edmund S. (March 28, 1914–March 27, 1996). When Muskie died, President Bill Clinton praised him for his involvement with environmental issues and noted that "generations to come will benefit from his steadfast commitment to protecting the land." While he served in the U.S. Senate from Maine (1959–1980), Muskie was especially committed to problems of air and water pollution. He helped draft and was floor manager of such legislation as the Clean Air Act of 1963, the Water Quality Act of 1965, and the appropriation in 1967 of more than $428 million for pollution control.

Muskie was born in the textile mill town of Rumford, Oxford County, Maine. His father, a tailor, had fled to the United States in 1903 to escape Czarist tyranny in Poland. Muskie attended the Virginia Primary School in Oxford County and Stephens High School in Rumford. He was his high school valedictorian in 1932. Muskie worked his way through Bates College in Lewiston, Maine, serving as a waiter in college and a bellhop and dishwasher at a nearby resort during the summer. He was elected to Phi Beta Kappa and graduated, cum laude, from Bates in 1936. He earned his LL.B. degree in 1939 from Cornell University Law School and was admitted to the Massachusetts bar in 1939 and to the Maine bar in 1940.

Following enlistment in the Naval Reserve (1942–1945), he served on destroyer escorts in the Atlantic and Pacific until his release from active duty as a lieutenant. He returned to Waterville, Maine, to practice law and to become active in the Democratic Party. He served as member and secretary of the Waterville Board of Zoning Adjustments (1948–1955); was appointed district director for the Maine Office of Price Stabilization (1951–1952); served as city solicitor of Waterville (1954); was elected to the state house of representatives (1946, 1948, 1950) and was Democratic floor leader (1949–1951); served as governor of Maine (1955–1959); and was elected the first Democrat to the U.S. Senate from Maine (1958) and served until his resignation on May 7, 1980, to enter President Jimmy Carter's cabinet as secretary of state (1980–1981).

Following his term as secretary of state, Muskie practiced law in Washington. At the time of his death, he was a partner in the Washington office of Chadbourne and Parks, a large New York law firm.

When Muskie entered the U.S. Senate in January 1959, he was placed on three committees: Banking and Currency, Public Works, and Government Operations. In 1963, he became chairman of the Public Works subcommittee on air and water pollution. Consequently, he became an expert on air and water pollution and the chief sponsor and floor manager of the Clean Air Act of 1963 and the Water Quality Act of 1965. To accomplish this task, Muskie had first created a national environmental constituency. He held hearings in major cities across the country in 1963, 1964, and 1965.

In 1955, the federal government had decided that the problem of air pollution needed action on the federal level since many state and local governments had already legislated concerning air pollution. The congressional action resulted in the Air Pollution and Control Act of 1955, the first piece of federal legislation on this issue. This act provided funds for federal research on air pollution. It was an act to make the nation more aware of the environmental hazard.

The Clean Air Act of 1963 dealt with reducing air pollution by setting standards for stationary sources such as power plants and steel mills. It did not take into account mobile sources of air pollution, which had become the largest source of many dangerous pollutants. Once these standards were set, the government also needed to determine deadlines for companies to comply with them. Amendments to the Clean Air Act were passed in 1965, 1966, 1967, and 1969. These amendments authorized the secretary of health, education, and welfare to set standards for auto emissions, to expand local air pollution control programs, to establish air quality control regions, to set air quality standards, to set compliance deadlines for stationary source emissions, and to authorize research on low emissions from fuels and automobiles.

The 1970 amendment to the Clean Air Act continued to set new and demanding standards and increased funds for air pollution research. Clean Air Act amendments of 1990 increased substantially the authority and responsibility of the federal government. This act authorized programs for acid disposition control; it authorized a program to control 189 toxic pollutants, including those previously regulated by the National Emission

Standards for Hazardous Air Pollutants; it established permit program requirements; it expanded and modified provisions concerning the attainment of national ambient (environment) air quality standards; and it expanded and modified enforcement authority.

Muskie's Subcommittee on Air and Water Pollution worked as aggressively on water quality as it had on clean air. Water pollution was viewed as a state and local problem by the federal government when the Federal Water Pollution Control Act of 1948 was enacted. The act provided state and local governments with technical assistance funds to address water pollution problems, including research. Federal involvement was limited to matters involving interstate waters and only with the consent of the state in which the pollution originated.

During the latter half of the 1950s and well into the 1960s, water pollution programs were shaped by four laws that amended the 1949 statute. They dealt largely with federal assistance to municipal dischargers and with federal enforcement programs for all discharges. During this period, the federal role and federal jurisdiction were gradually extended to include navigable intrastate, as well as interstate, waters. Water quality standards became a feature of the law in 1965, requiring states to set standards for interstate waters that would be used to determine actual pollution levels. By the late 1960s, there was a widespread perception that existing enforcement procedures were too time-consuming and that the water quality standards approach was flawed because of difficulties in linking a particular discharger to violations of stream quality standards. This dissatisfaction with existing regulations, coupled with increasing public interest in environmental protection, led to further amendments in 1972, 1977, 1981, and 1987.

The 1972 statute mandated that municipal and industrial wastewater be treated before being discharged into waterways. It also provided for increased federal assistance for municipal treatment plant construction. Additional amendments in 1977 and 1981 continued federal assistance for treatment plant construction and also required the use of "best practicable control technology" to clean up waste discharges. The 1987 amendments were devoted to nonpoint source pollution (storm water runoff from agricultural lands, forests, urban areas, and so on). Federal financial assistance would support demonstration projects and control activities.

Muskie's interest in environmental issues continued after he left public office in 1981. In a 1990 article published in the *Christian Science Monitor*, he urged the United States to support a multinational environmental policy and to endorse giving the United Nations more power to deal with environmental problems on a worldwide basis. A summer 1992 article by Muskie again stressed the need for the U.S. government to formulate policies to avert the global environmental crisis in dealing with the problems of population control, depletion of natural resources, and pollution.

Shortly after Muskie's death in 1996, the Edmund S. Muskie Foundation was organized to honor and continue his lifelong commitments to preserving the environment and fostering civic responsibility. The foundation funds programs that extend Senator Muskie's work by holding seminars and forums on environmental, foreign policy, public integrity, and intergovernmental relations issues. It also supports the Edmund S. Muskie Archive at Bates College in Lewiston, Maine, and contributes to the programs of the Edmund S. Muskie School of Public Service at the University of Southern Maine in Portland.

S. JOAN RYAN, C.S.J.

Copeland, Claudia. *Clean Water Act: A Summary of the Law.* Washington, DC, 2002.

Lippman, Theo, and Donald C. Hansen. *Muskie.* New York, 1971.

Muskie, Edmund S. "The Global Environmental Crisis." *Boston College Environmental Affairs Law Review* (Summer 1902): 19.

———. "More UN Clout on Environment." *Christian Science Monitor,* November 16, 1990.

Nevin, David. *Muskie of Maine.* New York, 1972.

U.S. Senate. "Memorial Tributes Delivered in Congress: Edmund S. Muskie, 1914–1996. Late a Senator from Maine." Washington, DC, 1996.

Nearing, Scott (August 6, 1883–August 23, 1983). The father of the back-to-the-land movement, Nearing was an environmentalist, radical economist, pacifist, self-subsistent organic farmer, popular lecturer, and prolific writer. An author of hundreds of articles and over fifty books, he is perhaps best known for *Living the Good Life: How to Live Simply in a Troubled World* (1954), which he cowrote with his wife Helen.

Nearing chose to end his life in his house on his Harborside, Maine, hundred-acre farm just weeks after celebrating his one hundredth birthday. A stoic and extremely hardworking man, very little of the time during that century-long life was not put to productive use. Although a political activist during his entire adulthood, Nearing did not truly find his place in the world until settling onto his first farm of sixty-five acres of land on a mountain slope in Pike Valley, Vermont. He and Helen wrote in *Living the Good Life*, "So we went back there on a chilly day in autumn of 1932, and signed an agreement to buy the place."

Nearing was born in Morris Run, Pennsylvania, to Louis Nearing, a merchant, and Minnie (née Zabriskie) Nearing. He had three sisters, Mary, Dorothy, and Beatrice, and two brothers, Guy and Max. Nearing attended school in Philadelphia and graduated from Central Manual Training in 1901. He went on to study oratory at what was to become Temple University and economics at the University of Pennsylvania, where he was awarded his B.A. as well as, four years later, his Ph.D. In 1906, he began to teach as a professor of economics at the University of Pennsylvania and Swarthmore.

In 1911 he turned his attention to one of the more pressing issues of the day, that of child labor. He wrote and had published *The Solution to the Child Labor Problem*. This work alienated him from the wealthy supporters of higher education who prospered from child labor. In 1912 he, along with his first wife Nellie Seeds, wrote *Women and Social Progress*. He and Nellie had two sons, John and Robert. By 1915 his liberal positions caught up with him, and he was dismissed from the University of Pennsylvania.

Nearing opposed U.S. involvement in World War I and wrote a thirty-two-page pamphlet titled *The Great Madness* that, in part, encouraged resistance to the draft. He was fired from the Wharton School of Business, and he was indicted by a federal grand jury, tried, and acquitted. Despite the fact that he was vindicated in a court of law, his reputation in the academic community was severally tarnished.

For a while he found solace in the Communist Party but was eventually expelled because of his opposition to Leninist imperialism. He remained, fundamentally, an adherent of Marxist philosophy for the rest of his life and a staunch anti-imperialist. One rationale for Nearing's anti-imperialist stance was rooted in environmentalism. He argued that the dominating country would strip the natural resources of a subjugated nation to the point of detriment to that country's, and the world's, ecology.

By the outbreak of the Great Depression, Nearing had left Nellie and was living with his future wife Helen. (Ultimately, he and Helen were formally married in 1947, not long after Nellie's demise.) Helen was an organist, flutist, and concert violinist when she met him. She embraced Nearing's radical politics, supported his outspoken individualism, and would go on to coauthor six books with him.

Nearing believed that the Depression marked the collapse of capitalism. He and Helen determined that they would stand a better chance of survival if they could achieve a greater degree of self-sufficiency in a rural setting. He disapproved of a social order "activated by greed and functioning through exploitation, acquisition and accumulation." Nearing believed:

> If profit accumulation in the hands of the rich and powerful continued to push the economy toward ever more catastrophic depression; if the alternative to depression under the existing social system, was the elimination of the unmarketable surplus through the construction and use of ever more deadly war equipment, it was only a question of time before those who depended upon the system for livelihood and security would find themselves out in the cold or among the missing.

The Nearings wanted to set up a "semi self-contained household unit, based largely on a use economy, and as far as possible, independent of the price-profit economy which surrounds us." They wrote in *Living the Good Life*: "We decided to liberate and disassociate, as much as possible, from the cruder forms of exploitation, the plunder of the planet; the slavery of man and beast; the slaughter of men in war, and of animals for food." They searched for a place to live: "Finally, we settled on Vermont. We liked the thickly forested hills which formed the Green Mountains." In 1932 they purchased the land in Pike Valley, Vermont, for $1,100. It was "[a] rundown farm, with a wooden house in poor repair, a good-sized barn with bad sills and a leaky roof."

Nearing neither smoked tobacco nor consumed alcohol. At five feet eight inches and 160 pounds, Nearing was lean, athletically built, strong, and incredibly fit. He was a tireless laborer on his farm where he and Helen built, by hand and without the aid of machinery, a fieldstone house. Scott believed, "Everybody ought to do something with their hands instead of becoming parasitic to a post-industrial society." They rejuvenated the soil through organic means.

> Each day brought garbage from the house, weeds and tops from the garden, grass cutting and trash from the flowerbeds. We used every bit of available organic matter. These materials went on the current compost pile until they filled 4 to 6 inches in depth. Top soil like every other aspect of nature can be plundered and depleted by wrong practice until it is all but sterile. Reverse these practices, build a living soil and vegetation flourishes as it is reported to have done in the Garden of Eden.

They also wrote, "There is an old saying that we reap what we sow. Nowhere is this more evident than in the treatment of the good earth."

They dealt with the fall and long winters by utilizing solar-heated greenhouses, to "collaborate with Mother Nature rather than having to deal with the electric and fossil fuel companies." In keeping with their rejection of the trappings of the modern capitalist society, the Nearings had no radio (later no television) or telephone. Amenities included a hand pump and a black-iron sink in the kitchen and an outhouse at one end of the woodshed.

Out of a reverence for life, Nearing and Helen were strict vegetarians who did not believe in hunting, fishing, or the use of animals for the purpose of work, for food production, or as pets. They regarded pets as creatures in bondage. Typically, they left their property when hunting season came because they were unable to stop poachers from killing deer and did not want to be present when it happened. Nearing said, "We don't like to stay and see our friends killed."

In his autobiography *The Making of a Radical* (1972) Nearing wrote: "I became a vegetarian because I was persuaded that life is as valid for other creatures as it is for humans." They were convinced, based on their own experience, that people of almost any age, "with a minimum of health, intelligence and capital, can adapt themselves to country living, learn its crafts, overcome its difficulties and build up a life pattern rich in simple values and productive of personal and social good."

Ski resort developers began to intrude on the area near their farm, and so the Nearings sold the place to novelist Pearl S. Buck in 1951. They moved to a hundred-acre piece of property, called Forest Farm, in Harborside on the Maine coast.

Nearing was a spellbinding speaker who traveled the world, lecturing on international affairs, social engineering, and humans' place in the environment. His travels were curtailed in 1957, however, when his passport was revoked for going to Communist China.

By the 1960s thousands of pilgrim visitors began to flock to Forest Farm to bask in Nearing's wisdom and to learn the subtle machinations of self-subsistent farming. Nearing's individualism, revolutionary ideology, and rejection of industrialized society appealed to the countercultural youth of the era who were among the greatest numbers of his visitors. As a result of this newfound popularity, *Living the Good Life* was reissued in 1970 and became the bible for the back-to-the-land movement. In addition, in response to the demand for knowledge, the Nearings established the Social Science Institute, which disseminated information on important common concerns.

Nearing remained active throughout the rest of his life. He lectured, wrote, worked on the farm, and even appeared, as himself, in Hollywood star Warren Beatty's Oscar-winning film *Reds* (1981). When at last, as he approached his one-hundredth birthday, he began to slow down significantly, and he decided that he was "not going to eat anymore." At first he abstained from all solid food and drank only fruit juice and water. After a few weeks he drank only water and soon thereafter passed away peacefully.

Nearing wrote, "Simplicity, serenity, utility and harmony are not the only values in life, but they are among the important ideals, objectives and concepts which a seeker after the good life might reasonably expect to develop in a satisfactory natural and social environment."

MICHAEL HAYES

Freeman, Joseph, and Scott Nearing. *Dollar Diplomacy.* New York, 1925.

Nearing, Scott, and Helen Nearing. *The Making of a Radical: A Political Autobiography.* New York, 1972.

Sherman, Steve, ed. *A Scott Nearing Reader: The Good Life in Bad Times.* Metuchen, NJ, 1989.

Nelson, Gaylord (June 4, 1916–July 3, 2005). A Wisconsin governor and senator, Nelson is chiefly known as the originator of Earth Day, though his other environmental achievements, both in and out of public office, also deserve recognition.

Nelson was born in Clear Lake, Wisconsin, a town of about 700 residents; where main street—three blocks long—ended, the wilderness began. His father, Anton, was a country doctor and served several times as village president; his mother, Mary (Bradt) Nelson, was a registered nurse and community activist. His grandfather was a founder of the Republican Party in Wisconsin. Nelson recalled that when he was ten years old, he became seriously interested in politics after his father took him to hear Robert La Follete, the Progressive Party leader, deliver a speech from the rear platform of a train.

As a youngster, Nelson spent a good deal of time in the wood marshes and lakes surrounding Clear Lake. With his friends, he enjoyed trapping snakes and gophers and camping out. He always had pets, including dogs, pigeons, and a skunk. In 1934 Nelson graduated from Clear Lake High School, where he was captain of the football and basketball teams. In 1939 he received a diploma from California's San Jose College and thereafter entered Wisconsin Law School from which he graduated in 1942. Subsequently, he spent four years in the U.S. Army, participating in the Okinawa campaign, before launching his legal and political careers. While serving overseas he met an army nurse, Carrie Lee Dotson, whom he married in 1947.

After the war, he encountered Aldo Leopold, author of *Sand County Almanac* (1948), which Nelson considered "probably the most impressive and influential environmental book of the twentieth century." The book doubtless helped to shape Nelson's environmental ethic.

Nelson had his first political experience when, as a teenager, he conducted an unsuccessful campaign to plant elm trees alongside roads running into Clear Lake. In 1946 he incurred another defeat running as a Progressive Republican for the Wisconsin legislature. But two years later, campaigning as a Democrat, he was elected to the state senate, representing the people of Dane County, and served them for a decade. In 1958 Wisconsinites elected Nelson governor.

Nelson entered office at a time when there was an increasing call for more out-of-doors recreational areas. In August 1960, for instance, he visited ten crowded state parks and announced that Wisconsin should "double" and "redouble" facilities and land acquisitions. As a result, the governor championed an audacious scheme to enlarge his state's publicly owned lands. He was successful, as two years after entering office Nelson secured legislation backing the Outdoor Resources Action Program (ORAP), a ten-year, $50 million program—underwritten by a penny-a-pack cigarette tax—to finance the state's purchase of a park land and wetlands. ORAP provided for the recreational sites, throughout Wisconsin, to function both as wildlife habitats and as public parks. Thousands of acres of land would be untouched by development; the acreage, said Nelson, would be places "where people would always be able to enjoy clean air and water," as well as "the beauty and peace of nature."

Wisconsin was the first state in the nation to regulate laundry detergents that polluted the water supply. Indeed, under Nelson his state became a trailblazer in environmental protection.

In 1962 Nelson was elected to the U.S. Senate, starting a career that endured eighteen years. In 1965, three years after the publication of Rachel Carson's *Silent Spring*—an eye-opening account of the perils of DDT—Nelson introduced the first legislation to bar the use of DDT. Evidence from *Silent Spring* and other sources showed that the chemical—developed during World War II—not only killed intended targets, harmful bugs, but also threatened birds, fish, and frogs, as well as bees and other beneficial insects. Congress eventually responded to Nelson's initiative with the Federal Pesticide Control Act. He also introduced the National Hiking Trails System, a scheme of hiking trails preserved by the federal government.

The first Earth Day—an event closely associated with Nelson—occurred in 1970. The Wisconsin senator was searching for some way to alter Americans' cast of mind about the environment. The year before, Nelson had read about antiwar students organizing teach-ins—including speeches and classes—about the Vietnam War. "Why not have a teach-in on the environment?" Nelson asked himself. He announced his idea on the Senate floor. "I am proposing a national teach-in on the crisis of the environment," he declared. "The idea took off," Nelson later reported. "I was getting calls from people all over the country asking what they could do." And ninety of his Senate colleagues requested copies of his speeches. "The reason that Earth Day worked," Nelson realized, was that "it organized itself. The idea was out there, and everybody grabbed it."

The first Earth Day, April 22, 1970, was a national affair. Over 20 million Americans took part in the day's events. Nelson wanted a demonstration by so many people that politicians could not ignore it. And, said Nelson, "that's just what Earth Day did." In New York City, schoolchildren cleaned streets, while community bodies planted trees in city parks. Elsewhere students picked up trash, and in New Orleans they christened the oil industry as the "Polluter of the Month."

The political establishment responded. In a number of states, including New Jersey, Massachusetts, Maryland, and Ohio, legislatures recognized Earth Day by passing regulations to safeguard the environment. "It seemed that everyone was for Earth Day," observed the *New York Times*.

In January 1970, three months before the first Earth Day, Nelson gave an important speech in the Senate offering his "environmental agenda." Among the items on his agenda was a constitutional amendment that gave "[e]very person . . . the inalienable right to a decent environment," guaranteed by the "United States and every state." He also recommended steps to rid America in the 1970s of the massive pollution "from the internal combustion engine, pesticides, detergent pollution and aircraft pollution." Moreover, he proposed the reduction of ocean contamination, the establishment of an environmental advocacy agency, and the adoption of a national land-use policy to diminish the random mixture of urban sprawl and air, water, and land adulteration. Finally, Nelson called for a new ecological ethic that heeded the well-being of both present and future generations.

Over the decade that followed, Nelson made headway on some of his "environmental agenda." For instance, the Water Quality Control Act sponsored by the Wisconsin senator called for the cessation of polluting the planet's oceans. Other legislation introduced and successfully battled for by Nelson included the Clean Air Act and the National Lakes Preservation Act, the intention of which was to renew endangered waterways. Among other measures he espoused were the Clean Water Act, which was the first comprehensive federal legislation to cope with water pollution; the federal Environmental Pesticide Control Act, which enhanced federal enforcement authority over pesticides; and the Marine Mammal Protection Act, whose goal was to halt the slaughter of ocean mammals. Of course, in each instance, Nelson required the support of Senate committee chairmen, House of Representative politicos, and the presidents (Richard Nixon, Gerald Ford, and Jimmy Carter).

Though he had achieved a great deal in the Senate, Nelson lost his bid for reelection in 1980. Friends and supporters were shocked at his defeat, but Nelson took it in stride. "When you're licked, you're licked," he said. "This is no job for crybabies."

Nelson remained active as an environmentalist, signing on as chairman of the Wilderness Society, a Washington-based organization whose mission is to safeguard public lands and keep them wild. In his new position, Nelson was the society's voice, delivering speeches on college campuses and elsewhere. Subsequently, he acquired the designation of counselor, a kind of "utility infielder or designated hitter." As such, he informed society on policy issues and offered his advice—from a long career in public office—to staff members.

The culmination of Nelson's career was the twentieth anniversary of Earth Day. On April 22, 1990, over 55 million Americans came together to demonstrate their concern for the beleaguered planet. Indeed, it was a day of apprehension and remonstrance in over one hundred countries, from Brazil to Kenya. Clearly Earth Day had become a worldwide affair.

Fittingly, Nelson was named honorary chairman of Earth Day 1990, and he spoke to large crowds. "Unless we change our ways," he warned, "our legacy will be one of pollution, poverty, and ugliness." "The test of our conscience," he continued, "is our willingness to sacrifice something today for the future." And he cautioned that this "is a test we must take—and pass." Moreover, he counted on the younger generation to carry the environmental burden. "You and the generation after you are going to have to do it," he said. "And I believe you will."

From his years as governor of Wisconsin, Nelson was an early advocate of conservationist practices. As a senator, the founder of Earth Day introduced and shepherded legislation to protect oceans, rivers, and lakes, as well as the air, and to outlaw DDT and other pollutants.

His hefty legacy was acknowledged with medals and honors, including the Presidential Medal of Freedom, America's loftiest civilian award. President William Clinton, in conferring the award in 1995, aptly remarked: "As the father of Earth Day, he is the grandfather of all that grew out of the event." The United Nations conferred on Nelson an Environmental Leadership Medal and an Only One World Award. A Wisconsin state park, near Madison is named after him.

Nelson passed away at his home in Kensington, Maryland, a suburb of Washington, at age eighty-nine. The cause of death, according to his family, was cardiovascular failure.

"He was just an incredible person: humble, funny, proud of his roots . . . and never changed by the power pomp of the offices that he held," said Wisconsin governor Jim Doyle.

RICHARD P. HARMOND

Christoferson, Bill. *The Man from Clear Lake: Earth Day Founder, Senator Gaylord Nelson*. Madison, WI, 2004.

"Gaylord Nelson." *Chicago Tribune*, July 4, 2005.

Shulman, Jeffery, and Teresa Rogers. *Gaylord Nelson: A Day for the Earth*. Frederick, MD, 1992.

Webber, David J. "Gaylord Nelson: The Founder of Earth Day." *MU Political Science*, January 1996.

Nice, Margaret Morse (December 6, 1883–June 26, 1974). One of the most important women in the history of North American ornithology, Nice was born in Amherst, Massachusetts. Her father, Anson D. Morse, was a professor of history at Amherst College. She lived in secluded, rural conditions during her youth and while attending Mt. Holyoke College in South Hadley, where she began undergraduate studies in 1901. Milton B. Trautman suggests that the pastoral environment of Margaret's youth "appears to have awakened her absorbing interest in nature," a suggestion corroborated by her posthumously published autobiography *Research Is a Passion with Me: The Autobiography of Margaret Morse Nice* (1979). Another factor undoubtedly contributing to her interest in nature was the Wednesday-off policy at Mt. Holyoke, which provided her with time for weekly, daylong rambles in the countryside around the college and allowed her to explore a myriad of natural habitats and their inhabitants. She graduated in 1906, having spent one of the previous five years in Europe, where she exercised her lifelong passion for languages, seven of which she read fluently.

In 1906 she began graduate work at Clark University in Worcester and was a fellow from 1907 to 1909. In the latter year she married Leonard Blaine Nice (pronounced like "niece"), a graduate student in physiology at Clark. After Blaine, as he preferred to be called, graduated in 1911, the couple lived in Boston for two years, after which a move to Oklahoma in 1913 ended the New England phase of Margaret's life, although she returned periodically during the ensuing sixty years, including a trip back in 1915 to receive an

M.A. in zoology from Clark for work done six years earlier on "The Food of the Bob-white."

The Nices resided in Norman, Oklahoma, until 1927. During this period in her life, she balanced the demands of her growing family, eventually to include four daughters, one of whom died at age nine, with the demands of her growing passion for research. Although she at first focused her research efforts on child psychology, publishing eighteen articles between 1915 and 1933 about her own children's maturation, her interest in ornithological research gradually took control of her time and energy, especially after 1919. Her first significant ornithological publication—*The Birds of Oklahoma*—appeared in 1924. Thereafter, hardly a year went by that she did not publish at least one article about birds; eventually she published 250 articles and seven books on matters avian.

During the Oklahoma period, Nice's concern for conservation also appeared. In particular, she took up the cause of the mourning dove and almost single-handedly prevented an extension of the open hunting season on doves into August. She accomplished this feat by collecting, with help from her husband and daughters, considerable field data demonstrating that doves continue to breed in August and, therefore, should not be hunted while still attending young in the nest.

In 1927 the Nices moved to Columbus, Ohio, initiating one of the most fruitful and famous periods ever experienced by an American studying birds. The family moved into a home situated on a bluff above the Olentangy River. To the weedy, floodplain-dominated area along the river and between bridges north and south of her home Nice gave the name "Interpont." On this site she commenced a study of the lives of song sparrows (*Melospiza melodia*), including famous "M4," whose career Nice followed on an almost daily basis for eight years. She amassed a collection of data about individually color-marked birds unlike any ever collected previously. From this immense data set came the material for her most famous work. First published in condensed form in the *Journal für Ornithologie* (1933–1934), perhaps the foremost ornithological journal of the time, *Studies in the Life History of the Song Sparrow* was later published in two volumes, the first (*A Population Study of the Song Sparrow and Other Passerines*) as volume 4 of the *Transactions of the Linnaean Society of New York* (1937) and the second (*The Behavior of the Song Sparrow and Other Passerines*) as volume 6 of the *Transactions of the Linnaean Society of New York* (1943). These volumes established her reputation as an ornithologist of international prominence; they remain valuable references. They may also be viewed as pioneering works in ecology and ethology. As Bruce Peterjohn (head of the U.S. Fish and Wildlife Service's Breeding Bird Survey) claims, Nice was the first true "ecologist," trying to understand how species interact among themselves, with other species, and with the physical environment in order to survive (pers. com.).

This enormously productive period in Nice's life ended in 1936 when the family moved to Chicago, Illinois, where she resided for the remainder of her life, cut off, as she often wryly noted, from almost all birds except for the occasional house sparrow. However, she continued regular publication of articles and books on birds. In addition, during this period she also served on the editorial staff of the ornithologicial journal *Bird-banding*, to which she contributed reviews of ornithological works, including many in languages other than English. In all, she wrote and published over 3,300 reviews.

Among Nice's books from this period, *The Watcher at the Nest* (1939; reprinted 1967 Dover Publicatons) is noteworthy not only for its ornithological content but also, as Marcia Myers Bonta notes, because it airs her concern about a multitude of environmental issues. These include preservation of roadside vegetation as a source of bird habitat; condemnation of poor timber practices and misuse of wildlife refuges; and opposition to poisoning prairie dogs, killing albatrosses on Midway Island, and—as early as 1944—using lead shot in waterfowl hunting. Lead shot was not entirely banned in the United States until the late 1980s, offering testimony to Nice's early prescience about this deleterious practice. During the remainder of her life, she was outspoken about these and other environmental issues.

Nice deserves a place among American environmentalists for several reasons. Foremost among these is the fact that she nearly single-handedly initiated the ecological study of animal species with her famous song sparrow work; testimony to the importance of this study comes from such noted scientists as Nikolaas Tinbergen, Ernst Mayr, and Konrad Lorenz. She also deserves credit for proving that women could perform superior research and for increasing the opportunities for women in a formerly male-dominated field. Finally, from the time of her effort to prevent August hunting of mourning doves in Oklahoma, she was an outspoken advocate of numerous environmental causes, many of which have only in recent years ended in success. Without the long years of preliminary advocacy by

Nice and many like her, issues such the banning of lead shot in waterfowl hunting might yet be unresolved.

STEPHEN J. STEDMAN

Bonta, Marcia Myers. *Women in the Field: America's Pioneering Women Naturalists.* College Station, TX, 1991.

Ehrlich, Paul R., David S. Dobkin, and Darryl Wheye. *The Birder's Handbook: A Field Guide to the Natural History of North American Birds: Including all Species that Regularly Breed North of New Mexico.* New York, 1988.

Nice, Margaret Morse. Papers. Department of Manuscripts and University Archives, Cornell Univ. Library, Ithaca, NY.

Parkes, Kenneth C. "Margaret Morse Nice." *Wilson Bulletin* 86 (1974): 301–302.

Trautman, Milton B. "In Memoriam: Margaret Morse Nice." *Auk* 94 (1977): 430–441.

Oberholtzer, Ernest Carl (February 6, 1884–June 6, 1977). Born in Davenport, Iowa, a town on the Mississippi River, as a boy Oberholtzer was impressed most by "the long rafts of logs that came down the Mississippi . . . out of that vast unknown North!" Though he wrote no books and published relatively few essays (including adventure stories for boys and tracts of conservation philosophy), his lobbying efforts on behalf of preserving the natural resources and cultures of the North Woods of Minnesota and Ontario made him one of the key figures in twentieth-century conservation politics. Oberholtzer's proposals for managing land use in the Boundary Waters area dominated political discussions for many years and transformed the Quetico Superior region into an international testing ground for the idea of wilderness preservation.

From 1903 to 1908 Oberholtzer attended Harvard, studying landscape architecture under Frederick Law Olmsted. In 1906 he made his first visit to the North Woods, embarking on a canoe trip out of Ely, Minnesota. Although informed by doctors a few years later that he had only a short time to live because of a heart condition brought on by rheumatic fever, he proved them wrong by paddling some 3,000 miles in the Rainy Lake watershed the summer of 1909, returning to the region that had enchanted him from a distance when a boy. During this trip, he developed a lifelong interest in North Country exploration and the lifeways of its Native peoples. His canoeing companion was Billy Magee, an Ojibwe twenty-three years his elder who retained much of the tribe's traditional knowledge. In 1912, when Oberholtzer proposed that they journey from Winnipeg to the Northwest Territories by canoe, Magee responded, "Guess ready go end earth." Neither man had ever been north of the Rainy Lake watershed.

Having read the works of J. B. Tyrell, one of the greatest of all modern geographers, Oberholtzer decided to approximate Tyrell's route in 1894 across the Barrens. Several times he tried to employ a guide but to no avail; he and Magee carried on anyway. Embarking from the Pas northwest of Lake Winnipeg, they paddled and portaged to Pelican Narrows and Reindeer Lake in Saskatchewan to the Hudson Bay Company outpost at Fort Hall, and Oberholtzer became the first white man to see Nueltin Lake since

Samuel Hearne in 1772. From there they ran the Thlewiaza River to Hudson Bay; this portion of their route was virtually terra incognita, and Oberholtzer realized his dream of experiencing the North in the manner of the classic explorers of the eighteenth and nineteenth centuries. Occasionally they encountered a hunting party of Chippewyan Indians. From the mouth of the Thlewiaza they followed an Eskimo party south to Fort Churchill, then proceeded on to York Factory, returning to Lake Winnipeg via the Hayes River. All told the journey took six months and covered some 2,000 miles; they barely won the race with winter.

Though he kept a journal and took many photographs, Oberholtzer never published an account of this epic journey, explaining, "It was my ambition at the time to return to the general locality for five years, but that could never be carried out and my life has been very crowded ever since." Indeed it was. He moved to the Rainy Lake area permanently in 1913 and lived among the Ojibwe for the next five years. Eventually he acquired several islands, on one of which, Mallard, he built cabins to accommodate himself, visitors, and his growing book collection (at the time of his death some 12,000 works were shelved in nine different cabins). He used Mallard Island as a base from which to travel to Ojibwe villages by canoe and snowshoe within a radius of 250 miles, his purpose being "to record in their own language, and from their own dictation, their old legends and songs."

He became a conservationist partly out of a desire to preserve Ojibwe culture and partly to thwart the efforts of local timber baron Edward W. Backus, who proposed dam construction on Rainy Lake and its tributaries to raise water levels so that logging could be facilitated. In response, Oberholtzer developed an international land management program that divided some 15,000 square miles of the Quetico Superior into three zones. The innermost core would consist of wilderness, "with no roads and no sign of human activities except such as pertain to the present life of native Indians." A buffer area would surround the wilderness, which in turn would be encircled by a larger area containing towns, highways, businesses, and homes. Forestry would be permitted, except along shorelines, as would fishing and hunting.

Oberholtzer's plan was never adopted in its entirety. But it was on the cutting edge of the regional planning movement in vogue in the United States at the time. Benton MacKaye, engineer of the Appalachian Trail, was a friend of Oberholtzer's, and the two exchanged ideas. The plan also anticipated the movement today known as bioregionalism: defining cultural boundaries by ecological systems rather than arbitrary political lines drawn on a map. Arthur Carhart, another important landscape architect of the period, also proposed a management plan for the Quetico Superior in 1922, but it featured a wilderness in which "motorboat highways" provided access to hotels and resorts. Carhart's plan was set aside in favor of Oberholtzer's.

Backus's plan for damming the Rainy Lake watershed was defeated by Oberholtzer and others through the efforts of the Quetico Superior Council, of which Oberholtzer served as president from 1927 to 1964. The organization then lobbied to create an international wilderness park straddling the border between the United States and Canada. With the help of several conservation groups, the Shipsted-Nolan Act was passed in 1930, which prohibited altering the water levels of Rainy Lake and banned shoreline logging. It was the first law ever enacted by Congress specifically to preserve wilderness.

With that law as a foundation, Oberholtzer joined with Bob Marshall, Aldo Leopold, Sterling Yard, MacKaye, and others to form the Wilderness Society in 1935; he served on its executive council until 1964. With another, more famous Minnesota conservationist, Sigurd Olson, he campaigned for the creation of an airspace reservation over the Quetico Superior region in the 1940s when the incursions of floatplanes threatened the area's wilderness character. In 1949 President Harry Truman, in response to their lobbying, signed an order reserving airspace over the wilderness to an altitude of 4,000 feet and limiting landings to government and rescue operations—the first such restriction imposed on aircraft in the United States for nonmilitary purposes.

In the 1940s Oberholtzer continued his study of Native American cultures, making numerous trips to Ojibwe villages in the border country. He set out for these communities motivated in part by the pastoral impulse: the work of a lobbyist was sometimes debilitating and discouraging, and a return to a "simpler," "purer" way of life that seemed more in tune with the workings of the natural environment served as a tonic. But ultimately Oberholtzer was never able to complete and publish his lifelong study of the Ojibwe because of the increasing demands made on him to defend and protect the integrity of the Quetico Superior wilderness.

The Wilderness Act was passed in 1964, Voyageurs National Park was created in 1977, and the Boundary Waters Canoe Area Wilderness (BWCAW) was established in 1978 (one year after Oberholtzer's death at age ninety-three). Ontario imposed wilderness restrictions in Quetico Provincial Park in 1972. In piecemeal fashion, then, Oberholtzer's vision of an international wilderness park was realized, albeit on a reduced scale. But as early as 1947 he foresaw yet another danger to wilderness unanticipated by most preservationists at the time. "You can lose it," he warned, "even if you keep out roads, planes, and human occupancy. You can lose it because you will be so overwhelmed with even the most innocent and warmhearted wilderness lovers that the old-timer who wants to escape crowds . . . will find few remaining satisfactions in the region." His prophecy has come true for the BWCAW: it is now the most popular wilderness area in the country, attracting more than 200,000 visitors annually. And efforts by the government to restrict use in the Boundary Waters continue to raise strong opposition from a variety of user groups.

Much of Oberholtzer's voluminous correspondence and other papers are held by the Minnesota Historical Society in St. Paul. The cabins he built on Mallard Island in Rainy Lake remain, along with his books and extensive artifact collection, now owned and protected by the Oberholtzer Foundation, an organization dedicated to promoting his role as one of Minnesota's leading conservationists. Among the Ojibwe, for his interest in their culture and his storytelling skills, he acquired the nickname "Atisokan"—Legend. As a legendary figure in the history and preservation of the North Woods, he deserves to be more widely known.

DON SCHEESE

Backes, David. *Canoe Country: An Embattled Wilderness.* Minocqua, WI, 1991.

———. "Wilderness Visions: Arthur Carhart's 1922 Proposal for the Quetico-Superior Wilderness." *Forest and Conservation History* 35.3 (July 1991): 128–137.

Cockburn, R. H. "Voyage to Nutheltin." *Beaver* 66.1 (January–February 1986): 4–27.

Munson, Marit. *Ernest Oberholtzer: A Conservationist of Unusual Vision.* [St. Paul?], MN, 1992.

Nash, Roderick. *Wilderness and the American Mind.* 3rd rev. ed. New Haven, CT, 1982.

Searle, R. Newell. *Saving Quetico-Superior: A Land Set Apart.* St. Paul, MN, 1977.

Odum, Eugene (September 17, 1913–August 10, 2002). A biologist, naturalist, ornithologist, educator, and author, Odum is one of the most influential figures in the history of ecology. He explored connections between plants and animals, climate and weather, and watersheds and water supply. For more than a half century, he challenged academic assumptions about the physical world. At first, his ideas found little support in the scientific community, but today he is adulated as "the father of modern ecosystem ecology."

Odum was born in Newport, New Hampshire, and grew up in Chapel Hill, North Carolina. His father, Howard W. Odum, was a professor of sociology at the university; his mother, Anna Louise (Kranz) Odum, was an urban planner. From his father, Odum learned to approach subjects holistically; from his mother, to set up modes of action. As a boy, he had a fascination for birds, and in his teens he wrote a column on birdlife for a local newspaper, the *Chapel Hill Weekly*. An enthusiastic bird-watcher, he had such a keen ear that he could identify almost every bird by its chirping or warbling, its call or song.

After he earned an A.B. and an A.M. in zoology from the University of North Carolina, Odum considered pursuing his doctorate at Cornell or Michigan. He rejected them and decided on the University of Illinois, one of the few schools, as he put it, "with something like ecology in its zoology department." Wanting to get beyond taxonomy into function, into the physiology of birds and how they relate to the larger natural environment, he wrote his dissertation on physiological ecology. Specifically, he wanted to concentrate on avian cardiovascular systems. To do so, he had to invent a device that would allow him to measure the heartbeat of small birds.

In 1939, Odum was awared his Ph.D. Shortly thereafter, he was appointed resident naturalist at the Edmund Niles Huyck Preserve in Rensselaerville, New York. He remained at the preserve for a year. In 1940, he became a member of the Biology Department at the University of Georgia. At Georgia, Odum taught mainly zoology courses and an occasional course in ecology. After a few years, at a departmental meeting about the core curriculum and what courses biology majors should be required to take, he suggested that ecology be included. His colleagues were somewhat amused or indifferent. They tended to look on ecology as little more than just going out and finding animals and collecting and describing them. As a subject, they complained, ecology lacked essential principles. One even carped: "It is just organized natural history, not an important subject." Odum quickly realized that ecology was basically unknown and misunderstood. No one had ever written a book on it, so with the help of his younger brother Howard, then a graduate student in the physical sciences at Yale, he determined to write one.

In 1953, they published *Fundamentals of Ecology*. They were pleased with how well their work was received. Their volume has been frequently revised, expanded, published in several editions, and translated into a dozen languages. What helped make the work such a great success was that instead of starting with the concept of organism and ending with an ecosystem, the text written by the Odum brothers began with the ecosystem and worked down, in what they dubbed "a top-down approach." As a result of the volume's popularity, *ecosystem* became a watchword of the emerging environmental movement in the 1970s and provided the movement with a lexicon, as well as a comprehensive vision of nature as a dynamic system.

In *Fundamentals of Ecology*, Odum, in presenting a broad view of the environment, pioneered ecosystem ecology as an integrative science. Among other things, he proposed that an ecosystem theory could provide a common denominatior for humans and nature and that the goods and services of both are coupled. "Without healthy, natural systems to support and buffer industrial, urban, and agricultural activities," he affirmed, "there can be no healthy economy or high quality of life." In short, he sought for a more harmonious relationship between humans and nature, and he postulated how the earth's ecological systems interact with one another.

Odum's name is so closely associated with *ecology* it is surprising to some that he did not coin the term. It is believed that German biologist Ernst Haeckel first used the term in 1869, taking its root from the Greek *oikos*, meaning "house, habitation." Ecology has come to denote a discipline of biology that treats the relationships between organisms and their environments. Sociologists frequently employ the term, or *bionomics*, to indicate the study of human populations in their physical environment, spatial distribution, and cultural characteristics.

Nor was Odum the first to use the locution *ecosystem*. Raymond Lindeman, who viewed nature as shaped more by physics than by biology, used the term in 1942 to describe nature as a vital flow of energy from recycled chemicals moving through a thermodynamic system. Before Lindeman, others had used the word in various contexts. Most likely, *ecosystem* first appeared in a 1935 publication by British ecologist Arthur Tansley, though the term was coined by one of his colleagues, Roy Clapham.

As interesting as the history of ecosystem may be, it is more important to stress that before Odum ecology had been studied on a small scale and only within limited disciplines. Odum viewed the entire universe as a series of interlocking ecosystems. Each one, he proposed, embraced "a unique strategy of development . . . directed toward achieving as large and diverse an organic structure as is possible within the limits set by the available energy input and the prevailing physical conditions of existence." Judith Meyer, a president of the Ecological Society of America, declared that Odum made "ecosystem" into a household word. "He began with the system as a whole," she explained, "and demonstrated the power of that concept."

Over the years, Odum published a dozen or so books and over 200 scientific papers on ecosystems. Never one to coast, however, he also devoted much time and energy to establishing an Institute of Ecology at the University of Georgia in 1960, and he served as its first director. The institute, as its name implies, was set up to encourage undergraduate and graduate programs in ecological studies. To promote scholarly activity, Odum contributed $150,000 to the University of Georgia Foundation, money that had been bestowed on him for his environmental achievements. In tribute the institute unveiled a bronze bust of him on September 17, 1984. It stands in the foyer of the institute, with an inscription on the base of the sculpture reading: "The Ecosystem is greater than the sum of its parts."

Faculty members affiliated with the institute conduct primary research at several sites: at the Coweeta Hydrology Laboratory in North Carolina; at the Okefenokee Swamp on the border between Georgia and Florida; and at Horseshoe Bend, located near the University of Georgia campus. In addition, Odum founded the Marine Institute, Sapelo Island, Georgia. He also set up the Savannah River Ecology Laboratory, a 300-square-mile marsh area dubbed "one of the largest outdoor science classrooms on Earth." The Savannah Laboratory was initiated through grant money Odum had received from the Atomic Energy Commission to study the environmental impact of nuclear weapons production at the Savannah River Plant, a large governmental installation located a short distance from Aiken, South Carolina.

Odum spent virtually all of his professional career at the University of Georgia, but he also assisted in establishing endowments at the University of North Carolina, the University of Virginia, and the University of Illinois. Over and above, he helped endow the Ecological Society of America's E. P. Odum Award for Excellence in Ecology Education. For his own contributions to environmental causes, for engaging in large-scale research projects, and for motivating interdisciplinary education, Odum was the recipient of numerous national and international awards. In 1956, the Ecological Society of America singled him out for its Mercer Award. He was elected a member of the National Academy of Sciences in 1972. In 1974, he was named an honorary member of the British Ecological Society. In the same year, he was nominated for the Ecological Society of America's Eminent Ecologist Award. In 1975, he was elected to the French Institut de la Vie.

In 1977, in ceremonies at the White House, President Jimmy Carter presented Odum with the Tyler Ecology Award. At the time, he summed up Odum's career with the words: "We cannot overestimate Dr. Odum's work in making spaceship Earth a better place for us all." Greatly impressed with Odum's contributions to environmental studies, at another time Carter said: "The work of Dr. Odum changed the way we look at the natural world and our place in it."

Over the years, Odum continued to receive such praise and multiple awards and medals. In 1987, he was presented with the Crafoord Prize by the Royal Swedish Academy of Science. Similar commendations and tributes were oft repeated. Though unassuming in manner and makeup, Odum was especially pleased to be chosen the 1991 recipient of the Theodore Roosevelt Distinguished Service Award and Gold Medal.

After his official retirement from Georgia in 1984, Odum continued to devote full time to his research, writing and lecturing. He never stopped teaching, it has often been said, whether with a farmer in a pasture or with a student on a campus sidewalk. Always an evangelist for the values of ecology, he was generous with his time and wealth. He donated much of his estate that accumulated from royalties and awards to the University of Georgia. In his will he stipulated that property he owned on the Middle Oconee River in Athens, Georgia, be developed according to plans he laid out before his death.

The plans included protected greenspace and walking trails. Profits realized through the sale of land should go into an ecology fund. A sum of $1 million would be set aside for a professorial chair at Georgia in his name. Vibrant and creative until overtaken by death in his eighty-ninth year, to sum up his legacy is well-nigh impossible. Perhaps it is enough to state that he was that rare individual who was both thinker and doer. All that he accomplished speaks for itself. Odum ranks high among the foremost American environmentalists.

GEORGE A. CEVASCO

Chaffin, Tom. "Whole-Earth Mentor: A Conversation with Eugene P. Odum." *Natural History*, no. 107 (1998).

Craige, Betty Jean. *Eugene Odum: Ecosystem Ecologist and Environmentalist.* Athens, GA, 2001.

"Eugene Odum." *Environment and History* 3 (1997).

"Eugene Odum." *Notable Twentieth-Century Scientists.* Vol. 3. 1995.

Goldman, Ari L. "Eugene P. Odum Dies at 88; Founded Modern Ecology." *New York Times*, August 14, 2002.

Okun, Daniel A. (June 19, 1917–December 10, 2007). Hailed worldwide for his groundbreaking work in integrated water management, pollution control, and water reuse, Okun's well-earned reputation is a consequence of his ability to interconnect his scientific activities with practical applicability. His work brought him to eighty-nine countries, in which he engaged in a wide variety of hydrological projects and set up training programs in environmental engineering. In 1999 he was named in the *Engineering News-Record* one of the top 125 engineers "who singularly and collectively helped shape this nation and the world."

Okun was born in New York City to Will Okun and Leah (Seligman) Okun, immigrants from Belarus in eastern Europe. At an early age he first learned that one out of six individuals throughout the world lacked access to safe drinking water. When he was only twelve years old, he was drawn to his future career while watching his father oversee the engineering of a massive tunnel to bring water from the Delaware River to New York City. Influenced by his father, he matriculated at the Cooper Union Institute to obtain a degree in civil engineering, which was conferred in 1937. He continued his studies at the California Institute of Technology. In 1938 he earned his master's degree from Cal Tech. He then entered the U.S. Public Health Service as an assistant sanitary engineer, where he served for two years.

Between 1942 and 1945 Okun served in the U.S. Army as a member of the Sanitation Corps. He was stationed in Newfoundland, Latin American countries, the Southwest Pacific, and the Philippines. To ensure the health of the troops he had them avoid polluted water, tainted food, and questionable women. At one time his duties included the inspection of brothels in Guatemala, where he had to put his knowledge of sanitation and hygiene to good use.

After military service Okun entered Harvard University to further his study of sanitary engineering. His research involved the measurement of dissolved oxygen in water treatment and the protection of source water for public water supply. Additionally, he worked on the reclamation of wastewater for urban nonpotable use to conserve limited resources of high-quality waters. In 1948 he was awarded a Doctorate of Science degree.

Engineers with his education and experience were in demand, and he was offered a position with Malcolm-Pirnie, Inc., an organization that specialized in municipal and industrial water and wastewater management. In 1952 he decided on an academic career when he was approached by the University of North Carolina to join its faculty. He accepted a position as associate professor of environmental engineering. An outstanding teacher, he was appointed chairman of his department in 1955 and served until 1973. Over the years, he created a multidisciplinary approach to sanitary engineering and increased the size of his division from three to twenty-five members.

Okun transformed his departmental traditional sanitary engineering program into a prestigious environmental science and engineering agenda, adding to its curriculum courses in pollution control, industrial and radiological hygiene, aquatic and atmospheric science, as well as majors in environmental management and policy. Today it is widely conceded that the University of North Carolina has one of the best programs of its kind in the world.

In addition to his teaching and research, Okun served as consultant to numerous companies and governments. He was also a major contributor to the development of environmental sciences and engineering courses for major universities in the Netherlands, England, China, Peru, Thailand, France, Finland, and Singapore. The World Bank, the World Health Organization, UNESCO (United Nations Educational, Scientific and Cultural Organization), and the United Nations were recipients of his solutions to serious environmental problems. A map of the world that hung in his office was covered with 149 pushpins to mark the places he worked. Closer to home, he established environmental programs in such major American universities as Duke, Vanderbilt, Rensselaer, the University of Michigan, and the Univerity of Georgia.

Okun's concentrated passion was water. Everyone in the world, he held, should have access to free freshwater. Toward that end, he directed all his energies. To prevent pollution of groundwater, rivers, streams, and lakes, he was always concerned with watershed protection. In North Carolina and several other states, he helped draft legislation to keep industry and construction out of water-supply watersheds.

Another of his goals was to work toward the establishment of dual systems to promote the availability of potable water. Most systems, however, were designed to carry water for multiple purposes, not exclusively for human consumption but also for industrial needs and even fire control. Reclaimation would be an obvious way to conserve potable resources, as well as a way to avoid possible future droughts. To install dual systems for existing communities he realized would be fiscally difficult, if not impossible, but should always be considered in plans for new developments.

In his multitudinous presentations, books, manuals, and articles, Okun explained the components of his proposals and programs. Between 1966 and 2002 he delivered over twenty keynote addresses before such groups as the American Water Works Association, the Water Science and Technology Board of the National Research Council, the Water Environment Federation, and the International Reuse Symposium of the Institution of Civil Engineers.

Among his more than a dozen books and manuals the more significant are the *Manual on Sewage Treatment Plant Design* (1959), *Water Systems* (1994), and *Regionalization of Water Management* (1977). Always a zealous researcher and prolific author, he wrote over 250 papers for refereed and scientific journals. They cover such subjects as "Drinking Water for the Future" (*American Journal of Public Health*, 1976), "From Cholera to Cancer to Cryptosporidiosis" (*Journal of Environmental Engineering*, 1996), and "Distributing Reclaimed Water Through Dual Systems" (*Journal of American Water Works Association*, 1997).

Though the expression has it that a prophet is without honor in his own land, Okun was always admired and appreciated by his colleagues, students, and administrators at Chapel Hill. In 1970, he was appointed chair of the faculty at the University of North Carolina (UNC). Three years later he was honored with the university's Thomas Jefferson Award, conferred on the faculty member "who through personal influence and performance of duty in teaching, writing and scholarship best exemplified the ideals and objectives of Thomas Jefferson." Such honors and awards were bestowed on Okun year after year. Capping them all, perhaps, was the establishment of the Daniel A. Okun Professorship in 1999. The following year he was awarded an honorary Doctor of Science degree, the first bestowed on a faculty member at UNC.

Okum's reputation extended far beyond UNC; in fact, it was international in scope. He served as a Visiting Professor at the Technological University of

Delft, the Netherlands (1960–1961); at the University College of London (1966–1967; 1973–1974), and at Tianjin University, China (1981). His contributions to water management were so diverse and farsighted that on September 10, 2006, he was the recipient of the prestigious International Water Association's Grand Award for his outstanding achievements as a water engineer and eminent scientist. The award was presented at the opening ceremony of the World Water Congress at the Beijing International Congress Center.

Okun's formal honors and fadeless laurels from professional associations dedicated to water resources and environmental engineering made him one of the most esteemed hydrologists in the world. Over the years, hundreds of his students, including many from third-world countries, have advanced to critical field research and policy-making positions across the globe. The chief legacy of Okun is that they are furthering scientific knowledge as they devise and implement solutions to serious environmental problems.

GEORGE A. CEVASCO

"Okun, Daniel A." *International Who's Who*. 67th ed. 2003.
"Okun, Daniel A." [obit.]. *New York Times*, December 12, 2007.
"Okun, Daniel A." *Who's Who in Engineering*. 7th ed. 1988.
"Okun, Daniel A." *Who's Who in Technology*. 7th ed. 1995.
"OWASA Adopts Resolution to Honor Dan Okun." *Chapel Hill News*, June 20, 2006.
Weigl, Andrea. "He Crusades for Safe Water Supply: Expert Points Way to Dual Systems." *News and Observer* (Chapel Hill, NC), August 6, 2006.

Olson, Sigurd F. (April 4, 1899–January 13, 1982). A lyrical conservationist, Olson is credited with saving much of the American wilderness. Born in Chicago, Olson moved to Door County, Wisconsin, at an early age with his father, a Baptist minister, and mother. The early life experience on Lake Michigan initiated Olson's deep love of nature.

Early influences on Olson were his father and his grandmother. His father believed that there were only two occupations to choose from: working for the welfare of humankind and tilling the soil. Olson chose the former and never waivered in his commitment to saving the wilderness for others to experience. His father's library, filled with poetry, novels, travel, history, and essays, provided the philosophical basis for Olson. The essays of the contemplative naturalists Henry David Thoreau, John Burroughs,

William Hudson, and John Muir particularly influenced Olson's intellectual development.

In *The Singing Wilderness* (1956) Olson celebrated the other early influence in his life: his grandmother. Her enthusiasm and appreciation for the spiritual aspect of nature transferred to the child whom she encouraged to venture forth alone into the wilderness. Her delight in his verbal reports after each foray, too, stimulated Olson's adventures. His descriptions for his grandmother of what he saw and felt and did prepared him for his role as a writer, which came much later in his life.

In his adolescent years, Olson spent as much time as possible in the outdoor world. He began college at Ashland, Wisconsin, majoring, of course, in the sciences. A transfer to the University of Wisconsin at Madison allowed him to continue studies in zoology, biology, geology, and botany. He received a B.S. degree in 1920. Studying a new field, ecology, Olson received an M.S. degree from the University of Illinois in 1931.

A teaching position at Ely (Minnesota) Junior College gave Olson the opportunity to teach biology and geology in a setting close to the wilderness he loved. As a teacher, Olson preferred to teach his science classes in the outdoors, not separating the information from the natural environment. He served as Biology Department chairperson and academic dean at Ely from 1922 to 1935. During the weekends and summers he began exploring the vast wilderness of the Quetico Superior area, making friends with many of the vintage wilderness guides. After many trips into the wilderness, Olson applied for and received a certificate to be a wilderness guide.

The more experience he had in the wilderness, the deeper brotherhood he felt with the voyageurs, those French trappers and fur traders who ventured far into the interior of the continent.

With each passing year, Olson became more convinced that the wilderness was a vital teacher to those who would listen to the silences and hear the vast message of nature. He, however, was realistic enough to know that commercial greed seriously threatened the wilderness. He joined a three-month-old organization in 1935, the Wilderness Society, and was an active member (president, 1968–1971; honorary president, 1974–1982) all his life. Other organizations he joined and worked industriously for included the National Parks Association (president, 1954–1960), the Ecological Society of America, Nature Conservancy, Sierra Club, and Cosmos Club. He served as ecological consultant to the Izaak Walton League and to the U.S. Department of the Interior.

Most of his activist work centered on the Quetico Superior problems and projects on the Minnesota-Canada border. He was a leader in the 1920s trying to convince the U.S. Forest Service to decree the Quetico Superior area a roadless area. That decree finally came after years of bitter struggles; the Forest Service established a roadless area of a million acres on the Superior National Forest. This became Olson's first step in protecting the Boundary Waters Canoe Area.

The next major controversy Olson fought concerned the proposed building of a huge power reservoir that would have seriously affected the border lakes and destroyed islands, rapids, and waterfalls. Olson, active in the Quetico Superior Council, not only lobbied against the reservoir but also supported the idea of an International Peace Memorial Forest uniting the American and Canadian interests in the wilderness area. Olson and the council successfully blocked the building of the reservoir. The efforts to have an international forest, however, failed. The U.S. and Canadian governments did exchange official papers agreeing to work closely together to protect the wilderness area. Olson was instrumental in the finalizing of those papers.

The third conservation project in the Quetico Superior involved Olson and others fighting for airspace control. Olson worked on this project from 1938 to 1949. Finally, the Air Space Reservation Act (1949) created an airspace reservation over the roadless areas of the Superior National Forest. This limited, again, the possibility of commercialization of the wilderness area by seaplane transports into wilderness resorts.

But a corollary problem had arisen for Olson to work on; funds were needed for private land acquisition to solidify the wilderness area. Olson was instrumental in getting governmental funds to buy the private lands in the Boundary Waters Canoe Area. By 1960, 98 percent of the million acres were government owned. Then Olson turned his attention to the issue of mining in the Boundary Waters Canoe Area. His efforts were paramount in the passing of the 1960 Multiple Use Act requiring that all national forest lands balance the use of the resources and reiterating that outdoor recreation was the goal of national forest administration. Olson's efforts also led to the Wilderness Act of 1964, which provided wilderness areas on all federal lands. Olson specifically worked in defense of the Grand Canyon and Potomac River conservation projects, as well as for the Alaskan wilderness.

In 1945 Olson had chosen to leave teaching and commit himself to these activist conservation efforts as participant, consultant, and writer. He began writing down his feelings about the wilderness. Olson wanted to use the poetical prose of Aldo Leopold, whom he had worked with, and help others understand Leopold's land ethic concept. He asked the publishing advice of Rachel Carson, a contemporary conservationist. Finally, in 1956 he published *The Singing Wilderness*, followed by *Listening Point* (1958), *The Lonely Land* (1961), *Runes of the North* (1963), *Open Horizons* (1969), *Wilderness Days* (1972), *Reflections of the North Country* (1976), and *Of Time and Place* (1982). He also authored over one hundred articles, political and philosophical, insisting that preservation of the wilderness remains critical to humankind.

Olson's legacy is vast in environmental studies. He firmly advocated educating the public on the importance of preserving the wilderness. Through his activist work and his writings, he lived and taught that philosophy; he never compromised. He is remembered, also, for his lyrical interpretations of the meaning and teachings of the wilderness. He celebrates the living silence in each of his books, and he links the reader with the wilderness experiences, as well as with the historical past. He recounts the tales of the voyageurs, looking to them as ones who felt and experienced life and the land deeply. With the voyageurs, too, he shared the belief of the wilderness as the mystical link with the past from which one can draw self-understanding and deep personal satisfaction. Because of this philosophical stance, combined with solid knowledge bases, Olson made conservation the powerful social force it is today.

Olson and his wife, Elizabeth, had two sons: Sigurd Thorn and Robert Keith.

KELLY A. FOTH

Backes, David. *A Wilderness Within: The Life of Sigurd F. Olson*. Minneapolis, MN, 1997.

Graham, Frank. "Leave It to the Bourgeois: Sigurd Olson and His Wilderness Quest." *Audubon*, November 1980, 28–39.

"Obituary: Sigurd Olson." *Audubon*, March 1982, 5.

Olson, Sigurd F. "Six Decades of Progress." *American Forests*, October 1962, 16–19.

"One Who Never Compromised." *Living Wilderness*, Spring 1982, 26–27.

Searle, R. Newell. *Saving Quetico-Superior: A Land Set Apart*. St. Paul, MN, 1977.

Osborn, Henry Fairfield (August 8, 1857–November 6, 1935). Born in Fairfield, Connecticut, Osborn was the second child and eldest of three sons of William Henry Osborn, a wealthy businessman, and Virginia Reed (Sturges) Osborn. His middle name was taken from his place of birth, and he was called "Fairfie" and later "Harry" as a child and in young manhood. His parents, and particularly his mother, were staunch Presbyterians, and this was to influence his scientific outlook all his life.

Having made a fortune in the shipping business as a young man, Osborn's father, with the aid of Jonathan Sturges, his father-in-law, took over management of the Illinois Central Railroad in the 1850s and handled its affairs with great success, becoming president in 1853. George B. McClellan, the future Civil War general, was one of the senior Osborn's Illinois Central colleagues. The Osborn family spent their winters in New York City and built a summer estate on the Hudson River in Garrison, New York, when Henry was two. As a young banker, J. P. Morgan, a business associate of Henry's father, occasionally played with young Henry while courting the boy's mother's sister. Morgan became Henry's uncle by marriage in 1861 and would later provide strong financial support for his nephew's scientific publishing ventures. On occasion young Theodore Roosevelt, a friend of Henry's younger brother Frederick, was also a visitor in the family home.

Osborn was educated at the Columbia Grammar School and Lyon's Collegiate Institute in New York City. In 1873, he entered the freshman class at the College of New Jersey (later Princeton), of which a maternal great uncle had been a founder. In 1875, the Osborn family suffered a double shock when Henry's brother Frederick died while swimming in the Hudson River, and then his older sister Virginia succumbed to illness while the family vacationed in Paris. Despite these heavy blows, Henry soldiered on with his studies. During his undergraduate years, two men exerted particular influence on his intellectual development. One was Andre Guyot, a native of Switzerland, who became the college's professor of geography and geology in 1854. The other was the college president, James McCosh, a Scots-born evangelical Presbyterian minister who had studied evolutionary theory and published a book on the subject.

In 1876, William Berryman Scott, a classmate, proposed that he, Osborn, and another friend, Francis Speir, follow the example recently set by several Yale students and spend the summer out West, collecting fossils. The three young men secured the support of the college administration and, with its help, the U.S. Army. Thus a group of eighteen students and two faculty members, dubbed the first Princeton Scientific Expedition, spent several months in parts of Wyoming, Colorado, and Utah. Osborn, along with Scott, Speir, and two other students, did most of the fossil collecting. The specimens they brought back to Princeton were valuable enough to become the nucleus of the college museum's natural history collection. This experience served to launch lifelong careers in paleontology for both Osborn and Scott.

Encouraged by President McCosh, who was seeking to build up the college's graduate program, Osborn and Scott unofficially became the college's first graduate students in the sciences. Osborn's training at Princeton was primarily in zoology, Scott's, in geology. After a postgraduate year with McCosh, the two young men were encouraged to continue more specialized studies elsewhere. Princeton lacked the faculty and facilities to offer more advanced work, and McCosh hoped to eventually appoint both of them to the college faculty. Scott headed off for graduate work in England and Germany, while Osborn published a paper on their western findings, then spent a year studying anatomy and physiology at Bellevue Hospital in Manhattan.

Osborn's father wanted his son to join him in the railway business in New York but agreed to finance his studies for a year at Cambridge University, then later with Thomas Henry Huxley, who introduced the young American briefly to the aging Charles Darwin. On his return to New York, Osborn went to work for his father, but McCosh and others badgered the senior Osborn, urging him to give Henry a chance for further scientific work. He was granted some time off to collect natural history specimens in the West. His father had originally expected his deceased son Frederick to become a scientist, while Henry would become a businessman. Reluctantly, he released him from business, built a library for his son's scientific books at the family home in Garrison, and began supporting his son's scientific work in other ways. In turn, Osborn helped his mother provide care for his father, whose health had declined owing to the pressure of business.

The American Museum of Natural History first offered Osborn a position as a geologist in 1880, but he declined, believing that this would be ill-advised while the two titans in American paleontology, Othniel Marsh at Yale and Edward Drinker Cope in Philadelphia, still dominated the field. Osborn had met both men and hoped to avoid getting caught up in the increasingly acrimonious paleontological quarrels between the two, which continued into the 1890s. Nevertheless, Osborn was greatly influenced by Cope's views and much later, in 1931, published a biography of the famed Philadelphian.

Instead of going to New York, the young scholar completed the requirements for his doctorate at Princeton and accepted McCosh's offer of an assistant professorship there in 1881, although his salary was underwritten by his father because the college did not have the resources to pay him. His Princeton classmate W. B. Scott soon began his tenure at Princeton and enjoyed a long and productive career there as a paleontologist.

In 1881, Osborn married Lucretia Thatcher Perry, a descendant of Commodore Matthew Perry, at Governor's Island in New York, where her father, an army general, was then stationed, and they set up housekeeping in Princeton. The couple would have five children. Osborn spent the 1880s teaching embryology and comparative anatomy at Princeton, but criticisms of his published work in embryology were instrumental in his shift to paleontology. Though he did little fieldwork himself, he analyzed specimens provided by Scott and his students. Osborn was also much influenced by E. D. Cope's Lamarckian view of evolutionary change as well as the work of his friend and colleague Scott.

In 1891, Osborn left Princeton for New York, accepting a joint appointment at Columbia University and the American Museum of Natural History. For a half-dozen years at Columbia, he served as dean of the university's Faculty of Pure Science, establishing the foundations for what soon became an outstanding graduate program in biology. Some years later, the department was split into zoology and botany departments. Osborn initiated research and publication programs, established a biology laboratory, and made it possible for Columbia students and faculty to participate in programs at the Marine Biological Laboratories at Woods Hole, Massachusetts. He also developed summer offerings on Long Island and a field program on the West Coast at Port Townsend, Washington.

Osborn was also very interested in providing opportunities for residents and visitors to New York to learn more about natural history. Named president of the New York Zoological Society's executive committee, he selected the location for the Bronx Zoo in 1896 and chose its first director.

At the American Museum of Natural History, Osborn served as curator of vertebrate paleontology, beginning a career that would last more than four decades. With financial assistance from his uncle J. Pierpont Morgan, a museum trustee, and Morris Jessup, a businessman and the museum's president when Osborn was appointed, Osborn was able to conduct most of his research, producing a large number of books and articles. He held a firm belief in the importance of explaining natural history to the general public.

For many years, the museum maintained its own printing shop and a corps of illustrators who created most of the artwork contained in Osborn's sumptuously illustrated books. From the beginning of his museum career, and increasingly after he became president of the museum's board of trustees in 1908, Osborn's assistants carried out much of the research on which he based his conclusions for publication. He remained board president for twenty-five years, until his retirement.

During his tenure at the museum, Osborn developed many museum galleries where the general public could view mounted specimens. These exhibits were created on a scale unmatched at most other institutions in the United States. Their superiority was impelled by Osborn's strong personal interest. An increasingly central figure among paleontologists and museum administrators, his outlook reflected the social and moral values of the New York elite to which he belonged. He was thus able to attract wealthy patrons from among this group in support of his projects. Many biologists at the time considered paleontology a field outside the mainstream of biology. They considered this specialty old-fashioned, impractical, and cost-intensive. Many colleagues felt that paleontology failed to meet the standards for systematic inquiry, which modern science required. Because their focus was primarily causal, rather than historical, many biologists increasingly favored research in genetics and experimental biology. Osborn, however, resisted this trend. Fortunately, his reputation and resources allowed him to keep paleontology in the mainstream of biological science.

Under Osborn's leadership, the museum undertook a number of expeditions that brought back substantial collections for research and display. Osborn was responsible for museum staff in various parts of the world, correlating fossil remains to their appropriate geological strata. Osborn was deeply involved in theorizing about the origin and geographical distribution of fossil vertebrates. He was one of the first museum administrators in the country to assemble representative dinosaur skeletons for public exhibition. Many conclusions reached by Osborn and his associates about the likely posture and movements of dinosaurs resulted from their work with modern reptiles. Nevertheless, Osborn's fame rested most heavily on his administrative skills and his many publications.

Among his earlier scientific works were *On the Structure and Classification of the Mesozoic Mammalia* (1888), *Evolution of Mammalian Molar Teeth to and from the Triangular Type* (1907), *The Age of Mammals in Europe,*

Asia, and North America (1910), *The Origin and Evolution of Life* (1917), and *Equidae of the Oligocene, Miocene, and Pliocene of North America* (1918). His semipopular works included *From the Greeks to Darwin* (1894), *Men of the Old Stone Age* (1915), *The Earth Speaks to Bryan* (1925), *Man Rises to Parnassus* (1927), and *Impressions of Great Naturalists* (1928). The mammalian paleontologist George Gaylord Simpson later wrote that Osborn "planned as if he were to live forever, and he laid out more work for himself than could have been completed in ten lifetimes."

Osborn's energies were not confined to the American Museum or to Columbia University. An early supporter of resource conservation, Osborn was active in a number of organizations such as the Audubon Society, American Bison Society, and Save the Redwoods League.

Osborn believed that evolution was a slow, gradual process that exhibited purpose and direction. His ideas reflected the religious views of his upbringing, his training at Princeton, and the conservative social and political convictions of his peers. An elitist uncomfortable with the flood of immigrants entering the United States, his outlook was tinged by racism. His views on this subject were influenced by his ideas on human evolution; he believed that civilization and the mingling of races had led to a lowering of humankind's condition. Osborn concluded that the Nordic peoples should shape human development, a theory he termed *aristogenesis*. Osborn was uncritical of anti-Semitic ideas prevalent among his social acquaintances. He favored immigration restriction, was active in the field of eugenics, and supported the work of nativist authors such as Madison Grant. Toward the end of his life, he was sympathetic with some of the objectives of Adolf Hitler, believing that the Germanic peoples were trying to overcome challenges thrown up in the aftermath of World War I and the Great Depression.

As he grew older, Osborn became more pompous and tactless in manner. To avoid the summer heat in New York, Osborn generally worked at his home in Garrison, so museum staff members, principally secretaries and women editors with documents requiring his approval, had to take a morning train upriver during the summer months to conduct necessary business. One woman staffer was obliged to carry a heavy typewriter with her on her frequent trips to Garrison. When lunchtime came, Osborn would go off and dine privately, while the secretaries were provided with sandwiches they ate by themselves.

Edwin H. Colbert, an assistant to Osborn during the last five years of his life and later an eminent paleontologist, recalled the occasion when a young office staffer, helping Osborn with his correspondence, prepared to use an office blotter on Osborn's "large, flourishing signature." Osborn held up his hand and in complete seriousness said, "Never blot the signature of a great man." Osborn arranged to have a bust of himself placed outside the museum building in a niche between two windows of the Osborn Library, where it was highlighted at night by a spotlight. The Paleontology Department's Christmas dinners were strictly segregated by gender and position. Osborn and the senior department professionals, all male, dined together in the Osborn Library. Women staff members, along with the female secretaries, were banished to an outer office, while specimen preparators, taxidermists, and others ate in a museum laboratory.

Colbert described Osborn, in his later career, as "a large man with a somewhat overwhelming physical presence. His face was distinguished by a rather prominent, almost bulbous nose, and he had a clipped mustache. [His] movements were ponderous in a sort of elephantine way. . . . He had a love of big projects, including the museum, big exhibition halls and big animals to go in them."

Though it took him many decades, Osborn eventually produced several two-volume studies about ancient mammals. One set, published in 1929, dealt with titanotheres (extinct elephant-sized ungulates related to horses), though much of the actual writing was done by William King Gregory, a younger colleague. Another set dealt with proboscideans (elephants and mastodons), on which he worked until the end of his life. Publication of both works was underwritten by J. P. Morgan at a cost of approximately $400,000. Osborn did not live to see his completed elephant and mastodon work in print. The first volume was published in 1936, and the second appeared in 1942.

In time, many conclusions Osborn reached in his paleontological studies were superseded. The next generation of specialists, some on the American Museum staff, produced new developments and interpretations. Younger administrators preferred broader, less personalized objectives for the institution, and eventually, their views prevailed. Osborn's board colleagues forced him to retire from the museum's board of trustees in 1933, but he continued to research and write and maintained a connection with the Columbia University Zoology Department. He died while working at his desk in his Garrison home at age seventy-eight.

KEIR STERLING

Gregory, William King. "Henry Fairfield Osborn." *National Academy of Sciences Biographical Memoirs*. Vol. 19. 1938.
Rainger, Ronald. "Henry Fairfield Osborn." *American National Biography*. 1999.
Regal, Brian. *Henry Fairfield Osborn: Race and the Search for the Origin of Man*. Burlington, VT, 2002.
Simpson, George Gaylord. "Osborn, Henry Fairfield." *Dictionary of American Biography*. Supp. 1. 1944.

Osborn, Henry Fairfield, Jr. (January 15, 1887–September 16, 1969). A naturalist, conservationist, and zoo administrator, Osborn was born in Princeton, New Jersey, to Henry F. Osborn, a distinguished vertebrate paleontologist and president of the American Museum of Natural History, and Lucretia (Perry) Osborn. Better known as Fairfield, the younger Osborn was reared in a world of science and nature. Certainly, his father's love of science influenced him, but Osborn himself was drawn to living nature. He once stated, "Ever since I was old enough to see or sense anything, the mysteries of the natural world have possessed my soul." After spending his childhood at their Garrison, New York, home overlooking the Hudson River, Osborn graduated from the preparatory Groton School in 1905. He then attended Princeton University (his father's alma mater), receiving an A.B. in 1909. Osborn spent the following year abroad, pursuing graduate studies at Cambridge University, England.

On returning to the United States, Osborn worked in the freight yards of San Francisco and laid railroad tracks in Nevada. He then moved back to New York and was employed as treasurer of both the Union Oil Company and a small label-making business. On September 8, 1914, he married Marjorie Mary Lamond (an artist) in London, England; they were blessed with three daughters: Nathalie Hazard, Shirley, and Josephine Adams. During World War I, Osborn served in Europe as a captain in the 351st Field Artillery, U.S. Army. After military service, Osborn's career took another turn. From 1918 to 1935, he worked in investment banking at Redmond & Company (in which he was a partner) and Maynard, Oakley & Lawrence.

Although he did not pursue a career in conservation until he was forty-eight years old, Osborn is best known for his natural preservation efforts and achievements.

Following in his father's footsteps, Osborn became secretary (1935–1939) and president (1940–1968) of the New York Zoological Society. During his term, Osborn instituted many new approaches to zookeeping, including the innovative technique of re-creating natural habitats for the animals. No longer were the animals merely locked in cages; instead, penguins lived in a refrigerated environment, the great apes house was constructed, and dwellings were built in such a way that birds could fly free in their replicated native settings.

His greatest legacy, however, was the creation of the Conservation Foundation, Washington, DC, in 1948. Its purpose was "to promote the conservation of the earth's life-supporting resources . . . and to advance, improve, and encourage knowledge and understanding of such resources." Through his affiliation as president (1948–1962) and chairman (1963–1969) of the foundation, Osborn gained prominence and influence. He was appointed to the Conservation Advisory Committee of the U.S. Department of the Interior (1950–1957) and served on the United Nations planning committee of the Economic and Social Council. Correlatively, he was selected as one of the main speakers for the United Nations Scientific Conference on the Conservation and Utilization of Resources in 1949. Osborn lectured extensively during the late 1940s at such seminars as the Inter-American Conference on Conservation of Renewable National Resources, the National Conference on Land-Use Policy, and before the U.S. Chamber of Commerce, at the Massachusetts Institute of Technology.

He authored *The Pacific World* (1944), a grand-scale description of the geography, flora, and fauna found in the Pacific Islands, to aid World War II servicemen stationed in the area. In March 1948, Osborn published *Our Plundered Planet*, which detailed the necessity for maintaining balance in the ecological web. He posited that "man's conflict with nature" could cause "more widespread distress to the human race than any that ha[d] resulted from armed conflict." The work was highly acclaimed and helped pave the way to stricter antipollution laws and the creation of the Environmental Protection Agency and the Council of Environmental Quality. He also wrote *The Limits of the Earth* (1953) and edited *Our Crowded Planet* (1962). In addition, beginning in the 1950s until his death, Osborn authored numerous articles and editorials for *Science*, *Atlantic Monthly*, *This Week*, and *Animal Kingdom* (the New York Zoological Society's [NYZS] publication).

Working in conjunction with Laurence Rockefeller, then vice president of the NYZS, and under the sponsorship of the Rockefeller Brothers Fund, Osborn established the Jackson Hole Wildlife Park and the Jackson Hole Biological Research Station, both in Wyoming.

His indefatigable efforts did not pass unnoticed. Osborn received seven honorary degrees from 1955 to 1967; he was also awarded the Medal of Honor of the Theodore Roosevelt Memorial Association (1952), the Medal of the City of New York (1960), the Medal of Honor of the Municipal Art Society of New York (1962), the first Gold Medal for conservation from the Zoological Society of San Diego (1966), the Audubon Medal (1968), and the American Society of Planning Officials Medal (1969). Posthumously, the Rockefeller Brothers Fund created a lecture series in his name at Rockefeller University, and the Fairfield Osborn Reserve was dedicated as a permanent memorial at Sonoma State College.

Through his innovations in zoo administration and his tireless preservation efforts, Osborn achieved national and international status in the field of conservation.

JULIETTE M. FERNAN

Bridges, William. *A Gathering of Animals: An Unconventional History of the New York Zoological Society.* New York, 1974.

Current Biography. 1949.

"Henry Fairfield Osborn, Jr." [obit.] *New York Times*, September 17, 1969.

"Osborn, Henry Fairfield, Jr." [obit.]. *National Cyclopaedia of American Biography.* 1984.

Rockefeller, Laurence. "My Most Unforgettable Character." *Reader's Digest*, October 1972.

Paul, Sherman (August 26, 1920–May 28, 1995). An author, cultural historian, teacher, and editor, Paul was born in Cleveland, Ohio, to Jacob, a merchant, and Gertrude (Levitt) Paul. His Russian immigrant parents, who were lifelong urban dwellers, having settled in the Cleveland area after leaving Kiev, were not known to be actively involved in the natural environment. But raised in the suburb of Lakewood, Paul and his brother spent time exploring on and around the nearby Rocky River. After he became an Eagle Scout and was interested in sailing, his mother sewed his first sail on her sewing machine. His love for sailing small craft and swimming in fresh water led him in later life to settle in Wolf Lake in the Minnesota North Woods, the place where much of his work in nature would occur.

Paul began his education at the University of Iowa, from which he received the A.B. in 1941. He spent the next four years in the U.S. Army Air Force, where he earned the rank of captain. He then continued studying at Harvard, where he received an A.M. in 1948 and a Ph.D. in 1950. He remained on at Harvard for two years as an instructor, returning to the Midwest, first at the University of Illinois at Urbana (until 1967), then at the University of Iowa, where he was named Carver Distinguished Professor in 1974 and from which he retired in 1988 to write full-time at his home in Minnesota.

Paul's departure from academia to a life, as he liked to say, on the "margin"—margin offering for him as it did for Henry David Thoreau "the theme of settlement, of kindness to the ground, of nurture"—was carefully planned, completing as it did a lifelong exploration of the importance to Americans of being in nature, an exploration announced in his first two books: *Emerson's Angle of Vision: Man and Nature in American Experience* (1952) and *The Shores of America: Thoreau's Inward Exploration* (1958). He and his wife, an amateur archaeologist, set up "camp" on Wolf Lake a quarter of a century before moving there full-time, over the years improving their residence from tent to hut to cabin to house and tending the land on which it sat. Early on, Paul and his wife were activists in an effort to clean up Wolf Lake (an integral link in the uppermost reaches of the Mississippi) from the effluent waste of Bemidji, the city just upstream. The effort was eventually successful, after others from the larger Cass Lake community became involved. One of his more astonishing achievements (in which he was sometimes aided by his son) is his having planted by hand on some ninety acres over 30,000 trees, including red pine, white pine, balsom fir, spruce, and some hardwoods. Those who walked the acreage with him were impressed by his seeming to know the character and welfare of each tree, in spite of the immensity of the population.

But the camp in the North Woods was more than the stage on which he could encounter the natural world and its contemporary challenges on a physical basis. Like Aldo Leopold, who dedicated years not only to reclaiming but also to thinking macrocosmically about the wasted Wisconsin land upon which his shack was built, Paul drew on his personal experience of reconstituting and preserving his natural surroundings to develop values and metaphors for his writing. His prose, heretofore of a scholarly hue, in books like *Louis Sullivan: An Architect in American Thought* (1963) and *Edmund Wilson: A Study of Literary Vocation in Our Time* (1965), took on the accents of the woods in which he worked, and he began placing accounts of his personal experiences into his essays on the writers he read. Increasingly, these were writers who, like Sullivan, measured the quality of American life and thought by its capacity for having a symbolic and organic relation to nature. His next books were intense personal readings of such works: *The Music of Survival: A Biography of a Poem by William Carlos Williams* (1968), *Hart's Bridge* (1972), *Repossessing and Renewing: Essays in the Green American Tradition* (1976; a summation of the work of writers from Thoreau to Gary Snyder and Alfred Kazin, who helped to foster the "Emersonian tradition"), *Olson's Push: Origin, Black Mountain, and Recent American Poetry* (1978), *The Lost America of Love: Rereading Robert Creeley, Edward Dorn, and Robert Duncan* (1981), and *In Search of the Primitive: Rereading David Antin, Jerome Rothenberg, and Gary Snyder* (1986).

With the publication of *For Love of the World: Essays on Nature Writers* (1992), Paul addressed the theme of nature writing itself, testing by closely reading seven writer-naturalists—Thoreau, Aldo Leopold, Barry Lopez, Henry Beston, Richard Nelson, Loren Eiseley, and John Muir—Lopez's thesis that nature writing will "not only one day produce a major and lasting body of American literature, but . . . provide the foundation for a reorganization of American political thought." Paul had found his conclusion in his earliest reading of Ralph Waldo Emerson. But in this latest book, he reaffirmed it: "Having been dispossessed of 'America' so many times," he writes of the cultural experience of this country beginning with the Indians, "it is no wonder that the Republic, as Edmund Wilson said, has had to be saved again and again. This has been an essential task of our literature and now, more urgently, of nature writing." This is Paul's most personal work, beginning as it does with a journal he titles "Thinking with Thoreau" and interspersed throughout with autobiographical reflections inspired by the writers he discusses.

Paul was a Fulbright professor at the University of Vienna, held Ford Foundation and Guggenheim fellowships, and served on a number of editorial boards. In 1943 he married Jim McDowell, who continues to live at Wolf Lake. Their children are Jared Lempert, Meredith Olsen, Erica, and Jeremy.

JAMES BALLOWE

Peck, Daniel H., ed. *The Green American Tradition: Essays and Poems for Sherman Paul.* Baton Rouge, LA, 1989. This is a festschrift of critical and creative works by former students and fellow writers, and it contains a useful summation by Peck of the main themes of Paul's writing.

"Sherman Paul, 74, a Literary Scholar" [obit.]. *New York Times*, June 4, 1995.

Pearson, T. Gilbert (November 10, 1873–September 3, 1943). A biologist, ornithologist, wildlife conservationist, and author, Pearson was born in Tuscola, Illinois, to Thomas Bernard and Mary (Eliott) Pearson as the youngest of five children by twelve years. The Quaker farm family in their search for a setting in which to prosper moved to Indiana, to the village of Dublin. A physician advised Thomas to seek a warmer climate, and in January 1882, they moved to Archer, Florida. He purchased a log house, cleared several acres of woodland, and built a new house. He then planted a garden and a grove of orange trees and opened a nursery.

Gilbert began his education in a school conducted in a merchant's buggy shed. Early on, he showed more interest in bird-watching than in going to school or working at the nursery. Due to his being so much younger than the rest of his siblings, he was not expected to help out with the family business very much. He made friends with other boys who liked wildlife, and two of them had a large influence on Pearson's developing interest in birds and their eggs. They showed Pearson how to blow eggs for collecting and introduced him to his first issue of *Oologist*, a monthly magazine designed for boys interested in birds, eggs, and taxidermy. Despite being much younger than the other two boys, Pearson convinced them to allow him to tag along on a five-mile expedition to Bird Pond on April 27, 1886, by promising them almost his entire egg collection. From this experience he submitted the first of many reports he was to write over his lifetime. This first was one of nine published in *Oologist*. He later recalled having "wanted to tell the world" about the hike and the birds he had seen and identified. The enthusiasm with which Pearson faced every observation helped him to become a skilled field ornithologist (which was not a regulated title at that time) and collector and also helped him later to become successful at gaining allies for help in protecting birds.

In Gainesville, while running errands for his mother in March 1891, Pearson met Frank M. Chapman, a member of the American Ornithologists' Union (AOU). This chance meeting was beneficial to Pearson, as Chapman encouraged his work and helped to promote his career. Pearson was later nominated into the AOU by Chapman and was accepted as a member on November 17, 1891. This group of men was working to establish ornithology as an orderly field of study and also to begin a movement to protect non-game birds.

For some time Pearson made a living as a naturalist-ornithologist and operated a museum while selling his taxidermy services. To improve his social graces as well as expand his knowledge of things unrelated to bird studies, he wrote to several colleges in the summer of 1891, offering his museum in return for a term's enrollment. He was delighted when Lewis Lyndon Hobbs offered him two years of tuition, including room and board, for his museum, plus services as a curator of the college cabinet at Guilford College in North Carolina. His museum was noted as being the "finest College Collection in the State," and two years later, his extended collection was described as the "largest scientific collection of bird-eggs in the South." Besides becoming a leader in many areas, including the YMCA and the football club, at this time he also contributed articles to the *Guilford Collegian* and gave speeches on the importance of bird preservation and especially on the discontinuance of bird feathers being used in women's fashion. He continued to conduct explorations and raised money for the museum by asking for voluntary donations. After Pearson's two-year agreement was over, the trustees voted to extend his position as curator in exchange for $50 a year, plus room, board, and tuition.

One Saturday morning, Pearson recorded, when he went into the woods to collect a pair of Blue-gray Gnatcatchers, he experienced an unwillingness to shoot the birds because their song was so beautiful. It was the first time that sentiment had prevented him from shooting a bird that was in his opinion needed for educational purposes. This marked the beginning of his conversion from bird collector to eventual bird protector.

A paper he wrote graphically describing the cruelties of plume hunting was presented at the Congress on Ornithology, and he became one of the greatest adversaries to this practice. He received a Bachelor of Science degree from Guilford College in 1897 and entered the University of North Carolina, working for North Carolina state geologist Joseph A. Holmes in Chapel Hill to pay his tuition and board. By the end of October he had been chosen for the Dialectic Literary Society, a member of the junior class football team, and an editor of the *North Carolina University Magazine*. He continued to contribute articles to various publications while teaching a noncredit course in ornithology and collecting information for a book on the birds of North Carolina. In April 1898 Pearson went with Holmes and two other members of his Geological Survey staff on a ten-day trip to Pamlico and Albemarle Sounds. This expedition turned into a five-month period of exploring the North Carolina coast for Pearson, during which time he withdrew from classes. This period of time left a great impression on him and also provided much of the data needed for his book on North Carolina birds. When he reentered the university in September of that year, he wrote articles encouraging the promotion of bird studies in the public schools for children to build an admiration for nature in them. The teaching of bird appreciation in the public schools became one of Pearson's main areas of contribution.

Pearson earned another B.S. from the University of North Carolina in 1899. Afterward he took a teaching position at Guilford and left the Geological Survey in Chapel Hill. At his request for money to equip a biological laboratory at Guilford College, the trustees appropriated $75, half of which bought one microscope. Again he found himself soliciting contributions, a task that was often necessary throughout his career. He taught a conventional course in biology during which he tried to "arouse a deeper interest in ornithological study." While teaching he continued as always to contribute articles to journals and also to address audiences whenever possible. He maintained the position that everyone should learn something about birds.

Pearson was placed by Chapman on a *Bird-Lore* advisory council in 1900, as the representative for North Carolina. This council was designed to assist students by responding to their requests for information, therefore bringing the students into direct contact with a bird expert in their area. Pearson had a strong theory that he continually wrote, taught, and spoke on: there is a great interdependence of the "bird world and the human world." This idea was summarized by his "Bird Study in Elementary Schools," which he presented to the elementary education department of the Southern Educational Association's Richmond meeting. As he often referred to the evolutionary process despite his Quaker background, here he discussed bird's having reached the stage whereby they "must depend for their very existence upon the favors of the coming generations." With Elsie Weatherly (his future wife) drawing the illustrations, Pearson combined published articles with new material to create a book.

He, then, in the hopes of securing a high-enough salary to marry Elsie, took a position as a teacher at the State Normal and Industrial College, where he was required to develop a museum of natural history. The State Normal and Industrial College was dedicated to preparing women for teaching, and at Pearson's request, they began a tradition of field trips whereby the ladies could become familiar with nature. Pearson had by now become more interested in preserving birds than protecting them, and therefore the museum never developed as was hoped; however, the students became well prepared to "make use of nature study as a means in their teaching." In November 1901 Pearson's book *Stories of Bird Life* was published and adopted for use in the Boston Public Schools.

William Dutcher, chairman of the Bird Protection Committee, asked Pearson to organize an Audubon Society that could persuade the North Carolina legislature to adopt the AOU model law protecting nongame birds. This would involve a drastic change for him, as he had always worked to spread his enthusiasm for birds rather than lobbying for their protection. Pearson planned a meeting to organize an Audubon Society that was set for March 11, 1901, which was attended by approximately 150 people at the State Normal Chapel. The Audubon Society of North Carolina was chartered with 147 members, with Pearson as vice president and all the officers working in the state's educational system. As vice president, Pearson bore most of the responsibility for the society's work. The constitution reflected Pearson's views of the role that society should play. Its proposed three major objectives were to disseminate information about the value of birds, to promote bird study in the schools, and to build public sentiment against bird destruction.

In 1910 when Pearson chaired the Fifth Convention of the National Association of Game and Fish Wardens and Commissioners, he was elected president by the convention and held the position for two years. When the Audubon committee incorporated itself as the National Association of

Audubon Societies and expanded its dedication of protection for wildlife to include animals, Pearson was named special agent. He agreed to split his time between his duties as Audubon Society secretary and, now, special agent to the National Association with the specific assignment of "awakening public interest and securing financial assistance." His effectiveness as a public speaker, as well as his earnestness for the cause, made him very well suited for the position. On June 22, 1911, Pearson resigned as North Carolina's Audubon Society secretary, after which he was officially elected executive officer of the National Association on January 4, 1911. He remained president until retirement in 1934.

Pearson went on to become the founder and president of the International Committee for Bird Preservation from 1922 to 1938 and acted as chairman of the U.S. section until his death. His lasting work can be seen in established bird refuges and the laws he lobbied to have passed. Among these are protective laws supporting the Audubon Plumage Bill and the Federal Migratory Bird Law of 1913. Pearson also served on President Herbert Hoover's Yellowstone Park Boundary Commission and was the national director of the Izaak Walton League of America.

Pearson's contributions reached beyond the boundries of the United States. His international work included visiting President Porfirio Díaz of Mexico in 1909; negotiating with Canada, resulting in the important Migratory Bird Treaty of 1916; and visiting Europe in 1922 and founding the International Committee for Bird Preservation at a meeting in London. He lectured in Europe, in North, Central, and South America, and in the West Indies and organized and presided over the World Conference for Bird Protection at Geneva, Switzerland, in 1928 as well as the International Bird Protection Conference in Amsterdam in 1930. He was also the U.S. delegate to the 9th International Ornithological Congress in Rouen, France, in 1938. He chaired both the U.S. and Pan-American sections of the International Committee for Bird Preservation in 1938 and founded and chaired the National Conference on Wild Life Legislation.

In addition to Pearson's numerous magazine contributions, he also published his autobiography, *Adventures in Bird Protection* (1937); edited *Portraits and Habits of Our Birds* (1920); was senior editor of *Birds of America* (1917); was coauthor of *The Birds of North Carolina* (1919); and was coeditor of the *Book of Birds* (1937). Books that he wrote for children include *Stories of Bird Life* (1901), *The Bird Study Book* (1917), and *Tales from Birdland* (1918).

Among the various honors that Pearson was given over his lifetime, two of the most distinguished were an honorary LL.D. degree in 1924 from the University of North Carolina and the medal of the Société Nationale d'Acclimatation of France in 1937. In the *Greensboro Record* in 1906 it was noted that the suggestion of Pearson's being some day memorialized with a monument was inferior to the fact that his work in protecting birds would "erect a monument of itself."

LAURIE ANN CROMPTON

Orr, Oliver H., Jr. *Saving American Birds*. Gainesville, FL, 1992.

"T. Gilbert Pearson." *National Cyclopaedia of American Biography*. Vol. 33. 1947.

"T. Gilbert Pearson" [obit.]. *New York Times*, September 5, 1943.

"T. Gilbert Pearson." *Who Was Who in America*. Vol. 2. 1950.

Peattie, Donald Curloss (June 21, 1898–November 16, 1964). A botanist, silviculturalist, novelist, and author of many popular almanac-style discussions of the interaction between humankind and nature, Peattie was the son of *Chicago Tribune* writer Robert Burns Peattie and novelist Elia Amanda (Wilkinson) Peattie. In *The Road of a Naturalist* (1941) he described his boyhood isolation in Tryon, western North Carolina, a period when he became aware of "the democracy of science" and of his love for botany ("my weedy self"). He learned to take comfort in his belief that "[science] will not tell even a small and very white lie. Not even to comfort the suffering." In 1918, while a reader for George H. Doran Company in New York, Peattie visited the Bronx Botanic Garden and decided to become a trained botanist. His first work was *Flora of the Tryon Region* (1928–1931) after his studies at Harvard. After two years with the Bureau of Foreign Seed and Plant Introduction, he became a freelance writer. He had his first major popular success with *An Almanac for Moderns* (1935), in which his style, a combination of austere, scientific observation presented through a lush, elliptical prose, struck his readers as a new, "modern" approach.

Peattie's major works fall into two categories: (1) taxonomic descriptions of plants and trees and (2) general discussions of the application of scientific thinking to human problems. They are *Cargoes and Harvest* (1926), *Flora of the Indiana Dunes* (1930), *An Almanac for Moderns* (1935), *Green Laurels* (1936), *A Book of Hours* (1937), *A Prairie Grove* (1938), *A Gathering of*

Birds (1939), *Journey into America* (1943), *A Cup of Sky* (1945), and *The Rainbow Book of Nature* (1957), for children. Due to ill health, only two volumes of *Trees of North America* (1948, 1950) were completed.

Writing in the *Nation* (April 24, 1937), Joseph Wood Krutch immediately located Peattie's importance as a naturalist who could create a new group of readers, a group that wanted the work of a trained scientist. Krutch said, "What has interested me most in the books . . . is an attitude toward the world of nature recognizably individual and unmistakably modern." He felt Peattie's work was "modern" because he conveyed "a detailed knowledge of the actual ways of life which many pantheists have actively avoided." Krutch wanted Peattie to develop a new modern subject, "a sort of history of the love of nature." This is what Peattie did, but his style became increasingly florid after the 1930s, and much of his later popular work now seems overwrought, overstated, and sentimentalized.

In *An Almanac for Moderns*, Peattie offered his readers his best blend of careful scientific fact and honest speculation on humans' "proper place" in nature. His prose surrounded a very austere, mechanistic view.

Unlike Walt Disney or John Burroughs, Peattie found "that there are attitudes, emotions even, toward nature, more real and vital and valuable than love. There is a great deal more stingo, more savor and bite in Nature if you do not try to love everything you touch, smell, hear, see or step on."

It may have been the social turmoil fostered by the Great Depression, or some realism gained while he was a failed novelist in Europe, that made Peattie view his modern 1930s: world as surrounded by circling wolves (Nazism, Stalinism): "At every moment in that destiny the beauty and terror of life confront a man—still more a woman—and round the circle of the days the eyes of Death move watchfully, pondering on children running in the light, on woman in the night or man at labor."

The extraordinary delicacy of perception Peattie possessed in the 1930s, and the powerful style that has always attracted other nature writers, can be found distilled in the following passage. It is very close to that great artistry that skirts on the very tiptop edge of emotion. Here Peattie creates in his 1930s readers a sense of shame for the beauty they were wantonly destroying both in nature and in human civilization:

> I remember the first baldpate duck I ever saw, floating upon a marsh, in a cold evening damp-floating motionless, with speckled and green head, and blue bill outstretched lovingly upon the water, the exquisite mantle

> of brownish gray laved by the wind-driven dark ripples, the green and black-bordered wings outspread as if in an ecstasy to catch the wind. So, like a lovely boat, this creature of beauty drove on before the breeze, toward open water, more graceful and more silent than a swan—and dead. Gone was the fowler who had wounded him, but failed to retrieve him. With a bullet in his body the wild thing had still fought for its life, got clear away—to die unconquered, its proud plumage still unplucked; to drift, like this, a Viking's funeral, between the water and the sky.

Peattie carefully put before sportsmen in pre– and post–World War II America the need for a decision: Would they *conserve* nature or destroy nature, represented by deer, birds, and fish? Peattie's influence has been far-reaching, as he helped to create the desire in modern Americans to preserve, then conserve, wildlife.

On May 22, 1923, Peattie married novelist Louise Heegard. They had four children—a daughter and three sons. Most of Peattie's work examines the nurturing of marriage and children, as he viewed people as dependent on nature's processes. During his last decade, his wife helped him extensively when his diabetes and liver ailments curtailed his research and writing.

RODNEY SMITH

"Donald Curloss Peattie Is Dead: A Leading Naturalist and Writer" [obit.]. *New York Times*, November 17, 1964.

Hendrick, Kimmis. "The Peatties Talk About Writing." *Christian Science Monitor Magazine*, September 23, 1950.

Krutch, Joseph Wood. "Communion with Her Visible Forms." *Nation*, April 24, 1937.

Peterson, Roger Tory (August 28, 1908–July 28, 1996). An ornithologist, naturalist, educator, writer, and environmentalist, Peterson is widely recognized as a bird artist. His *Field Guides* have become standard reference volumes. Before Peterson, bird-watching had been primarily a precursor to bird hunting, but he taught ecologists to appreciate the beauty of all winged creatures, their importance in the natural world, and that it is wrong to destroy wildlife.

Peterson was born in Jamestown, New York, a small factory city in the western part of the state populated by Swedish immigrants. His father Charles worked as a traveling salesman and cabinetmaker. His mother had emigrated with her family from Poland to the upstate town of Olean, where the couple met.

By all accounts a mischievous and unruly young man, Peterson was transformed by birds at an early age. Encouraged, along with the rest of his class, by his seventh-grade teacher Blanche Hornbeck to join the Junior Audubon Society, Peterson sent in his 10¢ entry fee and received a series of leaflets that contained descriptions and pictures by well-known bird artists. On his first birding expedition with his friend, Carl Hammerstrom, the two were able to pet the head of a sleeping flicker, which awoke and flew away. Peterson, Hammerstrom, and a third friend, Clarence Beal, were soon embarking on regular excursions throughout the area in pursuit of birds to spot, sketch, and photograph.

After graduating from Jamestown High School in 1925, Peterson took a job with the Union Furniture Company, painting Chinese landscapes on the cabinets of high-end lacquered furniture. He worked under the tutelage of the firm's chief decorative artist, Willem Dieperink Von Langereis, who encouraged him to pursue an artistic career. After two years, Peterson was able to save enough money to attend art schools in New York City, first the Art Students League, then the National Academy of Design. Upon completing his art studies, he took a position teaching science and art at the Rivers School in Brookline, Massachusetts, a prep school for the children of Boston's elite. There he taught future Attorney General Elliot Richardson, and they became lifelong friends. Richardson would eventually write the foreword to Peterson's authorized biography.

The idea for publishing a guidebook apparently was given to Peterson by his friend William Vogt, a Westchester drama critic, after the two completed a bird-spotting hike along the Hudson River. He published his first *Field Guide to the Birds* in 1934. The book had been rejected by several publishers, but Houghton Mifflin took a chance on it, publishing it during the height of the Depression. The inspiration, according to Peterson in his preface, was the character Yan, in Ernest Thomas Seton's *Two Little Savages* (1903). In that book Yan encounters stuffed ducks at a museum and, viewing them up close, is able to learn their identifying marks, allowing him to become adept at spotting them in the wild. Seton's book also included 200 illustrations that supplemented the textual descriptions. The first *Field Guide*, which followed Seton's basic format of illustration and description, with the aim of identification, was an immediate, if seemingly improbable, success, during the height of the Great Depression.

The first copies of the guide were sold for $2.60. It supplanted previous guidebooks that tended toward detailed scientific descriptions and analysis. Importantly, the Peterson guides were compact enough to be carried in a pocket. The book's initial success was not a lucky stroke, and it has sold steadily, more than 7 million copies being published in five editions over more than sixty years.

The first guide covered birds east of the Rocky Mountains; a subsequent edition was expanded to include western birds. Peterson eventually, with his spouse and collaborator Virginia, also wrote a *Field Guide to the Birds of Britain and Europe* (1953) and a *Field Guide to Mexican Birds* (1973). He became the education director of the National Audubon Society and the art director for the National Wildlife Federation, until, secure in the financial success achieved through the bird guides, he pursued his passion for birds and art as a full-time occupation. He was undoubtedly instrumental in increasing binocular sales as well.

One of Peterson's most significant contributions was the development of the Peterson Identification System, a simplified methodology for identifying plants and animals and one that was eventually extended to species other than birds. It involved the seemingly obvious technique of using arrows to highlight the identifying marks of different species. Peterson used a visual epistemology, grouping birds together that looked similar, making it easier for users of the guidebook to correct possible mistakes quickly. Previously, illustrations of birds and other species had been ordered phylogenetically, that is, according to evolutionary development, making direct comparisons more difficult. Simple descriptions in the field guides also aided in the identification and remembrance of distinguishing features. The introduction of the Identification System, along with the beautiful illustrations, encouraged the development of bird-watching as an amateur pastime. Now there are hundreds of bird-watching societies all over the globe and an estimated 70 million "birders" in the United States alone.

Peterson's trips to various places throughout the globe also became an inspiration to birders and are documented in his autobiographical book *All Things Reconsidered: My Birding Adventures*, published ten years after his death, which was drawn from a regular column that he contributed to *Bird Watcher's Digest*. On these trips, he was often accompanied by his friend James Maxwell McConnell Fisher, a highly regarded English ornithologist. Peterson and Fisher also collaborated on *Wild America* (1955), a book that tracked their 30,000-mile journey around the continent. Peterson's extensive field experience was crucial for refining the disciplined artistry that characterized his approach to painting and illustration. The Peterson Identification System became a model for guidebooks on other topics, seventy of which Peterson either wrote or edited. Interested amateurs were given entry into the worlds of insects, plants, minerals, weather patterns, and constellations. An estimated 18 million copies of these various guidebooks have been sold worldwide. Peterson eventually became something of a one-man bird industry. Calendars, Christmas cards, and even bird trading cards were published under his name, always with his supervision.

Peterson has been criticized by some ecologists for diverting attention away from broader issues of land and ecology toward the narrower one of plumage, identification, and the game of spotting various bird species. His guidebooks, nonetheless, helped initiate the modern environmental movement. Certainly, Peterson's books constitute one of the most important informal sources contributing to an understanding of the place of avian life in nature, and they support his belief that birds can serve as harbingers of the state of our environment.

Birds are, in Peterson's terms, "environmental litmus papers." Here he supports Rachel Carson, who in her *Silent Spring* (1962) concurred with Peterson that birds are important as indicators of environmental matters. He also testified at congressional hearings in 1964 on the subject of DDT's impact on birdlife since, as he put it, "they are the best known animals in the world," and they enable millions of bird-watchers to monitor the changing state of global environment.

As an artist, Peterson is often ranked as one of the two most important bird artists in history, the other James Audubon. Until the publication of Peterson's *Field Guides*, Audubon's *Birds of America* (1840) had been the standard reference guide. Audubon had provided the model for synthesizing scientific understanding with artistic representations. Peterson would eventually write an introduction for a 1978 edition of the book. Another of Peterson's predecessors was Louis Agassiz Fuertes, the renowned bird artist who, like Peterson, traveled the world in search of bird images to capture. Especially later in life, Peterson attempted to emulate Fuertes, whose works can be found at the New York Museum of Natural History, as a serious artistic presence.

Peterson's numerous awards included the Presidential Medal of Freedom, the Gold Medal of the New York Zoological Society, the Conservation

Medal of the National Audubon Society, and the Gold Medal of the World Wildlife Fund. He was nominated for the Nobel Peace Prize. He held twenty-two honorary degrees. The Roger Tory Peterson Institute of Natural History was established in Jamestown, New York, in 1993, combining artistic and natural history exhibitions, educational seminars, and outreach programs, thus permanently institutionalizing Peterson's legacies as naturalist, artist, environmentalist, and educator.

THOMAS SHEVORY

Devlin, John C., and Grace Naismith. *The World of Roger Tory Peterson: An Authorized Biography*. London, 1977.

Dunlap, Tom. "On Early Bird Guides." *Environmental History* 10 (January 2005).

Kaufman, Kenn. "Roger Tory Peterson: A Man of Many Talents." *Audubon* 100 (November–December 1998).

Line, Less. "He Transformed Us into a World of Watchers." *National Wildlife* 40 (February–March 2002).

Ripley, S. Dillion. "The Nation's Foremost Birder Has Perfected the Field Guides into Essential Tools for Understanding the Vagaries of Nature." *Smithsonian* 15 (April 1984).

Phillips, John Charles (November 5, 1876–November 14, 1938). A naturalist, conservationist, avianist, curator, and author, Phillips was born in Boston, Massachusetts, a son of John Charles Phillips and Anna (Tucker) Phillips. In 1895, he entered the Lawrence Scientific School at Harvard University (now the Graduate School of Applied Science); four years later he was awarded a Bachelor of Science degree. In 1904 he graduated from Harvard Medical School but did not practice medicine. Instead, he devoted himself to travel and came to develop an interest in wildlife.

When the United States became involved in World War I, Phillips had to postpone his concentration on wildlife. In November 1915 he joined the Second Harvard Surgical Unit and was assigned to serve in France as a surgeon in a British General Hospital. Later he served with the U.S. Medical Corps, from September 20, 1917, to July 22, 1919. He reached the rank of major and was placed in charge of a Field Hospital in the Fourth Division. Shortly after he resigned his commission, he once again devoted himself to travel and the study of wildlife. He authored numerous papers on birds, genetics, and experimental animal breeding.

As Phillips's reputation grew, he was offered various positions and accorded multiple honors. He was, for example, appointed chairman of the Massachusetts Conservation Council, associate curator of the Harvard Museum of Comparative Zoology, a trustee of the Peabody Museum in Salem, and a member of the Massachusetts Trustees of Public Reservations. He was also elected president of the Massachusetts Fish and Game Association. Though heavily involved with multiple activities, he still found time to complete several books. His two-volume study of ducks, *A Natural History of Ducks* (1923), is considered one of the finest of its kind. Among his other highly lauded books are *American Waterfowl* (1930), *American Game Mammals and Birds* (1930), and *Migratory Bird Protection in North America* (1934).

In 1932, as chairman of the American Committee for International Wild-Life Protection, he served as a delegate in Europe to the International Conference for the Protection of the Fauna and Flora of Africa. Concern for the environment and conservation causes were responsible for his turning over wide tracts of land in Boxford, Wenham, and Rockport to the state of Massachusetts for the preservation of wild birdlife.

Phillips died suddenly in his sixty-second year while on a trip to southern New Hampshire. He was mourned by his wife Eleanor Hyde Phillips (whom he had married on January 11, 1908); two sons, John Jr. and Arthur; two daughters, Mrs. Schulyler Watts and Mrs. Stanley Phillips; and innumerable colleagues and family friends.

GEORGE A. CEVASCO

"Dr. John Phillips, Noted Naturalist" [obit.]. *New York Times*, November 1, 1938.
"Phillips, John C." *Who Was Who in America*. 1981.
"Phillips, John C." *World's Who's Who in Science*. 1968.

Pinchot, Gifford (August 11, 1865–October 4, 1946). A forester, governor, politician, public official, and leader of the early conservation movement, Pinchot was born in Simsbury, Connecticut, the eldest son of second-generation Frenchman James W. Pinchot and Mary Pinchot. James was a wealthy merchant who ran a profitable dry goods business and owned a large, forested estate in Pennsylvania. Pinchot considered a career in medicine or the ministry, but before matriculating at Yale in 1885, he was encouraged by his father to become a forester. The elder Pinchot was familiar with forest management practice in Europe and had the foresight to see the need for such practice in the United States even though formal study of forestry was impossible in this country.

At Yale, Pinchot read as much as he could on the subject of forestry and took various courses that he thought would be the most useful. Upon graduation Pinchot decided he must go to Europe to acquire a real education in forestry. He left for Paris in the fall of 1889 and there, through acquaintances of his family, was referred to the forestry school in Nancy. Before deciding on a course of action, Pinchot visited Sir Dietrich Brandis, who had established a systematic forest management system in India and Burma for the British government. Brandis had a profound influence on Pinchot, and Pinchot corresponded regularly with Brandis on forestry matters. Brandis recommended Pinchot attend the school at Nancy. At Nancy, Pinchot embarked on an intense study of forestry, supplemented with regular excursions to view European forests and consult with the eminent foresters on the continent.

Pinchot returned to the United States at the end of 1890 and set out to establish sound forest management practice in the United States. He was the only American educated in forestry at the time and made presentations and wrote articles on how to apply European management practice to American forests. He acquired several jobs inspecting forests for both private companies and federal and state governments. During the early 1890s, Pinchot made several trips to the western United States.

Pinchot's first opportunity to put his forestry education to a practical test and apply forestry principles on a large scale in this country occurred when he was hired to manage 20,000 acres of land surrounding George Vanderbilt's Biltmore estate in the Carolina mountains. Pinchot's work at Biltmore demonstrated to others that a forest could be managed as a renewable resource and make an annual profit. His work at Biltmore gave Pinchot a platform on which to spread the gospel of forestry in the United States and was summarized in *Biltmore Forest: Account of First Year's Work* (1893).

Pinchot advised private companies and several states during the early 1890s as a private forestry consultant and wrote *The White Pine* (1896) and *The Adirondack Spruce* (1898). These publications outlined comprehensive plans for managing eastern forests and included provisions for cutting, replanting, fire suppression, and soon, to produce sustainable yields of timber.

In 1896 Pinchot was appointed to the National Forest Commission by the National Academy of Sciences. He was the only non-Academy member of the seven-person commission and was appointed secretary at its first meeting. The commission was charged with filing a report and making recommendations to administer public forests. Additionally, the commission was to recommend forests to be included in federal reserves and propose legislation to guide conservation of the national forests. As a member of the commission Pinchot made an extensive trip to the western United States to survey forest areas. The commission's most significant and lasting contribution was the recommendation to President Grover Cleveland to create thirteen new forest reserves totaling more than 21 million acres. The reserves from seven western states more than doubled the acreage of national forests at the time. Cleveland acted on the commission's recommendation on February 22, 1897, ten days before leaving office. A bitter battle, mounted largely by legislators from western states, ensued in Congress. Pinchot played a key role in saving Cleveland's withdrawal through artful compromise between the commission, the McKinley administration, and Congress. Pinchot was opposed to many of the commission's final recommendations and even considered filing a minority report. Eventually he signed the commission's report but felt much of the commission's work was ineffective. Pinchot fought against military control of the forest reserves. The commission's report recommended the creation of a forest service in the Department of the Interior and the appointment of a director of forests and other personnel to administer management of national forests.

After the commission was dissolved, Pinchot was asked by the secretary of the interior to examine the federal reserves and make recommendations on establishing a service. On July 1, 1898, Pinchot, after initially declining, assumed the position of head of the Forestry Division (chief forester) in the Department of Agriculture. This small division had been created in 1880 and dealt mainly as an informational bureau. Pinchot's predecessor Charles Sargent, a member of the Forest Commission, was skeptical of Pinchot's forest management ideas and believed many of his ideas were impractical. Pinchot was given free rein to organize the ten-person division, which at that time was in jeopardy of being eliminated. He immediately started to build the division as well as the forestry profession in this country by giving advice to private landowners on how to manage their timberland for profit. During this period federal lands and forest reserves were administered by the Department of the Interior and were not under Pinchot's authority. By providing consulting services to private landowners, he supplemented the division's congressional appropriation. Pinchot started to expand the division immediately after assuming control by hiring college graduates who were willing to make forestry a profession and who accepted meager wages in exchange for on-the-job training. A year after he assumed leadership the division numbered 123 and in the next year grew to 179. Pinchot lobbied Congress for an increased appropriation and was successful in acquiring increased funding for the division in subsequent years.

Pinchot's most significant contributions to forestry occurred during the Theodore Roosevelt administration. Pinchot had become acquainted with TR when the latter was governor of New York, and Pinchot provided advice to Roosevelt on forest management and stream reclamation. Pinchot and TR shared similar political philosophies and were mutual friends. Pinchot was one of Roosevelt's most trusted advisers. Since his appointment as chief forester, one of Pinchot's greatest goals was to have all federal forests managed by the Division of Forestry. He fought vigorously to have administration of federal reserves transferred from the Department of the Interior to the Department of Agriculture. With TR's backing and the division's demonstrated ability to manage forest lands, this goal was realized in February 1905. At this time the Division of Forestry became the Forest Service.

Upon acquiring administration of the national forests Pinchot immediately set out to apply the forest management principles developed from the division's previous experience. The overall goal of the Forest Service during Pinchot's tenure as chief was to put the nation's forests to productive use for all citizens through sound, systematic, and scientific management. These practices were outlined in the division's *The Use Book* (1906), which explained standard operating procedures on all aspects of forest management. Pinchot had critics who resented his tight control of federal reserves. Under the Department of the Interior administration, use of federal reserves had been very loose and highly politicized. Pinchot's close scrutiny of federal forests led to the term *Pinchotism* as representative of autocratic, arbitrary, and unreasonable control of forest reserves. As expected, Pinchot's most vocal critics were from western states, where he tended to alienate private interests in mining, ranching, and utility companies.

Pinchot was instrumental in advancing forestry education in this country. The first professional graduate school of forestry was started at Yale in 1898 with an endowment of $150,000 (subsequently increased to $300,000) from the Pinchot family. In 1900 Pinchot helped establish the Society of American Foresters as a professional organization to unite individuals in this new profession. In 1903 he was appointed a professor in the Yale forestry school and regularly presented short courses, gave lectures, and assisted in the management of the school on a limited basis.

Pinchot's forestry work naturally extended his interests into soil conservation, water, mining, grazing, wildlife, and comprehensive natural resource management. While his original contribution was to forestry during the TR administration, he expanded on his forestry experience and became one of the foremost leaders of the conservation movement. He believed conservation was "the key to the future": "the very existence of our Nation, and of all the rest, depends on conserving the resources which are the foundations of its life." The conservation movement called for natural resources to be used for the greatest good, for the greatest number, for the longest time. Pinchot's early work in conservation involved him as a member of the Public Lands Commission, created in 1903, and the Inland Waterways Commission, created in 1907. Pinchot, with the help of TR, was instrumental in convening governors, national political leaders, and prominent private citizens for a conference on the conservation of natural resources in 1908. This may be regarded as the first national conference to address natural resource issues comprehensively. The tone of the conference established natural resources as the source of the nation's future wealth and a treasure of all citizens of the United States. The Governor's Conference on Natural Resources established the National Conservation Commission, which Pinchot chaired. He also chaired the North American Conservation Conference held at the White House, and plans were made for a World Conservation Conference. The latter was canceled by TR's successor William Howard Taft. During this period Pinchot's *The Fight for Conservation* (1910) was published.

Pinchot was optimistic that Taft would continue the fight for conservation, but he became quickly disillusioned when Taft replaced several of TR's cabinet appointees with individuals who did not support conservation ideals to the extent Pinchot thought was necessary. Foremost among the appointees with whom Pinchot was at odds was Secretary of the Interior Richard Ballinger. Ballinger was a Seattle attorney who waged a bitter fight against Pinchot and the Forest Service over the Forest Service's withdrawal of water power sites and the issue of Alaska coal claims. Pinchot's battles with Taft's administration came to a head in January 1910 when he was fired as chief forester for insubordination. A Congressional Joint Committee was established to investigate the Forest Service and Department of the Interior and the issues surrounding the Ballinger-Pinchot controversy. Ballinger was exonerated of Pinchot's charges of improper handling of Alaska coal claims.

Pinchot's dismissal as chief forester marked his transition from forester to full-time politician. In 1909 he founded the National Conservation Association to advance the principles of conservation. The association was staffed and dominated by close former Forest Service associates. Pinchot lent great financial support to the association and was instrumental in the association's national publicity campaign in the name of conservation.

Pinchot's animosity toward Taft combined with his unfaltering loyalty to TR made Pinchot an ardent supporter of the Progressive (Bull Moose) Party in the 1912 presidential election. In 1914 Pinchot failed to win the Pennsylvania Senate seat as a Progressive candidate. Several months before the election Pinchot married Cornelia Bryce, a wealthy, outgoing New Yorker who shared many of Pinchot's Progressive principles. The 1914 election was a harbinger of subsequent attempts to win political support in Pennsylvania. These attempts were often thwarted by mainline conservative Republicans both within Pennsylvania and nationally. Pinchot abandoned the waning Progressive Party and returned to the Republican Party in 1916 and spent a brief stint during World War I working for Herbert Hoover in the Food Administration before resigning out of disgust for Hoover's policies. From 1920 to 1922 Pinchot served as Pennsylvania state forest commissioner but had higher political ambitions in mind. In 1921 Pinchot rallied support against the Department of the Interior's attempts to gain control of national forests in Alaska and open these to private development.

In 1922 Pinchot was elected governor of Pennsylvania. He was strongly supported in rural areas and built a coalition of prohibitionists, suffragettes, and farmers. As governor, Pinchot sturdily supported legislation regulating utility companies to prevent monopolistic control and exploitation of natural resources. He was opposed to electric companies gaining access to water sites in perpetuity at no cost. Pinchot's interest to regulate water rights associated with hydroelectric power originated from his work with the Forest Service and eventually contributed to passage of the Federal Water Power Act of 1920.

Pinchot lost a second attempt for the Republican Senate nomination in 1926 and ended his first term as governor in January 1927. He was then able to devote more time to forestry and the conservation movement. Pinchot strongly supported federal regulations making timber cutting on private lands conform to practices required for cutting on federal lands. In 1929 Pinchot and George Ahern, a longtime colleague from his Forest Service days, submitted a minority report in response to a report of the Committee on Forest Policy of the Society of American Foresters. Pinchot felt the committee's report did not come down hard enough on private timber cutting practice, and the minority report chastised lumber men for their reckless cutting of forests on private lands.

Pinchot won a second term as governor and served from 1931 to 1935. Pinchot lost in his third attempt to obtain the Republican nomination to the Senate in 1934. During this period, he found Franklin Roosevelt's New Deal ideas attractive and consistent with much of his own political philosophy but remained a moderate Republican.

Once again when his term as governor expired, he demonstrated his belief "I am a forester all the time." Until his death he advanced his ideas on forestry and conservation through public speeches, writing journal articles, and editorializing in the popular press. He continually fought periodic attempts to transfer the Forest Service to the Department of the Interior or a newly created Department of Conservation and Public Works. A principal proponent of the transfer proposal was FDR's secretary of the interior Harold Ickes. Ickes resented Pinchot's opposition and resurrected charges against Pinchot stemming from the Ballinger controversy of years before.

Pinchot spent his last years lobbying presidents to convene an international conference on conservation. This finally occurred in 1949, three years after his death, and fulfilled a dream unrealized during Pinchot's life. Pinchot's other significant contribution during this period was completion of his autobiography *Breaking New Ground*. This work, published posthumously in 1947, was dedicated to the men and women of the Forest Service and outlined the development of forestry in the United States. A large portion of the book was devoted to Pinchot's interpretation of the Ballinger controversy. As might be expected, the book received mix reviews, depending on whose point of view and political leanings one took.

Pinchot's lasting contribution was to bring the idea of conservation and wise use of natural resources to the forefront of national policy. In this respect, he outlined many of the guiding principles that are still accepted today concerning natural resources and their wise use and management.

With his conservation work he helped to change the mind-set of his contemporaries as well as future generations.

RICHARD MYERS

McGeary, M. Nelson. *Gifford Pinchot, Forester-Politician.* Princeton, NJ, 1960.
Pinchot, Gifford. *Breaking New Ground.* New York, 1947.
Pinkett, Harold T. *Gifford Pinchot, Private and Public Forester.* Urbana, IL, 1970.

Preble, Edward Alexander (June 11, 1871–October 4, 1957). A biologist and naturalist, Preble was born in Somerville, Massachusetts. An early interest in nature was fanned by a friendship with Frank Blake Webster, the noted taxidermist and ornithologist. After completing high school in 1889, he briefly worked in Boston before rejecting city life and returning home.

Through his acquaintance with naturalist Frank Hitchcock, he was appointed by C. Hart Merriam to the staff of the U.S. Biological Survey in 1892, where he was to remain until retirement in 1935. His first assignments, intended to train him in the procedures and standards used by the Survey, were with Vernon Bailey in Texas, followed by collecting and study periods in Georgia, western Maryland, Oregon, Washington, and Utah. These projects were examinations of specific life zones, a type of research that was to become a hallmark of his career. In 1900, Merriam assigned Preble the exploration of the flora and fauna of northern Canada, inaugurating some of the most extensive fieldwork ever done in the sub-Arctic. Two expeditions to the Hudson's Bay and Mackenzie River areas of Canada in 1901 and 1904 resulted in *A Biological Survey of the Hudson Bay Region* (1902) and *A Biological Investigation of the Athabaska-Mackenzie Region* (1908). This latter, considered the best of the North American Fauna series issued by the Survey, documents the physical geography and climatology of the area, summarizes previous explorations between 1770 and 1907, and presents the mammal, bird, reptile, and plant life in precise detail.

Late in 1907, Preble accompanied Ernest Thompson Seton in a survey of the region northeast of Great Slave Lake, followed in 1910 by a trip from Wrangell, Alaska, to attempt to determine the species boundaries of mountain sheep in the Canadian Rockies. Mounting controversy over the decimation of the elk herd in Yellowstone National Park (publicized by such popular writers as Emerson Hough) called for an objective assessment of their condition. Preble's review of the problem, published in 1912, evaluated the food habits of the herd (and the effect on their movements) and their relationship to commercial livestock and made suggestions for long-term management. In 1913, he returned to British Columbia to the Nass and Upper Skeena Rivers, with the objective of hunting mountain sheep. The following year he traveled with Wilfred Osgood of Chicago's Field Museum of Natural History and George H. Parker of Harvard as a federal commissioner to the Pribilof Islands off Alaska. Their charge was to examine the state of the fur seal herds and offer suggestions for their preservation.

Returning to the Survey in January 1915, he switched his focus to the bird population in the metropolitan District of Columbia, while reworking his Alaskan data. This focus on ornithology led to several articles for *Bird-Lore* and *Condor*, as well as a memorial to his mentor Webster. Beginning with an article on the great anteater in 1924, Preble made his appearance in the pages of *Nature Magazine*, a journal to which he contributed frequently until his departure from the Survey in 1935, when he became one of its associate editors. The ensuing thirty-two years of publication in popular journals ranged over a wide variety of topics, from the plight of endangered species such as the wolf, passenger pigeon, and musk ox through extensive investigation of domestic species of birds (and a call for the preservation of waterfowl) to historical essays on significant American naturalists. The northland was never far away, as evidenced by pieces on the history of Alaska and the Arctic caribou. While at the Biological Survey, he also frequently served as editor for many of the scientific papers and reports prepared by other members of the agency.

Following his death during his eighty-sixth year, his ashes were interred at his grandmother's farm in Ossipee, New Hampshire, in the foothills of the White Mountains, where he had established a wildife sanctuary.

Preble's legacy as an environmentalist lies in his investigations of large areas of the Alaskan and Canadian Arctic and sub-Arctic and his recommendations for the care and preservation of threatened species, many of which form the basis for current policies. His numerous articles brought and kept the state of America's wild birds and mammals before public attention, while his ecological research helped to delineate the boundaries of regional ecosystems over much of North America.

ROBERT B. MARKS RIDINGER

McAtee, M. L. "Memorial: Edward Alexander Preble." *Auk* 79 (1962): 730–742.

McAtee, M. L., and Francis Harper. *Published Writings of Edward Alexander Preble (1871–1957)*. Lawrence, KS, 1965.
Westwood, Richard. "Edward Preble: An Appreciation." *Nature*, December 1957, 537.

Puleston, Dennis (December 30, 1905–June 8, 2001). In the 1960s, when osprey populations were plummeting due to exposure to DDT, Puleston, a sailor, naturalist, adventurer, and war hero, turned his love for the birds into activism, banding together with several others to sue New York's Suffolk County Mosquito Control Commission. The suit was not only successful; it was the first of its kind and thus gave birth both to similar suits and to one of the most powerful environmental organizations in the United States—the Environmental Defense Fund, of which Puleston was the founding chairman.

Born at the turn of the twentieth century, Puleston grew up in a small fishing village called Leigh-on-Sea, located approximately thirty miles outside London at the mouth of the Thames River. He spent his boyhood consorting with fishermen, painting pictures of birds and flowers, and exploring the waterfront and marshes that surrounded his home. An uncle who was an avid ornithologist fostered Puleston's love of nature, while his father encouraged a love of the water. As Puleston later wrote, "[I]t was actually my father who started me off on the sea, in a small but stout sailing dinghy. When it was calm, or we were exploring the maze of creeks that wind through the sea-lavender marshes, my brother and I were allowed to sit side by side on the 'midships thwart and tug with all our puny might at the oars" (*Blue Water Vagabond*). Later, he further gained his sea legs as a member of the crew in a yacht club race.

As a young adult Puleston attended London University, where he studied biology and naval architecture. He then went on to work as a bank clerk in London, using the commute to sketch the birds he saw along his route—a hobby that his mother, a painter of miniatures, encouraged him in and one that would later play a role in his work against DDT. Meanwhile, the books he read on these commutes—by adventure and travel authors like Joshua Slocum, Joseph Conrad, and Alain Gerbault—instilled in him a restlessness to see the world. And so, in 1931, Puleston commenced a life of trailblazing, as he and a friend cashed in their savings and set off on an engineless, thirty-one-foot yawl for a 4,000-mile trip across the Atlantic to the West Indies. Between them they had only the fully provisioned boat,

Uldra, and a few hundred dollars. And yet, remarkably, it was six years before Puleston found himself back home in England.

Right from the trip's beginning, Puleston took an active interest in the natural world around him, observing sperm whales, sharks, and porpoises, even nicknaming one loyal fish who trailed the boat for weeks. His adventures continued as he and his friend vagabonded across the isles of the West Indies in the company of two Yale graduates and later settled down to manage a rundown coconut plantation on the island of Tortola, surviving the devastation of a hurricane in the process. Ever restless, Puleston and his traveling companion next found themselves on the crew of a gaff-rigged schooner heading up the eastern seaboard of the United States. Despite an initial shipwreck off the coast of Cape Hatteras, North Carolina, they eventually made it to New York City, where Puleston signed on with a crew that was to pick up a schooner in the chilly waters off the coast of Newfoundland. From there Puleston and his shipmates sailed down to Santo Domingo, where there were rumors of treasure buried within the wrecks of long-forgotten sunken ships. They came back empty-handed, and so upon his return to New York, Puleston spent the summer months living aboard *Uldra*, teaching sailing lessons to youngsters as an instructor at the American Yacht Club. Just as the summer holiday came to a close, he was offered the chance to join the crew of a boat heading through the South Pacific—of course he jumped at the chance.

After passing through the Panama Canal, Puleston and company stopped off at the Galápagos Islands, then headed to the Marquesas, and on to Tahiti, where they stayed for eight months reveling in the warm weather. When Puleston's pet boa constrictor, Egbert, slithered ashore, creating a local stir and causing Tahitian tongues to wag, he and his shipmates cast off again, stopping to dive for pearls in Tongareva, drinking kava in American Samoa, and encountering cannibals in the New Hebrides. Later, on their way to New Guinea, Puleston and the crew discovered a group of small, uninhabited islets, which they christened the Director Islands, in honor of their ship. By this point, the ship had become the home of quite a menagerie of wildlife, including tortoises from Galápagos, a monkey, a honey bear, a Timorese dog, Puleston's dedicated pup Tiger, and a variety of birds. As Puleston, the amateur ornithologist, put it, "the more we had the happier we were. So our ship soon became a floating aviary" (*Blue Water Vagabond*).

After bouts of malaria brought them down near Manila, the travelers wandered up to Hong Kong and Peiping, where they hoped the northern climate would improve their health. Unfortunately, their stay in Peiping was cut short when the Japanese invaded, leaving Puleston stuck for a route out of China. In fact, it was not until he pulled out a newspaper clipping about his donation of a cockatoo to the Emperor's Imperial Zoo that a Japanese officer granted him passage on the Trans-Siberian Railroad, which eventually deposited him back in London. The whole wild tale is charmingly recounted in *Blue Water Vagabond*, which Puleston published in 1939 to a positive review by the *New York Times*, which described the book as "likable" and "peopled with modest, genial, gay, and sportsman-like characters" (April 9, 1939).

That same year, Puleston married his wife, Betty Wellington. "I met her in the middle of the Long Island Sound," Puleston later told the *San Francisco Chronicle* (March 7, 1993), explaining how they first met—he rescued her after she fell overboard during a sailing competition. Puleston and Wellington eventually had four children—Dennis, Peter, Jennifer, and Sally. Predictably, though, Puleston did not readily settle into domesticity. After gaining his American citizenship in the early 1940s, Puleston began working as a naval architect, contributing to the design of an amphibious troop-landing vessel, the DUKW (popularly pronounced "duck"), which was used at Normandy during World War II. He later supported the war effort by running DUKW training schools in India and Oahu, seeing action in Iwo Jima and Okinawa. At one point, he was wounded in Burma when a Japanese shell splinter struck him in the back, but fortunately he recovered. In 1948, President Harry S. Truman honored him with a Medal of Freedom for his work on the DUKW.

With the end of the war, Puleston settled in Brookhaven, New York, where he was soon employed by Brookhaven National Laboratory as a technical officer—a post he held for twenty-two years until his retirement in 1970. With a full-time job to occupy him during the day, Puleston became a weekend naturalist, and in 1948 he participated in an inventory of the local osprey population. To Puleston, the osprey was a rather mythical bird because it was no longer found on the British Isles; however, in 1948 populations were still high on Gardiner's Island near Long Island, and the results of the count in which Puleston participated found 300 active nests. Several years later, in 1955, fifty-two of Puleston's paintings of these and other birds were displayed at none other than New York City's American Museum of Natural History, their three-dimensional style influenced by both his commuting-time sketches and the Chinese bird art he saw in Peiping.

Unfortunately, by the mid-1960s Long Island's osprey population had fallen significantly. Puleston, fresh from reading Rachel Carson's groundbreaking book *Silent Spring* (1962), noticed that the shells of the ospreys' eggs were increasingly so thin that they were cracking before the chicks had fully formed; he wondered if this was one of the unforeseen effects of DDT that Carson had described. At the time, Puleston was a member of the Brookhaven Town Natural Resources Committee (BTNRC), which was focused on local concerns like the protection of the Carmans River ecosystem. The group held a fair amount of local clout, and thus under the banner of BTNRC, Puleston and a few other scientists took their suspicions about DDT's impact on local wildlife to court in 1966, successfully suing the Suffolk County Mosquito Control Commission to stop the spraying of DDT. For the trial, Puleston contributed seven paintings (later published in *Scientific American*) depicting DDT's negative reverberations throughout the food chain. As he later recalled, these illustrated charts were designed to "make an ecologist out of the judge hearing the case" (*Acorn Days*). Indeed, on seeing one of the charts that demonstrated the way DDT could affect the blue claw crab, Justice Jack Stanislaw is reported by Puleston to have commented, "So that is why the crab has disappeared from the Great South Bay!" Decades later Puleston used a nautical metaphor to describe DDT, writing, "The use of a broad-spectrum, persistent chemical like DDT to kill mosquitoes can be compared to torpedoing an ocean liner to get rid of the rats on board" (*A Nature Journal*). Needless to say, the ban that Puleston and BTNRC won against DDT was quite a coup, and as a result, communities began barraging BTNRC with requests for help with similar local lawsuits.

Juggling an increasing number of requests for help, and flush with the success of their suit against Suffolk County, in 1967 Puleston and his fellow plaintiffs formed the Environmental Defense Fund (EDF), taking as their motto "Sue the bastards!" Puleston, with his experience and vigor, was readily nominated to become the founding chairman of the group. Charles Wurster, one of the four founding members of the group, commented about Puleston, "[H]is unequaled eminence, dignity and gray hair made him the unanimous choice to be the first Chairman of the Board of EDF" (*Acorn Days*). Of course, when the group was first formed, they had no office and little money; in fact, they jokingly referred to themselves as the "Fundless Environmental Defenders." Meetings were occasionally even held in Puleston's swimming pool. Nevertheless, in the initial years of Puleston's chairmanship, the group's initiatives included several ambitious projects, including a suit to prevent the use of dieldrin in a Japanese beetle control project on the shores of Lake Michigan (despite traveling to Michigan to testify, they lost the case) and a case to reduce air pollution in Montana's Bitterroot Valley.

One of EDF's first major victories was the case of the Cross-Florida Barge Canal. Back in 1942, Congress had authorized the U.S. Army Corps of Engineers to create a military waterway across Florida that would negatively impact the Oklawaha ecosystem. When the project was finally about to get under way, EDF filed a civil suit, which the government in turn attempted to have dismissed. However, the judge allowed a temporary injunction, then later decided that all work must be stopped on the canal. Days later, President Richard Nixon issued a permanent halt to the work, citing environmental concerns.

With this victory came an awareness of just how many other aquatic systems needed protecting, and so Puleston and EDF soon found themselves involved in river cases nationwide. Meanwhile, they continued to broaden the scope of their activism, successfully delaying work on the Trans-Alaska Pipeline and becoming involved in energy cases in the western United States. In short, EDF continued to grow. As Puleston recalled about this transition period, "We were no longer considered by some as a handful of wild-eyed, bomb-throwing radicals. Our child had grown into a mature, respected adult, a vitally-needed protector of an environment that was under ever-growing pressures" (*Acorn Days*).

Today, Puleston's child, Environmental Defense, as it is now known, has more than 500,000 members and is a significant player in the environmental movement. Puleston stepped down as chairman in 1972, the same year DDT was banned nationwide (New York State discontinued its use in 1970). During his tenure, EDF went from an idea born of a lawsuit to a full-fledged environmental nonprofit involved in a wide variety of campaigns. Even after he stepped down as chairman, Puleston stayed active with the organization, and at the age of ninety-four he succeeded in halting the construction of a mall.

Of course, even while Puleston was agitating for environmental protection, wanderlust continued to strike, and beginning in the 1970s, he began working as a naturalist guide for the National Audubon Society and a cruise company heading to exotic locales including Antarctica, the Amazon, and his beloved Galápagos. In this second career as a lecturer, Puleston made almost 200 voyages, including at least 35 to Antarctica. "Here's a man in his 80s who's always the first one to jump ashore, the first one to head off into unknown territory," Kevin Schafer, then the director of field operations for the cruise company that employed Puleston, commented to the *San Francisco Chronicle.*

This pioneering spirit is precisely what Puleston is to be lauded for; he trailblazed a legal approach to environmental issues in the same manner that he ventured into unknown regions as a sailor both young and old: with energy and courage. Recognizing his commitment to environmental protection, in 1995 the State University of New York, Stony Brook, awarded him an honorary Doctorate of Humane Letters. Reflecting on Puleston's many accomplishments after his death in 2001, Environmental Defense executive director Fred Krupp commented, "He cared so deeply that he inspired others to care." And indeed, Puleston is further honored by the Dennis Puleston Osprey Fund, a group that encourages osprey research on Long Island.

The author of several well-received books, founding chairman of Environmental Defense, winner of a Medal of Freedom, and well-traveled naturalist and lecturer, Puleston spent his life exploring the natural world and communicating its fragile beauty. As he eloquently stated in his last book, *A Nature Journal* (1992), "There are many battles waiting for us in the coming years, some we are still unaware of. But our canaries, be they osprey, crab, fish, polluted streams, filled-in marshes, or other signals, are there to warn us of environmental threats, and we must pay full attention to these warnings."

ERICA WETTER

Puleston, Dennis. *Blue Water Vagabond: Six Years' Adventure at Sea.* New York, 1939.

———. *A Nature Journal: A Naturalist's Year on Long Island.* New York, 1992.

Rogers, Marion Lane. *Acorn Days: The Environmental Defense Fund and How It Grew.* New York, 1990.

Reilly, William Kane (January 26, 1940–). During the administration of President George H. W. Bush, Reilly served as the Environmental Protection Agency (EPA) administrator from 1989 until 1992. Prior to his appointment, environmentalists were very skeptical about the president's commitment to their agenda. Reilly was the first professional environmentalist to hold the EPA's top position since its founding in 1970. In 1973 he became president of the Conservation Foundation, and in 1985, after a merger with the World Wildlife Fund (WWF), he was concurrently appointed to head both organizations. During his tenure as EPA administrator, Reilly worked assiduously for the passage of the 1990 amendments to the Clean Air Act of 1970, as well as the recognition of environmental concerns regarding the North American Free Trade Agreement (NAFTA) and the General Agreement on Tariffs and Trade (GATT).

Reilly was born in Decatur, Illinois. His parents, George and Margaret, raised him and his sister on the family farm, instilling in them both religious and conservative family values. George was self-employed, running a supply company that sold steel products and metal culverts. His wife assisted him by performing accounting and bookkeeping duties.

When the business stalled due to a steel strike, the family relocated to Texas. George eventually became a contractor, and although this lasted only two years, William learned a valuable lesson from his father that would ultimately shape his future. It was his first exposure to the multicultural and international aspects of environmental affairs. He observed his father's dealings with the hiring of undocumented workers from the Rio Grande Valley. The Mexican-American border workers were often seized by the Immigration Bureau and sent back to Mexico. Today they are commonly known as "colonias." As EPA administrator, Reilly lobbied for immigrant workers in NAFTA and, more important, secured $50 million for them in the federal budget.

At the age of fourteen, Reilly went to live with his aunt in Fall River, Massachusetts. It was a European immigrant town where many languages were spoken, particularly French. This town represented a stark contrast from his Midwest upbringing. Reilly credits this cultural experience as the catalyst for his interest in international affairs. He eventually became fluent in five different languages. He graduated from Durfee High School in 1958 and later attended Yale, receiving a B.A. in history in 1962. He spent his junior year in France, then went on to Harvard Law School, graduating in 1965. Reilly's senior thesis was about land reform in Chile.

Reilly practiced law briefly in Chicago, working for the firm of Ross and Hardies, specializing in land-use issues. He served in the U.S. Army during 1966 and 1967, attained the rank of captain, and was stationed in Germany, working with an intelligence unit that was planning for the evacuation of U.S. troops from France. He returned home briefly to marry Mary Elizabeth Brexton on the campus of Yale University at the St. Thomas More Chapel. After returning home from Germany, Reilly went to Columbia University in 1968 to pursue a master's degree in urban planning. It was at Columbia that he was first exposed to the field and study of environmentalism. He contemplated a career in international planning and consulting during a summer project in Turkey, but the anti-American sentiment due to the Vietnam War convinced him otherwise.

Reilly then went to work for Urban America Inc., which soon merged with the National Urban Coalition. By 1970 he was appointed to President Richard Nixon's Council on Environmental Quality, serving as a land-use lawyer. The council afforded Reilly the opportunity to work with Russell Train and William Ruckelshaus. Besides his father, they were the two most influential persons in his professional life. Train and Ruckelshaus subsequently became the first two EPA administrators. Train's and Reilly's career paths are somewhat similar, as both held the top posts at the Conservation Foundation, and the World Wildlife Fund, in addition to the EPA. Reilly's contributions and impact while at the council included drafting the framework of what would eventually become the National Environmental Policy Act, and the Environmental Impact Statement.

In 1972 he was appointed executive director of the Task Force on Land Use and Urban Growth. The next year they produced a report titled *The Use of Land: A Citizens' Policy Guide to Urban Growth.* The thrust of the report was that Congress impose a national land-use policy by the granting or withholding of federal funds to states that did or did not comply with congressional guidelines.

In 1973 Reilly was appointed president of the Conservation Foundation, an environmental research group in Washington, DC. Under his stewardship the group lobbied for environmental issues on a global scope. In 1985 the World Wildlife Fund merged with the Conservation Foundation, and Reilly was named president of both groups. With Reilly at the helm, this expanded group saw its membership rise to 600,000, employing a staff of over 200 persons and working with an annual budget of over $35 million. The World Wildlife Fund was, and still is today, an international organization that specializes in the protection of flora and fauna and the preservation of rain forests, particularly in Latin America and the Caribbean, from overdevelopment and pollution.

During the presidential terms of Nixon, Gerald Ford, and Jimmy Carter, Reilly did not comment on the environmental policies of the government. However, that changed with the Reagan administration. Reilly clashed openly with controversial Secretary of the Interior James Watt. In 1985 Reilly criticized the Reagan administration for their EPA budgetary constraints and their neglect of the National Parks Service. The Conservation Foundation report "National Parks for a New Generation" noted the decline of national parks development; it predicted that the dangerous trends of surges in overall park attendance and new commercial development would result in greater pollution. This report, commonly referred to as "Preservation 95," urged that over a ten-year period the Park Service's budget, training, and research efforts be expanded. Reilly also lobbied for controls over acid rain, chemical pollutants, toxic waste, and wetlands preservation.

Reilly's greatest asset was his ability to assemble diverse groups of state and local lobbyists with the environmental community to reach a consensus and compromise on pertinent policy decisions. In 1987, concerned about the loss of thousands of acres of wetlands, Reilly worked with these groups to set a "no loss policy" for wetlands. Another agreement he brokered between the World Wildlife Fund, American Bankers, and third-world officials was a debt for nature program, where the protection of endangered resources in developing countries resulted in foreign debt relief by the banking community.

In December 1988, with the recommendation of former EPA head Ruckelshaus, Reilly was chosen to succeed Lee Thomas as EPA administrator by President H. W. Bush. In 1989 the EPA proposed legislation to curb the greenhouse effect for the coming decade. Reilly also proposed additional research for solar power, with the goal of curbing the levels of fossil fuel emissions. In an unprecedented decision, Reilly overruled the EPA regional director by canceling plans by the Army Corps of Engineers and the Denver Water Board to build the Two Forks Dam. An assembly of Colorado environmentalists feared that the construction of a dam in that area would negatively affect fish and wildlife. Reilly agreed that the potential water shortage did not outweigh the long-range threats to the environment and revoked permits for the project.

In May 1989 Reilly reported to the president about the failures and poor responses of both the federal government and oil industry regarding the *Exxon Valdez* oil spill at Prince William Sound, Alaska. In June 1989, working with the president, he oversaw amendments to the Clean Air Act of 1970. Reducing emissions by utilities and the chemical industry that led to acid rain and smog was a key revision of the legislation. Reilly also fought against automobile emission levels worldwide by insisting that foreign auto manufacturers utilize catalytic converters, and he proposed bans on the use of chlorofluorocarbons (CFCs) and products made from asbestos.

Reilly returned to head the WWF when President Bush failed to gain reelection. He is currently chairman emeritus of the World Wildlife Fund. The EPA was founded in 1970; at the twenty-year period in its history Reilly restored credibility and morale to the department. During his tenure the budget allocations and operating funds for the department rose over 40 percent. More fines were imposed, and the conviction rates for crimes against the environment were higher than at any other time in EPA history. His mediation skills, knowledge of the facts, and lifelong commitment to environmental issues on a global scale made his term as EPA administrator very productive and useful. In spite of political obstacles, and philosophical differences with Vice President Dan Quayle, White House Chief of Staff John Sununu, the Office of Management and Budget, and the Congress during the second half of Bush's term in office, he raised the consciousness of the United States and the world about ecological and environmental issues.

ANTHONY TODMAN

Abelson, P. H. "Reflections on the Environment." [Views of W. K. Reilly.] *Science* 263 (February 4, 1994).

Luoma, J. R. "Bungee-jumping in Brazil." [EPA administrator, William K. Reilly.] *Discover* 14 (January 1993).

Moritz, Charles. "William K. Reilly." *Current Biography Yearbook*. 1989.

Reilly, William K. "National Parks for a New Generation." *Environment* 27 (1985).

Schneider, K. "U.S. Environment Negotiator in Rio Walks a Tightrope in Administration." *The New York Times Biographical Service* 23 (June 1992).

Reisner, Marc (September 14, 1948–July 7, 2000). An environmental journalist, Reisner is most known as the author of *Cadillac Desert*, a comprehensive tome detailing the development of water in the U.S. West through engineering projects such as dams, aquaducts, and groundwater storage facilities. Through this book and a subsequent four-part PBS miniseries (1997) based on the publication, he provides a compelling and cautionary account that exposed the general public to the tenuous water situation in the American West.

Reisner was born in Minnesota, where he graduated with a B.A. in political science from Earlham College in 1970. After college, he worked in Washington, DC, for Environmental Action and the Population Institute. Then he worked as a staff writer for the Natural Resources Defense Council (NRDC) in New York from 1972 to 1979, where he also served as the communications director. In 1979 he received an Alicia Patterson Journalism Fellowship to research water in the West. This research eventually served as the basis for *Cadillac Desert*, first published in 1986. *Cadillac Desert* was later named in the Modern Library's list of the twentieth century's best nonfiction books in English. *Library Journal* and *Publisher's Weekly* named it one of the best nonfiction books of the year. The subsequent television miniseries also won a Columbia University Peabody Award.

Cadillac Desert is the first and best known comprehensive history of water development in the U.S. West. In this work, Reisner weaves the concurrent tales of population growth and water demand in which the characters often defy logic and reason:

> In the West, of course, where water is concerned, logic and reason have never figured prominently in the scheme of things. As long as we maintain a civilization a semidesert with a desert heart, the yearning to civilize more of it will always be there. It is an instinct that followed close on the heels of food, sleep, and sex, predating the Bible by thousands of years. The instinct, if nothing else, is bound to persist.

Feuds between individuals and agencies proved to be the motivating force behind many of the water projects Reisner details. These projects have since allowed for the growth of the West, but the growth and dependence on engineering projects have created vulnerability. He traces the history of water development in the West—from the original westward expansion to the city of Los Angeles' seizure of the Owens Valley water and subsequent construction of a tremendous aqueduct to deliver water to the city. Reisner depicts the bitter battles between the Bureau of Reclamation and the Army Corps of Engineers over water resources and then the dam-building boom of the mid-twentieth century. He describes the formation of the Colorado River Compact—the interstate policy that governs the distribution of Colorado River water to Colorado, California, Nevada, New Mexico, Utah, Wyoming, and Arizona. Reisner creates a readable account by including detailed descriptions of the personalities—including John Wesley Powell, William Mulholland, John Muir, and Floyd Dominy—involved in western water battles.

Ultimately, this book is about the engineering and politicization of a crucial natural resource to develop the largely desert American West. Consequently, Reisner is a bit pessimistic about the future: as he points out, most of the desert civilizations throughout history fell due to water-related issues including salinization. In *Cadillac Desert*, Reisner expresses some sense of defeat on this issue:

> Perhaps, despite the fifty thousand major dams we have built in America; despite the fact that federal irrigation has, for the most part, been a horribly bad investment in free-market terms; despite the fact that the number of free-flowing rivers that remain in the West can be counted on two hands; perhaps, despite all of this, the grand adventure of playing God with our waters will go on.

Reisner's extensive research demonstrates that as a result of the manner in which Americans have populated the land and developed water resources, the West now faces both water-supply challenges and environmental problems.

As a follow-up to *Cadillac Desert*, Reisner coauthored *Overtapped Oasis* (1989) with Sarah Bates, a natural resources lawyer. In *Overtapped Oasis*, Reisner and Bates extend the issues covered in *Cadillac Desert* into the future and develop a set of recommendations for water management in the West. Reisner and Bates include three sets of recommendations—for the federal level, for the state level, and for "A Conserved Water Trust for the Environment." On the federal level, they recommend that the Bureau of Reclamation rely on new direction from Congress, that federal water subsidies be gradually lowered, and that environmental issues resulting from water projects be ameliorated. On the state level, *Overtapped Oasis* suggests a streamlining of water laws and policies, restructuring of policies to address efficiency, increased protection of in-stream flows, promotion of water conservation, and removal of interstate transfer restrictions. Finally, they recommend that the Bureau of Reclamation begin an extensive water conservation program. Reisner published his third book, *Game Wars: The Undercover Pursuit of Wildlife Poachers*, in 1991. In it, he follows the life of an undercover U.S. Fish and Wildlife Service agent as he fights illegal poaching. He describes three incidents involving alligator hides in Louisiana, ivory in Alaska, and sacalait fish in the South.

In addition to his journalistic accomplishments, Reisner remained an active environmental advocate until the time of his death. With the Nature Conservancy, he founded the Ricelands Habitat Partnership. This partnership aims to create wildlife habitat on private farmland and to make use of rice production by-products for wood alternatives. He also consulted for the Pacific Coast Federation of Fishermen's Associations and the American Farmland Trust. He received a Pew Fellowship in Marine Conservation, which would have allowed him to study salmon in the Pacific Northwest, a project he was unable to complete prior to his death.

Reisner was an unabashed environmental realist. He resisted black-and-white solutions to what he viewed as very complicated problems and was also willing to change his own perspectives as environmental situations changed: "I was one of the first to argue that irrigation is a very inefficient use of water on a acre per dollar basis, but if sprawl is what you get by moving water out of agriculture, I think I'll stick with alfalfa."

When diagnosed with cancer later in life, Reisner exhibited this flexibility by joining the water market business. He was a director for the Vidler Water Company, which worked on groundwater storage while supporting dam removal. When asked if he was more or less optimistic than when he wrote *Cadillac Desert*, he responded: "Well, my perspective has changed because my identity has changed. I wrote that as a journalist and as a card-carrying environmentalist. Today, I am an entrepreneur involved in efforts to buy and sell water. I think of it as green capitalism because all the ventures that I'm involved in would create some environmental benefits, at least as I see it."

When Reisner entered the water business, he supported water storage as a means to assure water supply, despite some concerns about facilitating growth: "I know that new water storage will cause growth. But if you view growth as inevitable, as I do, then you try to make new storage supplies in the most environmentally benign way possible without putting great farmland out of production." Reisner's works reflect this concern about the environment and people of the West.

Reisner died at the age of fifty-one from cancer. After his death, his wife Lawrie Mott worked to publish his final book, *A Dangerous Place*, in 2003; this work describes California's vulnerability to earthquakes. When he was writing the book, Reisner intended to include discussion of earthquakes, climatic variability, invasive species, and wildfire, but the posthumous volume focuses solely on earthquakes. After publishing *Cadillac Desert*—a book heralded by many as being comprehensive—Reisner was approached by Michael Finch, a geologist at the California Department of Water Resources. Finch suggested that Reisner had omitted one important aspect of western water—the potential impact of a Hayward Fault earthquake on the San Francisco water supply. In *A Dangerous Place*, he seeks to redress this shortcoming by describing the history of population development and earthquakes in California. The second half of the book consists of a terrifying, first-person description of the mayhem following a fictional but feasible 2005 earthquake on the Hayward Fault. This book is an effort to explore those impacts and to expose readers to the potential impact of a Hayward Fault earthquake, which could be much greater than the impact of another earthquake on the San Andreas Fault.

And so, through *A Dangerous Place* Reisner's legacy of literary warnings about environmental and population pressures in the West persists. The style of *A Dangerous Place* extends what is perhaps Reisner's greatest contribution as an environmentalist. He conducted extensively detailed research, which he then translated into readable, compelling books accessible to the general public. Because of the style and execution of his written works, his books continue to be read in college courses, cited in news articles and dropped on the desks of politicians across the West.

KATE DARBY

Edgar, Blake. "Interview with Marc Reisner." *California Wild: The Magazine of the California Academy of Sciences* 53 (Winter 2000).

Pace, Eric. "Marc Reisner, Author on the Environment, Dies at 51." *New York Times*, July 25, 2000.

Rockefeller, John D., Jr. (January 29, 1874–May 11, 1960). An industrialist, philanthropist, and public servant, Rockefeller, with his diversified activities, had the greatest impact in the social, industrial, educational, and cultural arenas. Lesser known but attributed to him today are his extensive contributions to ecology, environmentalism, and conservation, directly and through organizations he supported.

Born in Cleveland, Ohio, the only son of John D. Rockefeller Sr. and Laura (Spellman) Rockefeller, he attended Brown University, where he received an A.B. degree in 1914, an M.A. in 1914, and an honorary LL.D. in 1937. His father was among the most prominent industrialists in the world and belonged to the group with the derogatory nickname "the robber barons." While perhaps deserving this epithet in the public eye for predatory capitalistic practices in building an industrial empire at the detriment of workers and the public in the nineteenth century, John Sr., in the early twentieth century, launched the most enlightened philanthropist "empire"—second to none, and indeed unparalleled in the world, ever.

Though it has often been said that Rockefeller became a philanthropist as an act of penitence, feeling guilty or at least uncomfortable about how he accumulated his wealth, a closer, more objective evaluation of the information allows the inference that he, his children, and his grandchildren evolved deeply seated moral commitments to doing good for humanity and, in the larger scope, for the environment.

John Sr.'s son John Davison Rockefeller Jr.—or JDR Jr., as he was dubbed—was far more his partner or personification in the philanthropic arena than in business, where he had a lesser role in the Rockefeller business empire than had his father. The charge has often been made that he had neither his father's business acumen nor his interest in business, though JDR Jr. studied accounting, common law, economics, and industrial management. He attended numerous conferences and meetings and, in place of his father, signed agreements and documents, including those with the giant Standard Oil Companies, the United States Steel Corporation, and a host of major railroads.

During 1913–1915, miners at the Rockefeller-controlled Colorado Fuel & Iron Company and its management had violent confrontations over wages and working/living conditions, which led to a strike when workers occupied the mines and facilities in 1915. The Colorado State Militia, called in by management, machine-gunned the tent colony of the miners, causing the death of about forty men, women, and children. This incident was of such magnitude that a Senate investigation was held. JDR Jr. took the stand at the hearings as representative of the Rockefeller family and impressed the Senate and the public by announcing: "I am going to Colorado as soon as I can to learn for myself the true and full situation."

JDR Jr. conducted an evaluation of the wages, living conditions, and management problems at Colorado Fuel, including environmental and conservation aspects. Aside from his immediate instructions to improve these areas, he implemented "Republics of Labor" arbitration committees to settle future disputes. He extended the concept of arbitration to many other Rockefeller enterprises.

At the Colorado Fuel facilities, JDR Jr. directed management to improve housing, sanitary, medical, and environmental conditions. Under the advice of enlightened consultants, he also focused management's attention on a prudent conservation policy of natural resources. Among his consultants was the then-renowned Canadian W. L. Mackenzie King, who, with JDR Jr. himself, spent several weeks at Colorado Fuel to study the situation firsthand.

While at Colorado Fuel—with the national media paying close attention—JDR Jr. made several speeches, professing his and the Rockefeller family's philosophy on the relationship between labor, management, and the public. His major speech was published in a lengthy article, "Labor and Capital—Partners," in the *Atlantic Monthly* in early 1916. The epitome of prudent and socially conscious corporate management, it established enlightened principles on the rights and restraints and obligations for all sectors. His speech incorporated environmental and conservationist (non-exploitation of resources) principles.

Following his father's advice, JDR Jr. engaged Ivy Lee, a foremost publicist, to guide the Rockefeller public image and exposure, a step that was of great importance to the Rockefeller family. Although father and son had established such major philanthropic organizations as the Rockefeller Institute for Medical Research (1901), the general Education Board (1902), and the Bureau of Social Hygiene (1913), the Rockefellers were still remembered in 1915 for JDR Sr.'s callous-sounding earlier edict in which he stated that the survival of the fittest in business was "[m]erely the working out of a law of God and nature . . . at the expense of a few."

JDR Jr. did what he could to improve the Rockefeller image. Indeed, he attempted to change the public image of business and capitalists in 1925 when he published *The Personal Relation in Industry*, a compilation of his previous speeches and articles, in which he advocated that business should govern its policies with the tempered aims of sharing profits and social benefits with employees while charging moderate prices to consumers. He also advocated conservation and care of the environment, in contrast to its exploitation for maximized short-range profits.

Nothing can be more illustrative of JDR Jr.'s public persona and image than the fact that in 1919 President Woodrow Wilson convened a National Industrial Conference at the height of labor-industry turmoil to which he appointed seventeen representatives of employers and bankers, nineteen representatives of unions, and twenty-one individuals to represent the public—and he selected JDR Jr. as one of the public representatives.

Enlightened principles and policies were greatly expanded by JDR Jr. through the enormous activities of the Rockefeller Foundation, which was launched in 1913 with JDR Sr. as the chairman; later, JDR Jr. succeeded his father as chairman. The foundation has been active in research, training, and investigating experimental hazards. Additionally the Rockefeller Institute for Medical Research was founded in 1901, which became the Rockefeller University in 1954, an institution that researches both the natural sciences and the humanities.

JDR Sr. transferred most of the ownership and wealth of the Rockefeller family in 1921 to JDR Jr., who continued to expand contributions to philanthropies. By 1940, the Rockefellers' contribution to philanthropies was over $200 million, and in 1981 the Rockefeller Foundation had over $800 million in assets. On May 12, 1960, the *New York Times* reported that JDR Jr. and his father contributed nearly $3 billion in gifts to philanthropic organizations and causes!

Among JDR Jr.'s direct conservationist activities were the reconstruction of colonial Williamsburg, Virginia; improvements to the Cloisters at Fort Tyron Park, New York; and the preservation of the environmental parks at Tarrytown, New York. In 1923, he was awarded the National Institute of Social Sciences Gold Medal. In 1959, the same award was made to his son Laurance S. Rockefeller; in 1967, jointly to his five sons; and in 1977, to his grandson William Rockefeller. In 1978, when President Jimmy Carter established a "Commission on Coal" to examine the "health, safety and living conditions in the Nation's coal fields," he appointed as its chairman the grandson of John D. Rockefeller Jr., John D. Rockefeller IV, governor of West Virginia.

JDR Jr. married Abby Greene Aldrich, daughter of Senator Nelson W. Aldrich, in 1901; they had six children: Jean, David, John D. III, Laurance, Nelson, and Winthrop. All five sons involved themselves in philanthropic and beneficent activities, as channeled by JDR Jr., who apportioned areas among them. Laurance was especially involved with environmental, conservation, and public improvement; Winthrop, in developmental projects in Arkansas; and Nelson, as governor of New York and vice president of the United States, in even more enlarged arenas.

CLAUDINE L. BOROS

Current Biography. 1941.
"Gifts to Nation and World Noted; Prelates, Business Leaders and Statesmen Join in Praise and Mourning" [obit.]. *New York Times*, May 12, 1960.
Notable Americans: What They Did from 1620 to the Present, 1988.
Raddock, Maxwell C. *Portrait of an American Labor Leader: William L. Hutcheson.* New York, 1955.
Who Was Who in America with World Notables. 1968.

Rockefeller, Laurance S. (May 26, 1910–July 11, 2004). A philanthropist, ecologist, financier, and venture capitalist, Rockefeller pursued a number of successful careers over the ninety-four years of his life. His many accomplishments have been recognized both nationally and throughout the world. In 1991, President George H. W. Bush bestowed the Congressional Gold Medal on him in a ceremony in the Theodore Roosevelt room of the White House. The room, filled to capacity, was chosen as the site for the award in memory of the earlier president for whom the room was named because he had done so much to initiate the conservationist movement in America.

In a short address to political figures, invited dignitaries, the Rockefeller family, and friends, President Bush extolled Rockefeller as "a hidden national treasure." In a modest response, Rockefeller stressed for the benefit of all in the Theodore Roosevelt room that they, too, had an obligation to protect our land, water, and air. The Gold Medal, he implied, really honored all devotees of nature and environmental concerns, all those concerned about preserving and improving our nation's heritage. The environmental movement, he added, glancing at the medal, "was central to the welfare of the people," and it would remain so for "as long as this piece of gold glistens."

The middle brother of the five prominent grandsons of John D. Rockefeller, Laurance was less gregarious and less well known than his older brother Nelson, the four-term governor of New York State and vice president under Gerald Ford; more reserved than his rather flamboyant younger brother Winthrop, who was governor of Arkansas; quite distinct from his oldest brother John D. III, who devoted himself to maintaining and expanding the charitable efforts established by his father and grandfather; and not so deeply involved in issues of international finance as his youngest brother David, who headed the Chase Bank.

Like his brothers, Laurance, though reared amid advantage, believed that wealth brought with it responsibity. "My grandfather gave like a forward pass," he once said. "He threw the ball, and it was up to others to make a go of it." Referring to himself and his brothers, he added, "We feel that you must give with the heart as well as the head. There must be a deep personal commitment along with an intellectual understanding of a project." When asked what especially motivated him, he replied: "I profoundly feel that the art of living is the art of giving. You're fulfilled in the moment of giving, of doing something beyond yourself. The act of giving [allows you to realize] you're alive and fulfilled."

Born in New York City into one of the wealthiest families in America, he attended the Lincoln School, a preparatory academy affiliated with Teacher's College of Columbia University. He then went on to Princeton, from which he was graduated in 1932. Next, it was Harvard Law School, but after two years, he concluded he did not want a legal career. Business and finance seemed to hold more interest. He gave considerable thought to his future, and in 1934 he married Mary French, a granddaughter of Frederick Billings, president of the Northern Pacific Railroad.

Upon his grandfather's death in 1937, Laurance Spelman Rockefeller—or "LSR," as he was addressed by colleagues and friends—inherited a seat on the New York Stock Exchange. During World War II he served in the navy as a procurement officer. After the war, he began a forty-year career with the Rockefeller Brothers Fund. Willing to risk capital on scientific and technological developments, he began to invest in aviation, aerospace, optics, lasers, and thermionics. One of his earliest financial successes was in the realm of commercial air travel. With Captain Eddie Rickenbacker, they grew Eastern Air Lines into one of the most profitable airlines to emerge after World War II.

LSR's investment activities often involved new or young enterprises on the frontiers of technology. Desiring to do more with his money than just have it multiply, he committed himself to culturally significant enterprises where "venture captial" (a term, it is said, he coined) would have the greatest influence. As a "catalyst for conservation" he primed the pump for worthwhile projects through seed money, especially for national parks. Recreational developments and public service became a major concern.

In 1939, he accepted Governor Herbert Lehman's appointment to the Palisades Interstate Park Commission. Similar appointments followed, one of the more important as vice-chairman of the New York Council of Parks under Robert Moses. LSR also commissioned a series of studies that suggested New York's huge Adirondack Park be put under federal control as a national park to prohibit commercial incursions by timber companies.

Nationally, he served several presidents in various capacities, focusing on such topics as outdoor recreation, national beautification, and environmental policies and programs. Under Presidents Dwight Eisenhower and John Kennedy, he operated as chairman of an Outdoor Recreation Resources Review Committee, which charted the nation's outdoor recreational needs. Reports that the commission issued represented groundbreaking achievements that, it is held, laid the framework for virtually all the significant environmental legislation of the following three decades.

When President Lyndon Johnson set up a Citizens' Advisory Committee on Recreation and Natural Beauty in 1966, he selected LSR to be its chairman. Johnson specifically requested LSR to push forward a long-stalled plan for a Redwood National Park in California, a plan first suggested in 1917. The park was finally established in 1968, after all sorts of squabbles among lumber corporations, politicians, and naturalists. Had LSR not mediated among the warring factions, the park (the most costly in the National Park Service) would never have been established. The committee also made strong recommendations for changes in highway planning and the freeing up of federal surplus lands for park and recreational purposes.

As an advocate for alliances between commerce and conservation, in 1956 LSR built an environmentally focused resort at Caneel Bay on St. John, the smallest of the U.S. Virgin Islands. A commercial venture, it was the beginning of his interest in ecotourism that extended from the Caribbean to Hawaii to the state of Vermont. The resort had to provide profitable returns, but it had to encourage self-renewal for its guests. At first it was planned there would be no telephones or such amenities as air conditioning, but such restrictions were relaxed when Rockresorts, as the enterprise was named, went on to build and operate additional facilities.

In 1956, Rockresorts constructed the Dorado Beach Hotel in Puerto Rico in cooperation with the island's Operation Bootstrap to stimulate employment and increase tourism. Other resorts followed in the British Virgin Islands and Hawaii, as well as ski resorts in Vermont. Over the years the resorts were sold, and LSR moved on to new interests, in particular, the creation of state and federal parks. His gifts of land from Wyoming to the Virgin Islands amounted to hundreds of thousands of acres.

Among the more important parks he established are the Grand Teton National Park in Wyoming, the Virgin Islands National Park of St. John, Tallman Mountain State Park along the west bank of the Lower Hudson River, and the Marsh-Billings-Rockefeller National Historical Park in Vermont. During this period he was invited to serve on committees of various important nonprofit and philanthropic organizations. He funded the Memorial Sloan-Kettering Cancer Center at critical junctures in its expansion and modernization of its operations. A longtime trustee of the New York Zoological Society, he also functioned as its president and chairman.

LSR's commitment to external nature and environmental causes had many roots. In his boyhood he loved to ride horses on verdant terrain and to hike paths through national parks in the West. It was in Yellowstone that he met Horace M. Albright, the park superintendent and a force in conservationism, who became one of LSR's first mentors. Another important mentor was Fairfield Osborn Jr., of the New York Zoological Society, who sounded an alarm of ecological disaster in his book *Our Plundered Planet* (1948). LSR seldom had the time to put pen to paper, but in 1972 he composed a tribute for Osborn in the October 1972 issue of the *Reader's Digest* series, "My Most Unforgettable Character."

It is regrettable that he did not write more than the little he did, but, as the bromide has it, he was more a man of action, one whose largesse benefited countless thousands. His driving passion, it is not an exaggeration to claim, was the environment. A concomitant enthusiasm was the support of cancer research and treatment. In 1958, he was awarded the Clement Cleveland Medal "in recognition of . . . many outstanding and meritorious contributions to furthering the cause of cancer control."

In acknowledgment of LSR's work on matters of conservation and the environment, ecological concerns, and medical research, he was singled out for all sorts of additional honors, tributes, and awards. In 1959, he received the National Institute of Social Sciences' Gold Medal for distinguished service to humanity, an award previously bestowed on his father and grandfather, the only time in the institute's history that three generations of any family had ever been so honored. In 1969, he was awarded the

Medal of Freedom, and in 1971 he was named a Commander (Honorary) of the Most Excellent Order of the British Empire. In 1995, he received the Chairman's Award from the National Geographic Society and the Theodore Roosevelt National Park Medal of Honor. In 1997, he was the recipient of the Lady Bird Johnson Conservation Award for Lifetime Achievement. In 2003, he became the first individual to be declared an honorary citizen of the British Virgin Islands in recognition of his many contributions to the area.

Regardless of all the plaudits bestowed on him, LSR remained a modest man, and he always maintained that to be extravagant and improvident was to be guilty of reprehensible behavior. As far as the environment was concerned, he often stated: "The concept that we have boundless resources of materials, manpower and spirit, and therefore can waste, clearly is no longer true." Despite his unending wealth, he always advocated for all a simpler lifestyle, one that would result in contentment and happiness rather than one of selfish pursuits and meretricious goals. Such was his credo, and his dedication to humanitarianism and ecological ideals is responsible for his countless contributions to society and his promotion of conservationism and modern environmentalism.

CHRISTOPHER ZEPPIERI

Kaufman, Michael. "Laurance Rockefeller, Passionate Conservationist and Invester Is Dead at 94." *New York Times*, July 12, 2004.

Rockefeller, David. *Memoirs*. New York, 2002.

Winks, Robin W. *Laurance S. Rockefeller: Catalyst for Conservationism*. New York, 1997.

Rockefeller, Margaret (September 28, 1915–March 26, 1996). An agriculturalist, conservationist, and environmentalist, Rockefeller was born in New York City and grew up in the suburb of Mount Kisco. A private person all her life, she preferred to remain in the background while devoting her time and energy to causes that were more worthwhile than fashionable.

The daughter of Francis Sims McGrath and Neva (van Zandt Smith) McGrath, Peggy, as she was called, attended the Shipley School in Bryn Mawr, Pennsylvania, and the Chapin School in New York. Early in the 1930s she met David Rockefeller. They married in 1940. Over the years they had six children: David Jr., Abby, Neva, Peggy, Richard, and Eileen.

Between 1950 and 1970, Rockefeller made substantial contributions to the Lincoln Center for Performing Arts and to the New York Philharmonic. She also served as a member of the Board of Managers of the New York Botanical Garden. Moved by her love of flowers, she organized the National Committee for the Wildflowers of the United States, which raised $500,000 to finance a series of six illustrated volumes on American wildflowers. Several years later she established the Peggy Rockefeller Rose Garden, a two-acre enclave of almost 3,000 rose bushes at the New York Botanical Garden. Originally designed by Beatrice Jones Farrand in 1915, the Rose Garden was restored and supported by a gift of $1 million from David and Margaret Rockefeller.

In the mid-1970s, Rockefeller began to raise cattle, first at the family summer farm on Bartless Island in Maine and then at Pocantico, the Rockefeller estate in Tarrytown, New York. In 1980, she and her husband obtained four farms in Columbia County, New York, where Rockefeller became an accomplished farmer and breeder of Simmental cattle.

"I always loved the country, farming and animals," she said shortly after the purchase of the Columbia County property, and she began to read all she could find about farmland preservation. "Little by little," she added, "I was trapped." Especially concerned with agricultural issues and the environment, she served on numerous conservation and environmental boards. She became a founding member of the American Farmland Trust, and she devoted considerable time to the Nature Conservancy and the National Historic Trust. In 1922, Rockefeller and her husband announced an agreement they had arrived at with the American Farmland Trust and the Columbia County Land Conservancy. They would place under protection for perpetuity almost 3,000 acres of their Hudson Valley properties. Their purpose: to preserve the visual and agricultural character of their land.

Three weeks after Rockefeller's death, a memorial service was held on April 12, 1996, at St. James Episcopal Church in Manhattan. Over 2,000 family members and friends attended to pay tribute to her memory and accomplishments. All recalled her as a warm, caring, remarkable woman. Various speakers called attention to her dedication to various causes. One speaker noted that Rockefeller had established the Maine Coast Heritage Trust and that she had been instrumental in saving thousands of acres and miles of shoreline in Maine. As a conservationist, she did all she could to preserve farmland and to protect the rugged coastline of Maine.

Others who spoke recalled the innumerable cultural, social, and environmental activities she initiated and supported over the many decades of her life. Chief among those who commended Rockefeller was Ralph Grossi, president of the Farmland Trust. Rockefeller, he emphasized, should be honored for "a conservation legacy rooted in a deep feeling for the family farmer and the protection of this nation's best land."

JULIAN ZEPPIERI

Faber, Harold. "Rockefellers Move to Preserve 2,972 Acres." *New York Times*, June 24, 1992.

"Margaret Rockefeller, Backer of Farm and Conservation Causes" [obit.]. *New York Times*, March 27, 1996.

Nemy, Enid. "Relatives and Dignitaries Honor Margaret Rockefeller." *New York Times*, April 13, 1996.

Roethke, Theodore (May 25, 1908–August 1, 1963). An individual who won most of the major literary awards available to poets during his lifetime, Roethke would not have identified himself as an environmentalist. Although he came of age during a time in which conservationism, as epitomized by Aldo Leopold, was a growing movement and lived to see the publication of Rachel Carson's landmark *Silent Spring* (1962), Roethke was not involved with these efforts—indeed, he seems not to have been politically active in any particular direction. However, his sympathies and structures of thought were always environmentalist, in the purest of senses. His deep feelings of kinship with the nonhuman world influenced "nature" poets in the latter half of the twentieth century, who valued Roethke's abilities to channel the complex power of natural beauty and to express humanity's place in this beauty through his words.

Roethke's life was focused solely on the production and teaching of poetry. However, the poetry he published was imbued with a strong and unwavering personal environmental ethic. Through his writing, Roethke was an advocate for all "small things": birds, plants, animals, insects, and anything living what he would call a "wild" life. Moreover, when Roethke wished to express an inmost feeling or emotional state or describe the essential conditions of life, he always returned to images of natural processes, constantly situating humans within the biotic world.

Roethke was born in Saginaw, Michigan. His family was of Prussian descent and had immigrated to the area in 1872. His father, Otto, died of cancer in 1923, when Roethke was in high school. His presence (and later absence) was to be a major influence in Theodore's life. Otto owned large greenhouses and supplied flowers for most of Saginaw's florists. The family remained in Saginaw for the entirety of Roethke's life. However, after his high school years, Roethke himself lived there only during brief periods, mainly moving about the country to fill teaching positions at various universities.

Imagery associated with the plant life inside the greenhouse, and with the field and forest attached to the family home, was to return time and time again in his poetry, forming an integral part of his fellow feeling for nonhuman life forms. These floral and faunal lives formed a major part of Roethke's childhood memories. For example, the life of the field attached to his childhood home taught young Roethke about mortality. He wrote in "The Far Field" (from *The Far Field*) of his early experiences exploring the field near his family's house, positing that they taught him the meaning of death "in the shrunken face of a dead rat, eaten by rain and ground-beetles." Denizens of the greenhouse and the forest were observers of his steps toward personal independence. Roethke wrote in a poem called "Child on Top of a Greenhouse" (from *Open House*) of his experience (as the eponymous child) of being the center of attention after climbing out of the reach of adults. He characterized the freedom he felt by noting the presence of both cultivated and "wild" nature: "half-grown chrysanthemums staring up like accusers," "the few white clouds all rushing eastward," and "a line of elms plunging and tossing like horses."

The greenhouse also signified his father's care for him and for his family, as he wrote in "The Rose" (from *The Far Field*) about a memory of his father "lifting him high" over the flowers: "the four-foot stems, the Mrs. Russells, and his own elaborate hybrids,/And how those flowerheads seemed to flow toward me, to beckon me, only a child, out of myself." Exultant in his memory of his time with his father, Roethke ends the poem asking: "What need for heaven, then,/With that man, and those roses?" Roethke's sympathy for small things and animals derived directly from Otto's ethic of caring for the land. For example, Roethke remembered in the poem "Otto" (from *The Far Field*) an episode when his father slapped the faces of two poachers who had "slaughtered game, and cut young fir trees down" on his land. Although Otto's hand could, as Roethke wrote in the same poem, "fit in a woman's glove," he clearly embodied what Roethke saw as the proper masculine way of protecting and standing up for the powerless.

Unlike Otto's, Roethke's life was centered on indoor pursuits of the intellectual variety. Roethke spent most of his adult years either attending or employed by institutes of higher learning. In 1929 he earned a B.A. at the University of Michigan in Ann Arbor and spent 1930–1931 at Harvard, studying for, but never actually attaining, a master's degree. In 1931, he began his teaching career at Lafayette College (where he taught through 1935), then finally received a master's degree from the University of Michigan in 1936. After this accomplishment, he moved to a teaching job at Pennsylvania State (1936–1943). By 1943, he was employed by Bennington College, in Vermont, where he stayed through 1946.

From 1947 through his death, he taught at the University of Washington (UW) in Seattle, where he became a fixture of the English Department, mentoring such poets as Carolyn Kizer, Tess Gallagher, and Richard Hugo. Roethke seemed to feel truly at home among what he called, in an unpublished essay, the "riot of wonderful natural life" he found on the UW campus, rhapsodizing, "There is no need to barber and pamper the landscape; everything grows green and strange . . . even the stones and sides of trees bear a fine mossy sheen."

Roethke was famous for the quality, intensity, and commitment of his teaching; but his career was somewhat sporadic due to his periodic bouts of manic depression, for which he was hospitalized several times (he shared this affliction with his friend and fellow "confessional" poet Robert Lowell). He remained unmarried until 1953, when he wed a former student, Beatrice O'Connell. The two remained childless, perhaps due to Roethke's illness and his emotional needs—though Roethke often expressed a sympathy and solidarity with children. Roethke died of a myocardial infarction while swimming in a friend's pool on Bainbridge Island, Washington.

Roethke published books including *Open House* (1941); *The Lost Son and Other Poems* (1948); *Praise to the End!* (1951); *The Waking* (1953); *Words for the Wind* (1958); *I Am! Says the Lamb* (a book of verse for children, with many poems featuring animals, published in 1961); *Party at the Zoo* (also written for children, published in 1963); and *The Far Field* (published posthumously, 1964). A collection of prose called *On Poetry and Craft* was also published posthumously in 1965. And indicative of the respect his earthbound vision earned within his field, even during his own lifetime, Roethke won the Guggenheim Fellowship twice (1945, 1950); the Ford Foundation

Fellowship (1952); the Pulitzer Prize (1953, awarded for *The Waking*); a Fulbright, which allowed him to lecture in Italy in 1953; and the National Book Award, also twice (in 1959 for *Words for the Wind*; in 1965 for *The Far Field*).

Like previous generations of Romantic poets, Roethke's poetry often associates his innermost emotional states with natural forces, eschewing rationality for intuitive action derived from inherent qualities of the flesh. In the poem "I Cry, Love! Love!" he asks, rhetorically: "Reason? That dreary shed, that hutch for grubby schoolboys!/The hedgewren's song says something else" (from *Praise to the End!*). In perhaps his most famous poem, "The Waking" (from *The Waking*), Roethke writes: "We think by feeling. What is there to know?/I hear my being dance from ear to ear." And then adds:

Great Nature has another thing to do
To you and me; so take the lively air,
And, lovely, learn by going where to go.

Roethke also pays nature the ultimate compliment by constantly comparing his beloved wife Beatrice to natural forms. In various poems, these forms include the "beast" ("When, easy as a beast,/She steps along the street/I start to leave myself"—"All the Earth, All the Air," from *Words for the Wind*); a lizard (in "Wish for a Young Wife," from *A Far Field*); and a cluster of earthen and oceanic creatures:

The breath of a long root,
The shy perimeter
Of the unfolding rose
The green, the altered leaf,
The oyster's weeping foot,
And the incipient star—
Are part of what she is.
She wakes the ends of life. ("Words for the Wind,"
from *Words for the Wind*)

But perhaps the most indelible aspect of Roethke's poetic environmentalism lies in his intense sympathy for the members of the biotic community that are the least spectacular, least likely to end up as the subjects of a glossy photograph or as the heroes of an iconographic calendar: common insects, fishes, birds, and plants. After the publication of *The Lost Son*, poet Stanley Kunitz, one of Roethke's friends, wrote to him that he thought Roethke was "a great fellow with bugs and bogs"—a designation that Roethke wholeheartedly embraced. In his notebooks, Roethke once wrote, "I wish I could photosynthesize" and called himself "[t]he leading under-the-stone poet of our time." He diagnosed in himself this empathy for the "under-the-stone" world: "More than a feeling: a desire for the qualities in primitive forms of life: crabs, snails." On a different occasion, he wrote: "Nietzsche and Whitman my fathers: and yet I cannot worship power. I hate power: I reject it."

Roethke's feelings of kinship with small (powerless) things can be seen in poems such as "The Minimal" (from *Lost Son and Other Poems*). Here, Roethke describes "studying the lives on a leaf":

> The little
> Sleepers, numb nudgers in cold dimensions,
> Beetles in caves, newts, stone-deaf fishes,
> Lice tethered to long limp subterranean weeds.

In "Cuttings" (from *Lost Son and Other Poems*), Roethke imagines the attempts of cuttings to take root in the earth where they have been transplanted:

> I can hear, underground, that sucking and sobbing,
> In my veins, in my bones I feel it—
> The small waters seeping upward,
> The tight grains parting at last.
> When sprouts break out,
> Slippery as fish,
> I quail, lean to beginnings, sheath-wet.

In "Moss-Gathering" (from *Lost Son and Other Poems*), Roethke explores the dark side of the human relationship with these small things, writing of his feelings upon having harvested moss from a hillside: "Something always went out of me when I dug loose those carpets/Of green." In rhetoric quite similar to that of a confirmed conservationist or environmentalist, Roethke compares the action to "pulling off flesh from the living planet."

Roethke's sympathy for the "innocent, hapless, forsaken" is perhaps summed up in his oft-anthologized "The Meadow Mouse" (from *The Far Field*), in which he tries to save a baby mouse he found, motherless, in a field. When the mouse leaves the "shoe-box house" where Roethke has been keeping him, feeding him "three different kinds of cheese" and watering him from a "bottle-cap watering-trough," Roethke worries about his fate, articulating his care for

> . . . the nestling fallen into the deep grass,
> The turtle gasping in the dusty rubble of the highway,
> The paralytic stunned in the tub, and the water rising,—
> All things innocent, hapless, forsaken.

Several writers of prose cite Roethke's poems, including Kurt Vonnegut in *Slaughterhouse-Five* (1969) and Frank Herbert in *Heretics of Dune* (1984). Echoes of Roethke's confessional style can be found in the poetry of Robert Bly and James Dickey, as well as his contemporaries Robert Lowell and Sylvia Plath. It is interesting to speculate whether Roethke would have—had he not died in such a premature fashion—become involved with the growing environmentalist movement of the 1960s and 1970s. Regardless, his poetry, with its fine-tuned sensitivity toward the natural world, has no doubt influenced the environmental ethic of countless readers.

REBECCA ONION

Malkoff, Karl. *Theodore Roethke: An Introduction to the Poetry.* New York, 1966.
Roethke, Theodore. *Collected Poems.* New York, 1966.
Seager, Allan. *The Glass House: The Life of Theodore Roethke.* New York, 1968.
Wagoner, David, ed. *Straw for the Fire: From the Notebooks of Theodore Roethke.* New York, 1974.

Rogers, Pattiann (March 23, 1940–). Known for having "a scientist's curiosity" and "a poet's eye," Rogers possesses "a singular voice" in the world of contemporary poetry. She is a lyrical virtuoso who weaves her extensive knowledge of science and fascination with its discovery and mystery with a profound vision of the natural world. An author of eleven books and countless magazine and journal articles, and recipient of prestigious Guggenheim and Lannan Foundation Fellowships, as well as numerous other awards, she is a prominent and beloved modern poet who makes our natural environment wholly and unimaginably acceptable.

Rogers, a student of zoology, lover of astronomy, ecotheologian, environmental nurturer, is fascinated by "the cosmological story told by science." Early on she realized that the often privileged and isolated language of science is beautiful, lyrical, evocative: it ignited a conscious desire in her to create music by utilizing the enormous vocabulary that scientists invented. This vocabulary, at the root of her poetry, provides a portal for the stories of science and an entryway for exploration and contemplation about a universe "teeming with life." Her poetry has been a vital building block in the relationship between literature and the environment. As she wrote in *The Dream of the Marsh Wren* (1999): "Without artistic expression, any fact remains alien, unpossessed, with no reverberation in the soul; it seems to me impossible to live in our world, to survive—the split, the rendering, being too great—if a union could not be found and created between two ways of knowing, the artistic and the scientific, both so essential in our lives."

Rogers was born in Joplin, Missouri, to William Elmer Tall and Irene (Keiter) Tall. Her parents were active members of a Presbyterian community in which ritual and prayer were part of daily life. When she was thirteen, her parents joined a small ultraconservative sect that rigorously studied the gospels. Their religious experiences had a profound impact on Rogers throughout her early years and later in her writing.

During her years in high school, she volunteered afternoons at a nursery school, a nonprofit organization that her mother established for children with working parents. The experience of Rogers acting as a mother figure to the young children established a reciprocal devotion and dependency, one that she would carry with her when she reared her own children.

While studying at the University of Missouri, she met John Robert Rogers, and they married in 1960. One year later she was awarded a bachelor's degree with a major in English and a minor in zoology. Her awareness of the world, the role of science and nature, and a reaction to various expressions of spiritality and divinity laid the groundwork for her experimentation with poetry. She found employment as a kindergarten teacher in order for her husband to complete his Ph.D. in physics at the University of Missouri. In 1967, their first son was born, and the family moved to a small suburb just outside Houston, Texas. A second son was born, and shortly thereafter, Rogers began taking correspondence courses in the writing of poetry at the University of Washington. She then enrolled in the University of Houston's first Creative Program, from which she was their first graduate, receiving an M.A. in 1981. That same year she published her first book, *The Expectations of Light,* which received a Texas Institute of Letters Award.

Rogers's reputation grew steadily over the years, and she was invited to fill teaching positions at the University of Texas, the University of Montana,

Washington University of St. Louis, and Mercer University as the Ferrol Sams Distinguished Writer-in-Residence. She also taught from 1993 to 1997 in the M.F.A. Creative Writing Program at the University of Arkansas. In May 2000, she was in residency at the Rockefeller Foundation's Bellagio Study and Conference Center in Bellagio, Italy. She is currently on the faculty of the M.F.A. Program in Creative Writing at Pacific University, Forest Grove, Oregon, and makes her home with her husband, a retired geophysicist, in Castle Rock, Colorado.

In "The Rites of Passage," a poem that helped ignite her literary career, she successfully bridges science and the arts. She recalls in this work time spent as an undergraduate student in Missouri when she had to perform fieldwork involving the development of frogs eggs, having to tag them to estimate pond populations. Concerned with the miracle of life, she did not want simply to describe the eggs. "That's what science does," she states. "I wanted to do more than that. I wanted to say, 'What does this mean?'—That there is a moment, an actual, identifiable moment, when the heart begins to pulse." The very best scientists and poets revere and love the world, and they express their reverence and love by observing carefully, "by paying very, very close attention to what they see, feel, taste, hear, and sense in all ways."

Rogers's devotees claim that her profound knowledge of natural history and her meticulous use of language are virtually unmatched in the literary world, that her poetry is fundamental in our day for those who seek to bridge the gap between scientific research and advocating active involvement in our environment. "Every environmentalist," she holds, "needs an understanding of poetry and an acute awareness towards the wellbeing of humans and their natural environment." The narratives of evolutionary changes, human nature, flora and fauna, and life cycles all exist because of the vocabulary of science and the awesome, ever-changing power of nature.

The constant adjustments made in our cosmological story are often overwhelming and incomprehensible—doctrines and dogmas can limit potentialities we can find in the physical world. In "Suppose Your Father Was a Redbird," Rogers contemplates what life would look like and how one would define experiences depending on a redbird for protection, nourishment, and guidance as a human child depends on his father. "Suppose his body was the meticulous layering/ of graduated down which you studied early/ . . . before you could speak, you watched." By the isolating of such moments, she explains, the natural environment and human flesh are evident as one—"born of the earth."

Rogers's poetry creates an indirect effect on the reader through the power of language, rather than a direct effect by dictating views. Her words are never rigid, didactic, or uncompromising. She believes that "productive poets and scientists are humble in the face of the universe," and she approaches her writing as if she were "sewing a garment and holding it up to nature to see if it fits." "The garment may need to be taken apart and redesigned. Nature makes the judgment, and scientists are willing, as are poets, to adjust their work as necessary."

In 1996, Rogers was asked to participate in "Watershed Writers: Nature and Community," a six-day conference cosponsored by the Orion Society and the Library of Congress that brought together noted novelists, poets, and storytellers to discuss writing, nature, and community. Prominent among those who took part were Gary Synder and Joy Harjo. Rogers's contribution to the conference was "Animals and People: The Human Heart in Conflict with Itself," a poem that meditates on the complex relationship that exists in our environment.

In collaboration with Joellyn Duesbury, well known for her vibrant landscape paintings, Rogers published *A Covenant of Seasons* in 1998, which unites a literary and visual response to the recurring spectacle of "nature passing through the metamorphoses of the four seasons." In *The Dream of the Marsh Wren* (1999), she focuses on "rapidly changing technology, social structures, and environmental concerns."

Rogers's literary work includes numerous poems and essays that intersect science and nature, such as *Intimate Nature* (1998), *Verse and Universe* (1998), *The Measured Word: On Poetry and Science* (2000), and *A Chorus for Peace* (2002). She has also made important contributions to numerous environmental journals such as *Whole Terrain* (1995–1996), which explores ecological and social issues, and *EnviroArts*, an online review that promotes activism through the arts. In addition, she devotes much of her personal time and effort to various organizations that seek community activism toward the environment. She serves on the advisory board for the Orion Society and was an avid participant in their Forgotten Language Tour, a national tour in which top writers and poets promoted nature literature, meant to help individuals achieve a better understanding of their relationship to the natural world.

In 1998, she joined a group of Utah high school students and fellow poet Linda Hogan in the River of Words Project, an interdisciplinary expedition involving writing lessons, painting, photography, animal tracking, bird-watching, and river ecology. She also added her voice to the Writer's Residency Project, a one-week stint in the fall of 2005 to encourage long-term ecological reflections. Jim Sedell, a developer of the project, was moved to remark: "What are many advocates doing? They're reading Pattiann Rogers and other writers who can connect the public with science in many ways better than scientists can."

Rogers has bestowed on the environmental movement not a critique of the human condition but an endless pursuit of literary exuberance and exploration of the beauties of the world and what we must do to preserve them. "The more we observe," she stresses, "the more alert we are to the details of the physical world around us, and the more we must do to value and revere it."

During an interview with *Timeline Magazine*, she continued:

> We are physical creatures immersed in a physical world, a world we come to know through our bodies, through our senses. We were born from the Earth and have inherent connections with it. . . . We are surrounded by and within the physical world, sunlight and shadows, wind or the still lack of it, the motion of clouds in the sky, the fragrance of rain, the silence of snow, the sound of a river, a bird calling in the background, grasses covered in frost, grasses in the wind, a fly at the window, a flowery weed by the roadside, the outline of a familiar tree at dusk, the sound of a door slamming, a dog barking in the distance, frogs or chickens or locusts beginning their calls as night comes on.

REGINA CORALLO

"An Interview with Pattiann Rogers." *Timeline Magazine*, no. 78 (November–December 2004).

Rogers, Pattiann. "Small and Insignificant, Mighty and Glorious." *Spiritus: A Journal of Christian Spirituality* 2.2 (2002).

———. "Twentieth-Century Cosmology and the Soul's Habitation." *The Measured Word*, ed. Kurt Brown. Athens, GA, 2001.

Wile, Kristin. "River of Words." *River of Words Organization*, May 2007.

Roosevelt, Franklin Delano (January 30, 1882–April 12, 1945). Wholly aside from his other achievements, Roosevelt was known for his ardent support of the Civilian Conservation Corps (CCC), known as

"Roosevelt's tree army," an innovative program, and his signing of the Soil Conservation Act. Nor should we ignore the national parks and national monuments established while he was president.

Franklin, the son of James Roosevelt, a lawyer and railroad manager, and Sara Delano Roosevelt, was born on his father's Hyde Park estate. The Roosevelts were an old (the first Roosevelt arrived in America in 1650) Establishment family who were comfortably fixed. James was worth $900,000 in an era of multimillionaires. He practiced law briefly and subsequently was a railroad manager. His wife's father, Warren Delano, grew rich in the China trade (that is, the opium trade) and lived on an estate near Newburgh, New York. Despite his resistance to their union, his twenty-six-year-old daughter married James, a widower, who was twice her age. They were a close-knit family, protective of their son, and this surely contributed to the sense of confidence Franklin developed that saw him through both private and public crises.

At the same time, Franklin led a rather untraditional childhood for an American child. He was taught at home by a tutor, and as early as the age of three, he traveled with his parents to Europe. Indeed, these European sojourns became part of his education, and as a child, he learned German and French.

While his mother took charge of the boy's management and discipline, his father introduced him to the pleasures of the out-of-doors, taking him on hikes in the woods. James knew the variety of trees on his estate (he owned a thousand acres of farmland and forests) and insisted that only diseased trees be cut down. Doubtless, Franklin's lifelong regard and respect for forests was due to his father's influence. He also acquired an interest in birds and became an avid bird-watcher. (It should be noted that historically, and with few exceptions, birders and sports fishermen have been conservationists.) In fact, presented with a gun when he was eleven, Franklin shot and, for a time, stuffed birds, developing quite a collection.

In 1896, having been tutored at home until then, Franklin was enrolled at the newly opened and exclusive Groton School. The instruction was rigorous and thorough. Four years later he went on to Harvard, where few of the professors interested him—though Harvard at the time had an impressive faculty—and was content to maintain a "gentlemanly C" average. He received his A.B. degree in 1903 and entered Columbia Law School; here he did fairly well. Passing the bar exam in the spring of the third year, he decided not to finish his courses and take a law degree.

In 1904, Franklin became engaged to Eleanor Roosevelt, a niece of then-president Theodore Roosevelt. They married in 1905, TR giving away his niece. Franklin spent the next three years with a Wall Street law firm. He entered politics in 1910, running from Dutchess County (where Hyde Park is located), a heavily Republican county. But Roosevelt campaigned energetically, and to his advantage, the Republicans were split that year between conservatives and progressives, and he was elected. In the state senate, he assumed the leadership of the progressive wing of his party and became enmeshed in a struggle between upstate Democrats and Tammany Hall over the election of a U.S. senator. Eventually, a compromise candidate was chosen, and Roosevelt went on to espouse the interests of upstate farmers and workingmen, becoming a spokesman for good government groups and conservationists. He was able to further the goals of the latter when he was named the chairman of the Committee on Fish, Forest and Game, and he attempted, vainly, to push through legislation that would have placed limits on tree cutting on private land.

He ran for reelection in 1912 but soon was involved in presidential politics, becoming the leader of the progressive anti-Tammany Democrats who backed Woodrow Wilson's nomination for the presidency. When Wilson won, Roosevelt was awarded the post of assistant secretary of the navy. After seven years in office, he established himself as one of the few young (he was thirty-one when he was appointed assistant secretary of the navy) bright lights among the progressive Democrats. Busy in his post, in 1918 he made a wide-ranging tour of navy facilities in Europe.

In 1920, he accepted his party's nomination to run on the ticket with the Democratic nominee for president, James M. Cox. Franklin campaigned vigorously throughout the nation, but the voters had wearied of the Democrats, and Republican candidate Warren G. Harding won handily.

So after ten years in elective and appointive office, Roosevelt was a private citizen. In New York, he became vice president of the Fidelity and Deposit Company, a bonding business, and returned to the practice of law. Then in August 1921 he was struck down with infantile paralysis. It was a severe attack that left him a lifelong cripple. In the years that followed he struggled to regain the use of his legs, but to no avail. Still, he remained active in party politics, and in 1928 he ran for governor of New York.

As a governor (he served two terms between 1928 and 1932) he molded his legislature agenda to appeal to as wide a swath of the electorate as possible. Thus he focused on public electric power development, a popular